Handbook of Ecology

Volume I

Handbook of Ecology
Volume I

Edited by **Jeffery Clarke**

R CALLISTO REFERENCE

New York

Published by Callisto Reference,
106 Park Avenue, Suite 200,
New York, NY 10016, USA
www.callistoreference.com

Handbook of Ecology: Volume I
Edited by Jeffery Clarke

International Standard Book Number: 978-1-63239-387-6 (Hardback)

Printed in the United States of America.

Contents

Preface

The study of interactions between organisms and their environment is known as Ecology. The study of interaction between two organisms also falls under the category of ecology. The concepts which interest ecologists are distribution, biomass, population and diversity of organisms. Competition and survival theories of an organism also are studied under ecology. The term 'Ecology', which was coined by German scientist Ernst Haeckel (1834–1919), can also be described as a science that includes concepts of biology and earth science. The field of ecology not only discusses environment, and environment sciences but it is also very closely related to evolution theories of biology, genetics, and the understanding of the impact of biodiversity on ecological functions. The main emphasis of ecologists are life processes, the movement of materials and flow of energy amongst living communities, development of ecosystems, and distribution of organisms and biodiversities in a particular environment.

Research in ecology is focused to study the influence of plant diversity on the numerical response of eriopis connexa to changes in cereal aphid density in wheat crops, plankton resting stages in the marine sediments, development of allometric equations for estimating above-ground liana biomass in tropical primary and secondary forests, farmers' interest in nature and its relation to biodiversity in arable fields, ant-related oviposition and larval performance in a myrmecophilous lycaenid, etc.

I invite students, researchers and teachers interested in environmental issues to read this book which presents different methodologies and studies on ecology. I would like to thank all the contributing authors who have shared their knowledge through this book. I would also like to thank my publisher for considering me worthy of this opportunity. Lastly, I would like to thank my family for their continuous support at every step.

Editor

Plankton Resting Stages in the Marine Sediments of the Bay of Vlorë (Albania)

Fernando Rubino,[1] Salvatore Moscatello,[2] Manuela Belmonte,[1,2] Gianmarco Ingrosso,[2] and Genuario Belmonte[2]

[1] Laboratory of Plankton Ecology, IAMC CNR, UOS Talassografico "A. Cerruti," 74123 Taranto, Italy
[2] Laboratory of Zoogeography and Fauna, CoNISMa U.O. Lecce, DiSTeBA University of the Salento, 73100 Lecce, Italy

Correspondence should be addressed to Genuario Belmonte; genuario.belmonte@unisalento.it

Academic Editor: Sami Souissi

In the frame of the INTERREG III CISM project, sediment cores were collected at 2 stations in the Gulf of Vlorë to study the plankton resting stage assemblages. A total of 87 morphotypes were identified and produced by Dinophyta, Ciliophora, Rotifera, and Crustacea. In 22 cases, the cyst belonged to a species absent from the plankton of the same period. The most abundant resting stages were those produced by *Scrippsiella* species (Dinophyta). Some calcareous cysts were identified as fossil species associated with Pleistocene to Pliocene sediment, although they were also found in surface sediments and some of them successfully germinated, thus proving their modern status. Total abundance generally decreased with sediment depth at station 40, while station 45 showed distinct maxima at 3 and 8 cm below the sediment surface. The depth of peak abundance in the sediment varied with species. This paper presents the first study of the plankton resting stages in the Bay of Vlorë. The study confirmed the utility of this type of investigation for a more correct evaluation of species diversity. In addition, the varying distribution with sediment depth suggests that this field could be of some importance in determining the history of species assemblages.

1. Introduction

Resting stages produced by plankton organisms in temperate seas accumulate in the bottom sediments of confined coastal areas [1]. Their assemblages represent reservoirs of biodiversity which sustain the high resilience of plankton communities, providing recruits of propagules at each return of favourable conditions, in accordance with the so-called Supply Vertical Ecology model [2]. The existence of benthic stages in the life cycles of holoplankton provides a new key for understanding the role of life cycles in the pelagic-benthic relationship in coastal waters [3, 4]. Consequently, assessments of biodiversity at marine sites should take account of the unexpressed fraction of the plankton community contained in the bottom sediments by performing integrated sampling programs [5, 6]. Despite the proven importance of resting stage banks in coastal marine ecology, the issue of "resting *versus* active" plankters has commonly been considered for single *taxa* and only rarely from the whole-community point of view. This is probably due to the great complexity (compositional, functional, and distributional) of resting stage banks. Indeed, it has been demonstrated that at any given moment the species assemblages in bottom sediments (as resting stages) are quite different from the species detectable in the water column (as active stages) [6, 7]. However, the study of such marine "seed banks" (as understood by [3], analogous to terrestrial seed banks in forest soils) is complex on many levels. Resting stages share a common morphological plan despite belonging to organisms from different kingdoms [8]. Consequently, resting stage morphology differs sharply from that of active stages and in some cases their identification is highly problematic. However, it is also true that for some naked dinoflagellates or for thecate ones with a similar thecal plate pattern, cysts are quite different, allowing correct identification without the use of SEM or molecular techniques. Cyst-producing dinoflagellates differ in the length of their life cycle and/or the timing of cyst production, and the rest capacity of resting stages also varies. They are generally programmed to rest for the duration of the adverse period, but fractions of them can also rest for longer periods, allowing the population to reappear decades later, ([9–11], for copepods, [12, 13], for dinoflagellates). Hairston et al. [14] reported a rest of more than 300 years for a calanoid resting egg, albeit of a freshwater species.

The scarcity of literature on whole resting stage communities encouraged us to describe situations in various parts of the Mediterranean, in order to obtain a rich data set useful for building models and in experimental situations. Here a detailed description of the structure of the marine "seed bank" produced by plankton in the Bay of Vlorë is reported.

The present study also focuses on an Albanian bay that has not been extensively studied from the marine biodiversity point of view. The data from the benthos are compared with analyses of the phytoplankton and microzooplankton [15] in the water column to assess with more precision the biodiversity of the plankton in the Bay of Vlorë.

Moscatello et al. [15] reported that the microzooplankton community of the Bay of Vlorë was composed of more than 200 taxa, of which 97 were classified as "seasonally absent." The aim of the present paper is to determine whether these absences in the water column correspond to resting stages in the sediments.

2. Materials and Methods

2.1. Study Site. An oceanographic campaign was carried out in the Bay of Vlorë from 17th to 23rd of January 2008 aboard the oceanographic vessel "Universitatis". This survey was conducted as part of the PIC Interreg III Italy, Albania Project for providing technical assistance for the management of an International Centre of Marine Sciences (CISM) in Albania. The sampling period of the present study (January 2008) coincided with that of Moscatello et al. [15] who investigated active plankton in the water column in the same area.

In order to investigate the presence and distribution of resting stages produced by plankton species in the area, 2 stations were chosen, representing 2 different types of environment: a deep zone (station 40, depth: 54 m), with sediments of terrigenous mud dominated by *Labidoplax digitata* (Holothuroidea) and a shallower site (station 45, depth: 28 m), with sediments of terrigenous mud dominated by *Turritella communis* (Gastropoda) (Figure 1) (for the classification of mud biocenosis in the Bay of Vlorë, see [16]).

2.2. Sampling Procedure. Samples of bottom sediments were collected in three replicates (named 40 a, b, c and 45 a, b, c) using a Van Veen grab with upper windows that allowed the collection of undisturbed sediment cores. At each station, 2 different PVC corers (h: 30 cm, inner ⊘ 4 and 8 cm) were used in order to obtain 2 different sets of samples. The smaller core was processed to obtain cysts of protists; the larger core was processed to obtain resting eggs of metazoans. This differentiation was necessary because metazoan resting

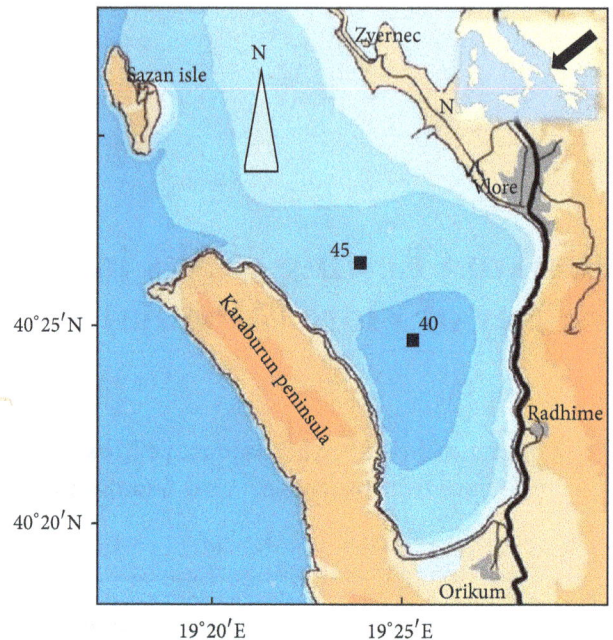

FIGURE 1: Map of study area showing location of two sampling stations (40, 45) in Bay of Vlorë (Albania).

stages are less abundant, so a greater amount of sediment is required. Moreover, their walls are only organic, allowing the adoption of a centrifugation method coupled with filtration to obtain a "clean" sample from a relatively large quantity of sediment. In contrast, protistan cysts are more abundant and have different types of walls (calcareous, siliceous, organic), which complicates the procedure when the whole cyst bank is studied. Thus, the most fruitful method of separating cysts from sediment is filtration through meshes of different sizes.

After extraction, sediment cores were immediately subdivided into 1 cm thick layers, until the 15th cm from the sediment surface. The thickness of 15 cm was chosen because in previous studies we noted that abundances diminished significantly at depths of more than 7–10 cm below the sediment surface [1, 12]. The outer edge of each layer was discarded to avoid contamination from material from the overlaying layer during the insertion of the corer into the sediments. Once obtained, the samples were stored in the dark at 5°C until treatment in the laboratory.

2.3. Protistan Cysts (20–125 μm). In the laboratory the small-core samples were treated using a sieving technique consisting of the following steps.

(i) The entire sample is homogenized and then subsampled, obtaining 3–5 mL of wet sediment which is passed through a 20 μm mesh (Endecott's LTD steel sieves, ISO3310-1, London, England), using natural filtered (0.45 μm) seawater (Gulf of Taranto).

(ii) The retained fraction is ultrasonified for 1 min and again passed through a series of sieves (125, 75, and 20 μm mesh sizes), obtaining a fine-grained fraction (20–75 μm) containing most of the protistan cysts and a 75–125 μm fraction with the larger ones (e.g., *Lingulodinium* spp.) and the zooplankton resting eggs. The material retained by the 125 μm mesh was not considered.

No chemicals were used to disaggregate sediment particles, in order to avoid the dissolution of calcareous and siliceous cyst walls.

Qualitative and quantitative analyses were carried out under an inverted microscope (Zeiss Axiovert S100 equipped with a Nikon Coolpix 990 digital camera) at 200 and 320 magnifications. Both full (i.e., probably viable) and empty (i.e., probably germinated) cysts were considered. At least 1/5 of the finer fraction and all of the >75 μm fraction were analyzed.

All the resting stage morphotypes were identified on the basis of published descriptions, ([17–19], for dinoflagellates, [20, 21], for ciliates and germination experiments).

Identification was performed to the lowest possible taxonomic level. As a rule, modern, biological names were used. The paleontological name is reported only for morphotypes whose active stage was not known.

A fixed aliquot (≈5 g) of wet sediment from each sample was oven-dried at 70°C for 24 h to calculate the water content and obtain quantitative data for each *taxon* as cysts g^{-1} of dry sediment.

2.4. Metazoan Resting Eggs (45–200 μm). For the analysis of the large-core samples the Onbè [22] method was used, slightly modified by using 45 and 200 μm mesh sizes to obtain a size range typical of mesozooplankton resting eggs.

For each sample a fixed quantity of wet sediment was treated (≈45 cm^3).

Only full (i.e., probably viable) resting eggs were counted and quantitative data for each *taxon* are reported as resting eggs × 100 g^{-1} of dry sediment.

2.5. Germination Experiments. To achieve germination, all putative viable cysts of protists isolated from the sediment were individually positioned in Nunclon microwells (Nalge Nunc International, Roskilde, Denmark) containing ≈1 mL of natural sterilized seawater. Cysts were incubated at 20°C, 12 : 12 h LD cycle, 100 μE m^{-2} sec^{-1} irradiance, and examined on a daily basis, until germination up to a maximum of 30 days. The incubation conditions were chosen on the basis of previous studies [5, 6, 23]. They have proved to be effective for a large number of species.

2.6. Data Analysis. For the 1st cm of the cores, data on resting stage abundance from the 2 sampling stations were obtained by merging the data from the 3 replicates of the 2 sets of samples (those for protistan cysts and metazoan resting eggs). For samples below the first cm, only the 45–200 μm fraction was used, in order to facilitate and accelerate the analysis. From the abundance matrices (*taxa versus* stations and *taxa versus* station and cm respectively) of both surface and deeper sediments, the Bray Curtis similarity measure was calculated

after 4th root transformation in order to allow rare species to become more evident.

The PRIMER "DIVERSE" function (Primer-E Ltd, Plymouth, UK) was used to calculate the taxonomic richness (*S*), *taxon* abundance (*N*), Margalef index (*d*), Shannon-Wiener diversity index (*H'*), and Pielou's evenness index (*J'*) for each sample.

The relationships between the samples collected at the 2 stations were analyzed by means of nonmetric multidimensional scaling (nMDS) with superimposed hierarchical clustering with a cutoff at 60% similarity (for surface sediments) and 70% (for the sediment core as a whole), while the SIMPER routine was used to identify relative dissimilarity and the *taxa* that contributed most to the differences.

The statistical significance of the differences between the 2 stations was calculated by means of a 2-way crossed analysis of similarities (ANOSIM) on the Bray-Curtis similarity matrix based on the stratigraphy.

All univariate and multivariate analyses were performed using PRIMER v.6 (Primer-E Ltd, Plymouth, UK).

3. Results

3.1. Total Biodiversity. Resting stages were found at all levels of the 15 cm sediment core columns from the 2 investigated sites in the Gulf of Vlorë.

Merging the data from the 2 sets of samples (20–125 μm and 45–200 μm) and considering both full (probably viable) and empty (probably germinated) forms from each station, 87 different resting stage morphotypes produced by plankton were recognized (Table 1). Most of them (59, belonging to 20 genera) were dinoflagellates, 16 were ciliates (9 genera), 4 rotifers (2 genera), and 5 crustaceans (4 genera), while 3 (1 protistan cyst type and 2 resting eggs) remained unidentified. Station 40 showed higher biodiversity, with 79 morphotypes, 35 of them exclusive to the site. At station 45, 52 morphotypes were observed, 8 of them being exclusive.

Moreover, analysis of the empty forms found among the 20–125 μm fraction led to the recognition of 11 morphotypes, all dinoflagellates.

A total of 36 cyst types were identified as *taxa* missing from the plankton list of the same period (January 2008; [15]). Partly due to nomenclature problems, uncertainty of identification, and differences in examined periods, it was possible to ascertain the contemporaneous presence of species in both pelagic and benthic compartments only in a very few cases.

Identification was frequently impossible due to the presence of previously unreported resting stage morphologies. In such cases, germination experiments allowed the cysts to be attributed to a high level *taxon* at least, as with a *Strombidium* (Ciliophora) cyst, whose morphology is reported here for the first time (see Figure 2).

3.2. Surface Sediments. The analysis of surface sediments (the 1st cm of the cores), that is, those most affected by cyst deposition and resuspension/germination, revealed sharp differences between the 2 analysed stations. In total, 36

TABLE 1: List of resting stage (cyst) morphotypes recovered from sediments of Bay of Vlorë (Albania).

Taxon	St.40		St.45	
Dinoflagellates				
Alexandrium minutum Halim	●		●	○
Alexandrium tamarense (Lebour) Balech	●			
Alexandrium sp. 1	●			
Alexandrium sp. 2			●	
Bicarinellum tricarinelloides Versteegh	●	○	●	
Calcicarpinum perfectum Versteegh		○		
Calciodinellum albatrosianum (Kamptner) Janofske and Karwath	●	○	●	○
Calciodinellum operosum (Deflandre) Montresor	●	○		○
Calciperidinium asymmetricum Versteegh		○		
Cochlodinium polykrikoides Margalef type 1	●			
Cochlodinium polykrikoides Margalef type 2		○		
Diplopelta parva (Abé) Matsuoka		○		
Diplopsalis lenticula Bergh	●			○
Follisdinellum splendidum Versteegh		○		
Gonyaulax group	●	○	●	○
Gymnodinium impudicum (Fraga and Bravo) G. Hansen and Möestrup			●	
Gymnodinium nolleri Ellegaard and Möestrup	●			
Gymnodinium sp. 1	●	○	●	
Lingulodinium polyedrum (Stein) Dodge	●	○	●	○
Melodomuncula berlinensis Versteegh	●	○	●	
Nematodinium armatum (Dogiel) Kofoid and Swezy	●	○		
Oblea rotunda (Lebour) Balech ex Sournia	●		●	
Pentapharsodinium dalei Indelicato and Loeblich type 1	●	○	●	
Pentapharsodinium dalei Indelicato and Loeblich type 2	●		●	
Pentapharsodinium tyrrhenicum Montresor, Zingone, and Marino type 1	●	○	●	○
Pentapharsodinium tyrrhenicum Montresor, Zingone, and Marino type 2		○		
Polykrikos kofoidii Chatton				○
Polykrikos schwartzii Bütschli		○		
Protoperidinium compressum (Abé) Balech		○		
Protoperidinium conicum (Gran) Balech	●	○		
Protoperidinium oblongum (Aurivillius) Parke and Dodge	●			
Protoperidinium parthenopes Zingone and Montresor			●	
Protoperidinium steidingerae Balech	●			
Protoperidinium subinerme (Paulsen) Loeblich III		○		
Protoperidinium thorianum (Paulsen) Balech	●	○		○
Protoperidinium sp. 1	●	○		○
Protoperidinium sp. 5			●	○
Protoperidinium sp. 6	●			
Pyrophacus horologium Stein	●		●	
Scrippsiella cf. *crystallina* Lewis		○		
Scrippsiella lachrymosa Lewis	●	○	●	○
Scrippsiella ramonii Montresor	●	○		○
Scrippsiella trochoidea (Stein) Loeblich rough type	●		●	
Scrippsiella trochoidea (Stein) Loeblich smooth type	●		●	○
Scrippsiella trochoidea (Stein) Loeblich large type	●	○	●	

TABLE 1: Continued.

Taxon	St.40		St.45	
Scrippsiella trochoidea (Stein) Loeblich medium type	●	○	●	○
Scrippsiella trochoidea (Stein) Loeblich small type	●	○	●	
Scrippsiella sp. 1	●	○	●	○
Scrippsiella sp. 4	●		●	
Scrippsiella sp. 5	●		●	
Scrippsiella sp. 6	●		●	
Scrippsiella sp. 8	●	○	●	
Thoracosphaera sp.	●		●	
Dinophyta sp. 2	●		●	
Dinophyta sp. 7	●		●	
Dinophyta sp. 17	●			
Dinophyta sp. 26	●			
Dinophyta sp. 30	●			
Dinophyta sp. 33	●		●	
Ciliates				
Codonella aspera Kofoid and Campbell	●			
Codonella orthoceras Heackel	●			
Codonellopsis monacensis (Rampi) Balech	●			
Codonellopsis schabii (Brandt) Kofoid and Campbell	●		●	
Epiplocylis undella (Ostenfeld and Schmidt) Jörgensen	●		●	
Rabdonella spiralis (Fol) Brandt	●			
Stenosemella ventricosa (Claparède and Lachmann) Jörgensen	●		●	
Strobilidium sp.	●		●	
Strombidium cf. *acutum* (Leegaard) Kahl	●		●	
Strombidium conicum (Lohman) Wulff		○	●	
Tintinnopsis beroidea Stein	●			
Tintinnopsis butschlii Kofoid and Campbell	●			
Tintinnopsis campanula Ehrenberg	●			
Tintinnopsis cylindrica Daday	●		●	
Tintinnopsis radix (Imhof)			●	
Undella claparedei (Entz) Daday	●			
Rotifers				
Brachionus plicatilis Müller	●			
Synchaeta sp. spiny type	●			○
Synchaeta sp. rough type	●			
Synchaeta sp. mucous type	●			
Crustaceans Cladocerans				
Penilia avirostris Dana	●		●	
Crustaceans Copepods				
Acartia clausi/margalefi	●			
Acartia sp. 1	●	○	●	○
Centropages sp.	●		●	○
Paracartia latisetosa (Krizcaguin)	●	○	●	
Unidentified				
Cyst type 1			●	
Resting Egg 1			●	
Resting Egg 9	●			

●: cysts observed as full (i.e., probably viable). ○: cysts observed as germinated (i.e., empty).

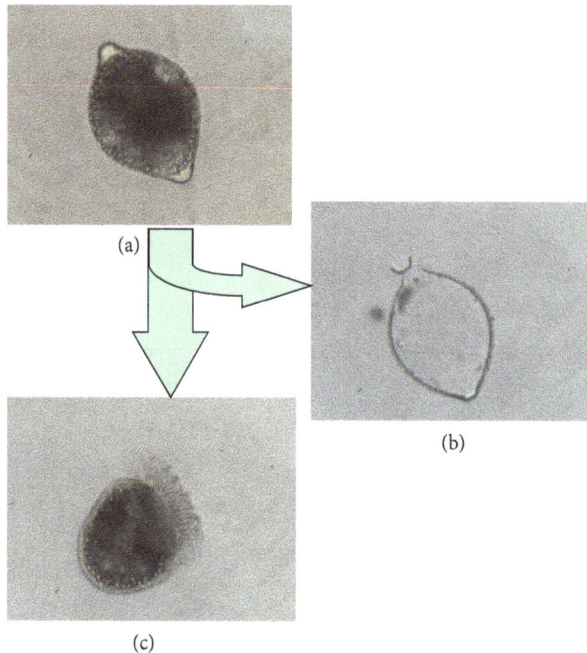

FIGURE 2: Photographs of Ciliophora cyst, with two opposite papulae (a). Its empty shell (hatch occurs from one of two papulae) (b). Germinated active stage, *Strombidium* ciliate (c).

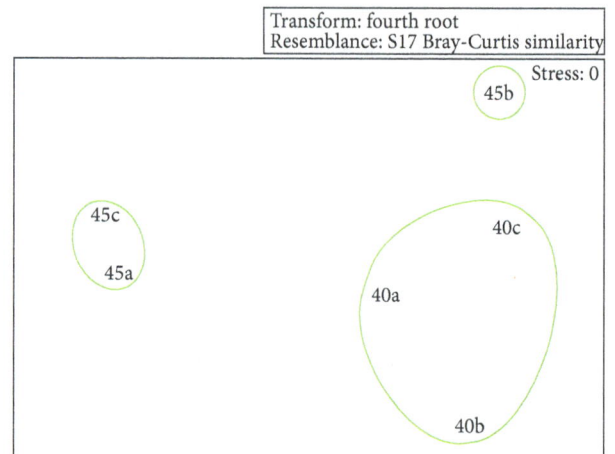

FIGURE 3: nMDS plot of surface sediment samples collected at stations 40 and 45 in Bay of Vlorë. Hierarchical clustering superimposed with cutoff at 60% similarity.

different resting stage morphotypes were observed in this first layer (Table 2), 23 produced by dinoflagellates, 6 by ciliates, 2 by rotifers, 4 by crustaceans, and 1 undetermined. Even considering the small amount of available data, station 40 showed higher biodiversity, in terms of both number of *taxa* and diversity indexes (see Table 3). Total abundances were comparable, however, with 389 ± 127 cysts g^{-1} (average ± s.d.) at station 40 *versus* 329 ± 123 cysts g^{-1} at station 45. SIMPER showed 58% dissimilarity between the assemblages of the two sites (Table 4).

The most abundant cyst morphotypes in the surface layers were calcareous cysts produced by species of the Calciodinellaceae family (Dinophyta). At station 40, five cyst morphotypes of this family accounted for 95% of total abundance, while at station 45, 99% was accounted for by just one cyst morphotype, *Scrippsiella trochoidea* medium type, confirming the lower evenness at this station.

The nMDS ordination (Figure 3, stress = 0), with the hierarchical cluster superimposed with a cutoff at 60% similarity, clearly reflects the separation between the samples from stations 40 and 45. Among these, due to its higher diversity, sample 45b is farther from samples 45a and 45c than it is from the samples of station 40.

3.3. Whole Sediment Cores. At both the investigated stations, a general decrease in total abundances was observed with depth along the sediment columns. At station 40, higher total abundance and diversity values than station 45 were registered (Figure 4), with a sharp decline between the 6th and 7th centimetres. Beyond this depth, total abundance remained below 100 cysts $100 \, g^{-1}$. In terms of species, *Codonellopsis*

schabii cysts and *Synchaeta* sp. and *Acartia clausi/margalefi* resting eggs were continuously observed along the whole sediment column at both stations. The ciliate *C. schabii* was by far the most abundant *taxon* at station 40 (43% of total abundance), with density highest in the 2nd cm (342 ± 192 cysts $100 \, g^{-1}$); as with total abundance, a sharp decrease was observed between the 6th and 7th centimetres. Other important species were the copepods *Centropages* sp. (181 ± 50 resting eggs $100 \, g^{-1}$ at 4th cm) and *Acartia* spp. (67 ± 32 resting eggs $100 \, g^{-1}$ at 1st cm). At station 45 the most abundant type was *Acartia* spp. (86 ± 57 resting eggs $100 \, g^{-1}$ at 1st cm) followed by *C. schabii* (70 ± 60 cysts $100 \, g^{-1}$ at 4th cm) and *Synchaeta* sp. (41 ± 23 resting eggs $100 \, g^{-1}$ at 5th cm).

In the nMDS ordination (Figure 5, stress = 0.12) with superimposition of the hierarchical cluster with a cutoff at 70% similarity, all the samples from station 45 cluster together, while the samples from station 40 were widely dispersed, a sign of greater variability at this site.

The assemblage structure of the two stations differed significantly at all layers (ANOSIM $R = 0.655$; $P = 0.001$), showing 59% dissimilarity (SIMPER, Table 5).

3.4. Germination Experiments. All putatively viable (i.e., full) protistan cyst types observed were isolated and incubated under controlled conditions to obtain germination. Successful germination generally allowed us to confirm the cyst-based identification, but in some cases it enabled us to go beyond this and discriminate between cysts sharing similar morphology. For example, *Alexandrium minutum* and *Scrippsiella* sp. 1, both have a round cyst with a thin and smooth wall with mucous material attached, *Protoperidinium thorianum* and *Protoperidinium* sp. 1 cysts are both round-brown and smooth, and *Gymnodinium nolleri* and *Scrippsiella* sp. 4 both produce round-brown cysts with a red spot inside. The germination of all these cyst types allowed us to correctly identify these species.

TABLE 2: Abundance (cysts g^{-1} dw) of probably viable resting stages (cysts) observed in surface sediments of two stations in Bay of Vlorë (Albania). Values from three replicates are reported.

	40a	40b	40c	45a	45b	45c
Calciodinellum albatrosianum	20.1	18.3	35.1	0.0	0.0	0.0
Calciodinellum operosum	0.0	0.0	11.7	0.0	0.0	0.0
Gonyaulax group	20.1	0.0	0.0	0.0	0.0	59.6
Gymnodinium sp. 1	20.1	9.2	0.0	0.0	22.2	0.0
Lingulodinium polyedrum	40.2	0.0	0.0	0.0	0.0	0.0
Melodomuncula berlinensis	40.2	0.0	0.0	0.0	0.0	0.0
Oblea rotunda	0.0	0.0	11.7	0.0	0.0	0.0
Pentapharsodinium dalei type 1	20.1	0.0	0.0	0.0	11.1	0.0
Pentapharsodinium tyrrhenicum type 1	40.2	18.3	23.4	0.0	11.1	0.0
Protoperidinium sp. 1	0.0	9.2	0.0	0.0	0.0	0.0
Protoperidinium sp. 5	0.0	0.0	11.7	0.0	11.1	0.0
Scrippsiella ramonii	0.0	9.2	0.0	0.0	0.0	0.0
Scrippsiella trochoidea rough type	40.2	18.3	46.8	0.0	0.0	0.0
Scrippsiella trochoidea smooth type	0.0	9.2	11.7	0.0	11.1	0.0
Scrippsiella trochoidea medium type	181.0	73.3	105.4	173.1	111.1	238.4
Scrippsiella trochoidea small type	80.5	64.2	58.5	230.8	0.0	0.0
Scrippsiella sp. 1	20.1	0.0	11.7	0.0	11.1	0.0
Scrippsiella sp. 4	0.0	9.2	0.0	0.0	0.0	0.0
Thoracosphaera sp. 1	0.0	0.0	11.7	0.0	11.1	0.0
Dinophyta sp. 2	0.0	0.0	23.4	0.0	0.0	0.0
Dinophyta sp. 17	0.0	18.3	0.0	0.0	0.0	0.0
Dinophyta sp. 26	0.0	18.3	0.0	0.0	0.0	0.0
Dinophyta sp. 33	0.0	0.0	0.0	0.0	11.1	0.0
Codonellopsis schabii	1.0	0.0	0.9	0.6	0.3	0.5
Stenosemella ventricosa	0.1	0.0	0.0	0.0	0.0	0.0
Strobilidium sp.	0.1	0.0	0.1	0.0	0.0	0.0
Strombidium acutum	0.0	0.0	0.0	0.0	11.1	0.0
Tintinnopsis cylindrica	0.0	0.0	0.0	0.0	0.1	0.1
Undella claparedei	0.1	0.0	0.1	0.0	0.0	0.0
Brachionus plicatilis	0.2	0.0	0.1	0.3	0.0	0.1
Synchaeta sp spiny type	0.3	0.2	0.0	0.2	0.0	0.1
Penilia avirostris	0.0	0.0	0.1	0.0	0.0	0.0
Acartia clausi/margalefi	1.0	0.3	0.7	1.5	0.3	0.8
Acartia sp. 1	0.1	0.0	0.1	0.0	0.0	0.0
Centropages sp.	0.3	0.2	0.0	0.2	0.1	0.2
Cyst type 1	0.0	0.0	0.0	57.7	0.0	0.0

TABLE 3: Abundance and diversity indices calculated for resting stages in surface sediments at two stations investigated in Bay of Vlorë.

	Abundance cysts g^{-1} dw	Total density cysts g^{-1} dw	S	d	H'	J'
Station 40	389 ± 127	1167	18 ± 2.7	2.9 ± 0.3	2.2 ± 0.1	0.7 ± 0.1
Station 45	329 ± 123	987	11 ± 4.4	1.8 ± 0.9	0.5 ± 0.2	0.5 ± 0.2

Abundance: average ± standard deviation from three replicates. Total density: sum of cyst abundances observed in three replicates from each station. S: number of *taxa* identified (average ± standard deviation). d: Margalef diversity index. H': Shannon diversity index. J': Pielou's evenness index.

TABLE 4: Results of SIMPER analysis for resting stages from surface sediments at stations 40 and 45 in Bay of Vlorë.

Taxa	Av. Abund	Av. Sim	Sim/SD	Contrib%	Cum.%
Station 40					
Average similarity: 56.81					
Scrippsiella trochoidea medium type	119.92	21.61	7.45	38.05	38.05
Scrippsiella trochoidea small type	67.73	15.81	6.14	27.83	65.87
Scrippsiella trochoidea rough type	35.13	6.44	2.78	11.34	77.21
Pentapharsodinium tyrrhenicum type 1	27.33	5.18	8.96	9.13	86.34
Calciodinellum albatrosianum	24.53	4.94	7.24	8.69	95.03
Station 45					
Average similarity: 40.37					
Scrippsiella trochoidea medium type	174.21	40.03	5.87	99.16	99.16

Stations 40 and 45
Average dissimilarity = 58.20.

Cysts ascribed to the paleontological *taxa Bicarinellum tricarinelloides* and *Calciperidinium asymmetricum* both germinated, thus confirming that they belong to modern *taxa*. The active stages obtained were tentatively identified as scrippsielloid dinoflagellates.

An unknown ciliate cyst, with a papula at both extremities, produced an active stage identifiable as belonging to the *Strombidium* genus (Figure 2).

4. Discussion

The total number of resting stage morphotypes recognized in the present study is particularly high compared with other studies in the Mediterranean. None of these studies gave a number higher than the one reported here, despite being based on a larger geographical area (the whole North Adriatic, in [24]) or a higher number of samples (157 sediment samples in [5]). This richness could be due to our enhanced ability, with the passage of time, to identify cysts from different species, but it could also depend on the consideration of different depths in the sediments. Indeed, the other mentioned studies only reported cysts from the sediment surface, while in the present case the type list grew by more than 60% when below-surface layers were considered.

As a consequence of its richness, the reported list adds 42 morphotypes to the Albanian list and 13 alternative morphotypes to already known *taxa*. This fact clearly demonstrates that the description of cyst assemblages in coastal Mediterranean areas is still far from being exhaustive.

The discovery of differences in the benthic species assemblage with respect to the plankton is partially due to the use in cyst studies of a terminology derived from paleontological studies which has yet to be standardised with reference to modern terminology. However, it is evident that the active stages in the water column assemblage of the Bay of Vlorë [15] differ in number and quality from that of the bottom sediments reported in the present study. By way of example, and only considering the surface sediment layer (i.e., the most affected by recent sinking and/or re-suspension), 4 different

species of *Scrippsiella* (Dinophyta) were isolated as cysts, but only 2 were reported [15] as active stages in the water column for the whole bay. Moreover, in this study 5 different cyst types for the single species *S. trochoidea* were identified, differing in terms of size and wall. This is evidence of great intraspecific diversity, but it could be also a sign of the presence of cryptic species, as discussed by Montresor et al. [25], differing in cyst morphology but not in that of the swimming stage.

The rotifer *Synchaeta* sp. was not found in the water column, but its resting eggs were easily recognizable and abundant, in the sediments.

While the case of *S. trochoidea* confirms that much remains to be discovered about the morphological variability of cysts produced by the same species (see [26], for Dinophyta or [27], for Calanoida), *Synchaeta* sp. is a clear case of a species not detected in the active plankton assemblage but waiting in the sediments for a favourable moment to rejoin the water column.

Also worthy of attention is the observation of a Ciliophora cyst with two papulae on opposite sides (Figure 2), which has never been reported before.

A study of plankton composition was carried out in the same site during the same scientific cruise (January 2008) as the present study [15]. In January 2008, the phytoplankton and the microzooplankton included a total of 178 categories. Considering only the main cyst producers (dinoflagellates and ciliates), examination of the water column at 16 stations gave a total of 76 *taxa* (48 dinoflagellates, 28 ciliates). The present analysis of sediments, from just 2 stations, gave a total of 75 *taxa*. This striking similarity of values was not, however, reflected in the taxonomic composition of the 2 compartments. Indeed, 36 cyst types were identified as *taxa* not present in the plankton list for the same period (January 2008). This number would be even higher if we considered only plankton from stations close to the two used here for the sediments.

It was not possible to correlate cyst abundance along the sediment column with age of deposition, which would require dating of the sediment layers. In any case, our results showed that the total abundance of cysts in the upper layers

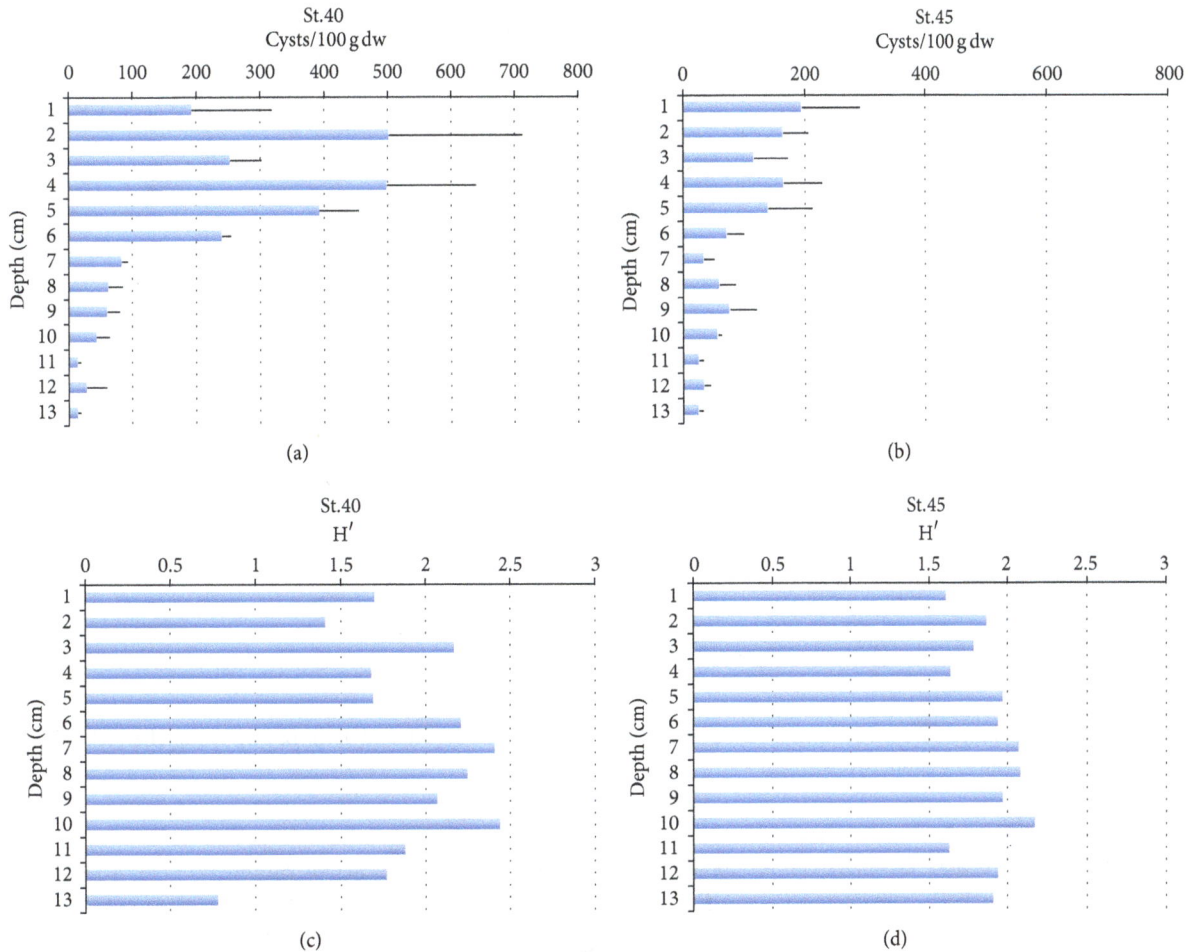

FIGURE 4: Resting stage abundance (average ± standard deviations) and Shannon's index (H') values recorded for each cm layer of sediment cores collected at two investigated stations in Bay of Vlorë (Albania).

was up to 10 times greater than in lower ones. The sharp decrease in abundance below the 5th cm of depth, at least at station 40, does, however, suggest that an event occurred at a certain moment in the history of the plankton in the Bay of Vlorë, a suggestion that clearly requires further study. Indeed, due to its position, station 40 is a candidate for studies of the history of cyst production (and their arrival in the sediment). Located in a depression on the seabed, the depth of St. 40 (−54 m) probably favours the sedimentation of fine particles and the depletion of oxygen content, and the deposition and accumulation of sinking resting stages can thus be considered undisturbed. In addition, the observed fall in diversity from lower to upper layers could be correlated with the growth of cultural eutrophication (i.e., urban development), as proposed for Tokyo Bay and Daja Bay [28].

This situation at St. 45 (depth 28 m) is not completely identical. It is near the slope of a detritus cone where materials from river Vjosa accumulate and marine currents possibly act at a different rate from those acting on St. 40.

Incubation of encysted forms under controlled conditions to obtain germination is a useful tool for confirming the identification made by observation of the cyst. In some cases, different species produce very similar cysts, especially when the morphology is very simple, that is, spherical, without processes or wall structures. In the present study, we observed many Dinophyta cysts with the same basic morphology, that is, round body and smooth brown wall with no apparent signs of paratabulation or spines or processes. Their germination allowed us to classify this basic type into at least 6 species. Round brown cysts are typical of *Protoperidinium* species [29, 30], but we also recognized *Diplopsalis lenticula*, *Gymnodinium nolleri*, and *Oblea rotunda*, as well as 3 additional *Protoperidinium* species. In the same way, it was possible to distinguish between *Alexandrium minutum* and *Scrippsiella* sp.1, whose cysts are very similar and whose distinctive features are recognizable only after germination.

Conversely, analysis of cysts may allow us to identify species whose active stages are indistinguishable, at least by optical microscope. This is the case in the present study for the *Scrippsiella* group, which produces active cells that are very difficult to distinguish, although their cysts differ in terms of the type of calcareous covering, colour, and the presence of spines [31, 32].

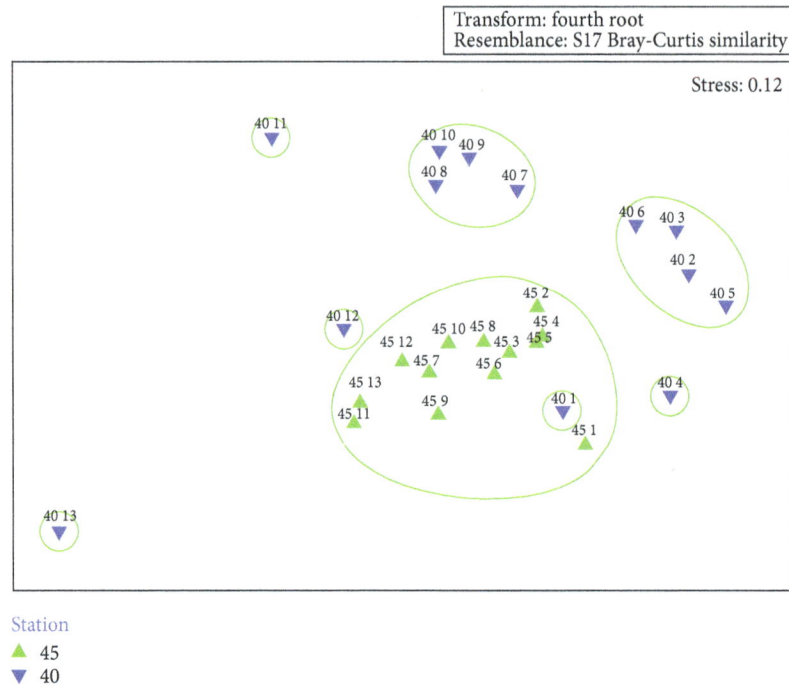

FIGURE 5: nMDS plot of samples from each cm of sediment cores collected at stations 40 and 45 in Bay of Vlorë. Hierarchical clustering superimposed with cutoff at 70% similarity.

TABLE 5: Results of SIMPER analysis for resting stages in sediment cores collected at stations 40 and 45 in Bay of Vlorë.

Taxa	Av. Abund	Av. Sim	Sim/SD	Contrib%	Cum.%
		Station 40			
		Average similarity: 44.16			
Centropages sp.	1.77	8.77	0.86	19.86	19.86
Codonellopsis schabii	2.10	7.00	1.31	15.86	35.72
Acartia clausi/margalefi	1.56	6.21	1.22	14.06	49.78
Synchaeta sp. spiny type	1.47	5.31	1.10	12.02	61.81
Penilia avirostris	1.14	4.47	0.98	10.13	71.93
Brachionus plicatilis	0.96	2.90	0.81	6.56	78.49
Stenosemella ventricosa	0.78	1.47	0.55	3.32	81.81
Strobilidium sp.	0.73	1.43	0.51	3.23	85.04
Scrippsiella spp.	0.57	0.95	0.36	2.14	87.18
Gonyaulax spp.	0.54	0.71	0.34	1.60	88.79
Strombidium conicum	0.44	0.69	0.33	1.57	90.36
		Station 45			
		Average similarity: 52.61			
Acartia clausi/margalefi	1.88	12.13	2.08	23.05	23.05
Synchaeta sp. spiny type	1.73	10.91	1.59	20.74	43.79
Codonellopsis schabii	1.43	6.73	1.12	12.80	56.59
Strobilidium sp.	1.12	6.05	0.92	11.51	68.10
Centropages sp.	1.20	6.05	1.02	11.50	79.60
Brachionus plicatilis	0.84	2.73	0.63	5.19	84.79
Acartia sp.1	0.75	2.57	0.58	4.89	89.67
Lingulodinium polyedrum	0.74	2.39	0.59	4.54	94.22

Groups 40 and 45.
Average dissimilarity = 58.63.

Worthy of special attention here is the recovery during the present study of Dinophyta cysts whose active stages have yet to be identified. As cysts, they are still classified with a pale-ontological name in accordance with their description from Pleistocene to Pliocene sediment strata in the Mediterranean [33]. Two of these cyst types (*Bicarinellum tricarinelloides* and *Calciperidinium asymmetricum*) were successfully germinated, producing motile forms recognisable as belonging to the Calciodinellaceae family. In any case their frequent observation in surface sediments in other Mediterranean areas [23, 34] and in sediment traps [35] is a clear sign that these species are present in the water column today and need to be better identified.

Conflict of Interests

The authors declare that there is no financial interest or conflict of interests involved.

Acknowledgments

The present study was funded by CoNISMa in the framework of the INTERREG III Italy, Albania Programme, CISM Project (Technical Assistance for Establishing and Management of an International Center for Marine Studies in Albania). The authors thank the crew of the Research Vessel Universitatis (CoNISMa) for the valuable field assistance, Professor Sami Souissi (Université de Lille, France) who took care of the paper editing, and two anonymous referees who were extremely helpful in the improvement of the paper.

References

[1] G. Belmonte, P. Castello, M. R. Piccinni et al., "Resting stages in marine sediments off the Italian coast," in *Biology and Ecology of Shallow Coastal Waters*, A. Elefteriou, C. J. Smith, and A. D. Ansell, Eds., pp. 53–58, Olsen & Olsen, Fredensborg, Denmark, 1995.

[2] N. H. Marcus and F. Boero, "Minireview: the importance of benthic-pelagic coupling and the forgotten role of life cycles in coastal aquatic systems," *Limnology and Oceanography*, vol. 43, no. 5, pp. 763–768, 1998.

[3] A. Giangrande, S. Geraci, and G. Belmonte, "Life-cycle and life-history diversity in marine invertebrates and the implications in community dynamics," *Oceanography and Marine Biology*, vol. 32, pp. 305–333, 1994.

[4] F. Boero, G. Belmonte, G. Fanelli, S. Piraino, and F. Rubino, "The continuity of living matter and the discontinuities of its constituents: do plankton and benthos really exist?" *Trends in Ecology and Evolution*, vol. 11, no. 4, pp. 177–180, 1996.

[5] S. Moscatello, F. Rubino, O. D. Saracino, G. Fanelli, G. Belmonte, and F. Boero, "Plankton biodiversity around the Salento Peninsula (South East Italy): an integrated water/sediment approach," *Scientia Marina*, vol. 68, no. 1, pp. 85–102, 2004.

[6] F. Rubino, O. D. Saracino, S. Moscatello, and G. Belmonte, "An integrated water/sediment approach to study plankton (a case study in the southern Adriatic Sea)," *Journal of Marine Systems*, vol. 78, no. 4, pp. 536–546, 2009.

[7] F. Rubino, O. D. Saracino, G. Fanelli, G. Belmonte, A. Miglietta, and F. Boero, "Life cycles and pelago-benthos interactions," *Biologia Marina Mediterranea*, vol. 5, pp. 253–259, 1998.

[8] G. Belmonte, A. Miglietta, F. Rubino, and F. Boero, "Morphological convergence of resting stages of planktonic organisms: a review," *Hydrobiologia*, vol. 355, no. 1–3, pp. 159–165, 1997.

[9] N. H. Marcus, R. Lutz, W. Burnett, and P. Cable, "Age, viability, and vertical distribution of zooplankton resting eggs from an anoxic basin: evidence of an egg bank," *Limnology and Oceanography*, vol. 39, no. 1, pp. 154–158, 1994.

[10] X. Jiang, G. Wang, and S. Li, "Age, distribution and abundance of viable resting eggs of *Acartia pacifica* (Copepoda: Calanoida) in Xiamen Bay, China," *Journal of Experimental Marine Biology and Ecology*, vol. 312, no. 1, pp. 89–100, 2004.

[11] H. U. Dahms, X. Li, G. Zhang, and P. Y. Qian, "Resting stages of *Tortanus forcipatus* (Crustacea, Calanoida) in sediments of Victoria Harbor, Hong Kong," *Estuarine, Coastal and Shelf Science*, vol. 67, no. 4, pp. 562–568, 2006.

[12] G. Belmonte, P. Pirandola, S. Degetto, and F. Boero, "Abbondanza, vitalità e distribuzione verticale di forme di resistenza nei sedimenti del Nord Adriatico," *Biologia Marina Mediterranea*, vol. 6, pp. 172–178, 1999.

[13] S. Ribeiro, T. Berge, N. Lundholm, T. J. Andersen, F. Abrantes, and M. Ellegaard, "Phytoplankton growth after a century of dormancy illuminates past resilience to catastrophic darkness," *Nature Communications*, vol. 2, no. 1, article 311, 2011.

[14] N. G. Hairston Jr., R. A. Van Brunt, C. M. Kearns, and D. R. Engstrom, "Age and survivorship of diapausing eggs in a sediment egg bank," *Ecology*, vol. 76, no. 6, pp. 1706–1711, 1995.

[15] S. Moscatello, C. Caroppo, E. Hajdëri, and G. Belmonte, "Space distribution of phyto- and microzooplankton in the Vlora Bay (Southern Albania, Mediterranean Sea)," *Journal of Coastal Research*, no. 58, pp. 80–94, 2011.

[16] P. Maiorano, F. Mastrototaro, S. Beqiraj et al., "Bioecological study of the benthic communities on the soft bottom of the Vlora Gulf (Albania)," *Journal of Coastal Research*, no. 58, pp. 95–105, 2011.

[17] C. J. Bolch and G. M. Hallegraeff, "Dinoflagellate cysts in recent marine sediments from Tasmania, Australia," *Botanica Marina*, vol. 33, pp. 173–192, 1990.

[18] J. A. Sonneman and D. R. A. Hill, "A taxonomic survey of cyst-producing dinoflagellates from recent sediments of Victorian coastal waters, Australia," *Botanica Marina*, vol. 40, no. 3, pp. 149–177, 1997.

[19] A. Rochon, A. De Vernal, J. L. Turon, J. Matthiessen, and M. J. Head, "Distribution of recent dinoflagellate cysts in surface sediments from the North Atlantic Ocean and adjacent seas in relation to sea-surface parameters," *AASP Contribution Series*, vol. 35, pp. 1–152, 1999.

[20] P. C. Reid and A. W. G. John, "Tintinnid cysts," *Journal of the Marine Biological Association of the United Kingdom*, vol. 58, pp. 551–557, 1978.

[21] P. C. Reid and A. W. G. John, "Resting cysts in the ciliate class Polymenophorea: phylogenetic implications," *Journal of Protozoology*, vol. 30, pp. 710–712, 1978.

[22] T. Onbè, "Sugar flotation method for sorting the resting eggs of marine cladocerans and copepods fron sea bottom sediments," *Bulletin of the Japanese Society for the Science of Fish*, vol. 44, p. 1411, 1978.

[23] F. Rubino, M. Belmonte, C. Caroppo, and M. Giacobbe, "Dinoflagellate cysts from surface sediments of Syracuse Bay

(Western Ionian Sea, Mediterranean)," *Deep-Sea Research Part II*, vol. 57, no. 3-4, pp. 243–247, 2010.

[24] F. Rubino, G. Belmonte, A. M. Miglietta, S. Geraci, and F. Boero, "Resting stages of plankton in recent North Adriatic sediments," *Marine Ecology*, vol. 21, no. 3-4, pp. 263–284, 2000.

[25] M. Montresor, S. Sgrosso, G. Procaccini, and W. H. C. F. Kooistra, "Intraspecific diversity in *Scrippsiella trochoidea* (Dinophyceae): evidence for cryptic species," *Phycologia*, vol. 42, no. 1, pp. 56–70, 2003.

[26] A. Rochon, J. Lewis, M. Ellegaard, and I. C. Harding, "The *Gonyaulax spinifera* (Dinophyceae) "complex": perpetuating the paradox?" *Review of Palaeobotany and Palynology*, vol. 155, no. 1-2, pp. 52–60, 2009.

[27] G. Belmonte, "Diapause egg production in *Acartia (Paracartia) latisetosa* (Crustacea, Copepoda, Calanoida)," *Bolletino di Zoologia*, vol. 59, pp. 363–366, 1992.

[28] Z. Wang, K. Matsuoka, Y. Qi, J. Chen, and S. Lu, "Dinoflagellate cyst records in recent sediments from Daya Bay, South China Sea," *Phycological Research*, vol. 52, no. 4, pp. 396–407, 2004.

[29] R. Harland, "A review of recent and quaternary organic-walled dinoflagellate cysts of the genus *Protoperidinium*," *Paleontology*, vol. 25, pp. 369–397, 1982.

[30] J. Lewis, J. D. Dodge, and P. Tett, "Cyst-theca relationship in some *Protoperidinium* species (Peridiniales) from Scottish sea lochs," *Journal of Micropalaeontology*, vol. 3, pp. 25–34, 1984.

[31] M. Gottschling, R. Knop, J. Plötner, M. Kirsch, H. Willems, and H. Keupp, "A molecular phylogeny of *Scrippsiella sensu lato* (Calciodinellaceae, Dinophyta) with interpretations on morphology and distribution," *European Journal of Phycology*, vol. 40, no. 2, pp. 207–220, 2005.

[32] H. Gu, J. Sun, W. H. C. F. Kooistra, and R. Zeng, "Phylogenetic position and morphology of thecae and cysts of *Scrippsiella* (Dinophyceae) species in the East China Sea," *Journal of Phycology*, vol. 44, no. 2, pp. 478–494, 2008.

[33] G. J. M. Versteegh, "New Pliocene and Pleistocene calcareous dinoflagellate cysts from southern Italy and Crete," *Review of Palaeobotany and Palynology*, vol. 78, no. 3-4, pp. 353–380, 1993.

[34] K. J. S. Meier and H. Willems, "Calcareous dinoflagellate cysts in surface sediments from the Mediterranean Sea: distribution patterns and influence of main environmental gradients," *Marine Micropaleontology*, vol. 48, no. 3-4, pp. 321–354, 2003.

[35] F. Rubino, S. Moncheva, M. Belmonte, N. Slabakova, and L. Kamburska, "Resting stages produced by plankton in the Black Sea—biodiversity and ecological perspective," *Rapport Commission International pour l'Exploration Scientifique de la Mer Mediterranee*, vol. 39, p. 399, 2010.

Modeling Impacts of Climate Change on Giant Panda Habitat

Melissa Songer,[1] Melanie Delion,[1] Alex Biggs,[1] and Qiongyu Huang[2]

[1] *Conservation Ecology Center, Smithsonian Conservation Biology Institute, National Zoological Park, Front Royal, VA 22630, USA*
[2] *Geography Department, University of Maryland, College Park, MD 20742, USA*

Correspondence should be addressed to Melissa Songer, songerm@si.edu

Academic Editor: A. E. Lugo

Giant pandas (*Ailuropoda melanoleuca*) are one of the most widely recognized endangered species globally. Habitat loss and fragmentation are the main threats, and climate change could significantly impact giant panda survival. We integrated giant panda habitat information with general climate models (GCMs) to predict future geographic distribution and fragmentation of giant panda habitat. Results support a major general prediction of climate change—a shift of habitats towards higher elevation and higher latitudes. Our models predict climate change could reduce giant panda habitat by nearly 60% over 70 years. New areas may become suitable outside the current geographic range but much of these areas is far from the current giant panda range and only 15% fall within the current protected area system. Long-term survival of giant pandas will require the creation of new protected areas that are likely to support suitable habitat even if the climate changes.

1. Introduction

Giant pandas (*Ailuropoda melanoleuca*) are endangered [1] and attract great popular attention, scientific interest, and conservation dollars. The species' historic range encompassed most of southeastern China, northern Myanmar, and northern Vietnam. Climate changes during the late Pleistocene and millennia of agricultural conversion and human settlement have dramatically reduced the geographic distribution of giant pandas and populations are now scattered across six mountain ranges between the Sichuan plain and Tibetan plateau [2–4].

One of the greatest threats to giant panda survival is habitat loss [2, 5, 6]. The species is limited to montane deciduous and coniferous forests with bamboo understories. During the twentieth century, giant panda habitat steadily and rapidly declined [3, 5, 7]. Driving forces of habitat loss are agricultural conversion, and large-scale activities such as road construction, logging, mining, and hydroelectric development. Habitat loss has led to a highly fragmented range; many giant panda populations are small and isolated, resulting in limited gene flow and risks from inbreeding [2, 4, 8, 9].

Climate change may pose a significant threat to giant panda survival. Current climate models estimate a 1.4–5.8 degree Celsius increase in temperature during this century [10–13]. Past and recent changes in climate have been shown to cause range shifts and contractions in plant and animal distributions [14–21]. Whether a species can survive changes in their environment is dependent on various life history characteristics. Characteristics that make a species more likely to be negatively impacted by disturbance include having a limited geographic range, poor ability to disperse, low rates of reproduction, and highly specialized habitat requirements [14, 22–24]. Giant pandas have a narrow range, do not disperse over large distances, produce one cub every 2-3 years, and depend on bamboo for 99% of their diet [25]. These traits suggest they will be highly susceptible to climate change. In addition to the limitations resulting from life history characteristics, species' response is also limited by the spatial configuration of habitat in the landscape. Species may have the capacity to shift as vegetation regimes shift; however, distance or other barriers may limit movement. Given the giant panda's restricted and montane geographic range, climate change may significantly reduce and isolate already fragmented giant panda habitats, decrease gene flow, and thereby substantially increase the species' extinction risk.

Extrapolating known suitable climate envelopes into future climate scenarios is one of the best approaches for predicting effects of climate change on species' geographical distributions [26, 27]. Based on the current giant panda

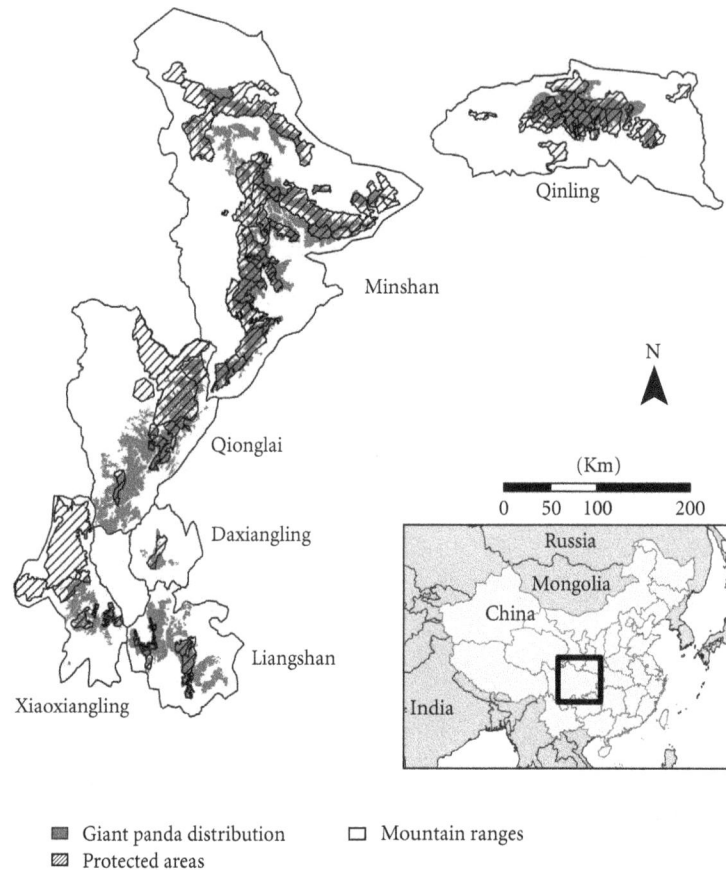

FIGURE 1: Current giant panda distribution, protected areas, and mountain ranges.

distribution and available general climate models (GCMs), we present a range-wide estimate for how climate change may affect giant panda habitats by the year 2080 and assess projected changes in fragmentation and protection levels. These data provide the most recent and informed estimate on how climate change will affect one of the most endangered and charismatic megavertebrates in the world. It also provides useful information as conservation organizations assess how to invest in giant panda conservation in the future.

2. Methods

2.1. Study Area. Our study encompassed six mountain ranges that constitute the extant geographic distribution of giant pandas: the Qinling, Minshan, Qionglai, Xiaoxangling, Daxiangling, and Liangshan (102°00′-108°11′E and 27°53′-33°55′N; Figure 1). Habitat types transition vertically through elevational changes within the giant panda distribution, from subtopical evergreen broad-leafed forest at lower elevations, to evergreen and deciduous broad-leafed forests, to mixed coniferous and deciduous broad-leafed forests, up to subalpine coniferous forests. There is a lot of variation in temperature and precipitation within the giant panda distribution and this, along with variation in soils, hydrology, slope, and aspect, have resulted in diverse plant and tree

species [28]. Baseline data on giant panda distribution is from the most recent national survey for giant pandas. This data is not available to researchers outside China, making direct modeling of giant panda locations impossible [29]. However, habitat associations and models derived from the data have been made available. Our study assesses the effects of climate change on giant pandas indirectly, by measuring how climate change will alter the geographic distribution and extent of giant panda habitat.

The current distribution is primarily above 1,200 m elevation; however, giant pandas were found at elevations as low as 500 m during the last century [30]. As our study area we used the distribution of giant pandas from the national survey as a baseline and extended it to include contiguous areas down to 500 m elevation.

2.2. Climate Change Data. We obtained future climate projections from the WorldClim database [31] at 30′ resolution for the year 2080. We included two general climate models; one described by the Canadian Centre for Climate Modeling and Analysis Coupled Model, version 3 (CGCM3) [32] and one from the Hadley Center for Climate Modeling Coupled Model, version 3 (HadCM3) [33]. Both are commonly used atmosphere-ocean coupled models and data is available for download (http://www.worldclim.org/).

TABLE 1: Bioclim variables and their percent contribution and percent permutation importance reported by Maxent. Variables are in order of highest to lowest permutation importance.

Variable	Description	Permutation importance (%)	Variable contribution (%)
Bio10	Mean temperature of warmest quarter	19.6	36.4
Bio15	Precipitation seasonality (coefficient of variation)	18.0	9.7
Bio7	Temperature Annual Range (Bio5-Bio6)	9.5	5.7
Bio12	Annual Precipitation	8.3	2.9
Bio4	Temperature seasonality (standard deviation *100)	8.0	6.1
Bio6	Min temperature of coldest month	7.6	17.1
Bio3	Temperature change/no change (Bio2/Bio7) (*100)	6.5	4.6
Bio11	Mean temperature of coldest quarter	5.0	1.6
Bio2	Mean diurnal range (mean of monthly (maximum–minimum temperature)	4.6	5.1
Bio17	Precipitation of driest quarter	2.8	6.2
Bio14	Precipitation of driest month	2.8	0.2
Bio18	Precipitation of warmest quarter	1.7	0.7
Bio9	Mean temperature of driest quarter	1.5	0.2
Bio1	Annual mean temperature	1.4	0.1
Slope	Angle of slope	1.2	0.5
Aspect	Direction of slope	0.8	0.8
Bio8	Mean temperature of wettest quarter	0.5	0.1
Bio5	Max temperature of warmest month	0.1	1.9
Bio13	Precipitation of wettest month	0.1	0.2
Bio19	Precipitation of coldest quarter	0.0	0.0
Bio16	Precipitation of wettest quarter	0.0	0.0

We restricted our study to models constrained by the conditions outlined in the A2 scenario of the Special Report on Emissions Scenarios [34]. A1 and A2 families assume more rapid economic development than B1 and B2 families, which also assume more ecologically responsible societies by the year 2100 [35]. The A2 family assumes more heterogeneous future societies with regionally divergent economic growth and more fragmented growth in technological changes while the A1 scenario assumes the world will be more homogeneous with similar standard of living levels and technological progress among various regions. Population is assumed to continually increase in the A2 scenario, but the A1 scenario assumes population will decline after reaching 9 billion. We chose the A2 scenario based on current trends in China where fossil-fuel CO_2 emissions have doubled since 2000 [36], population has doubled since 1960 [37], GDP has grown nearly 40-times since 1960 [38], and other environmental indicators have shown steady declines in recent years [39]. IPCC author Richard Tol has asserted that the A2 family is by far the most realistic [40]. The A2 is a strong scenario and should help us identify patterns and trends in predicted changes to giant panda habitat.

For each climate dataset, bioclimatic parameters from monthly precipitation and minimum and maximum temperatures were interpolated using BIOCLIM [41]. We selected 19 of the 35 BIOCLIM variables which seemed most relevant (Table 1). These bioclimatic parameters are calculated across the entire year to offer a wider range of climatic variables for analysis.

We used Maxent to relate current giant panda distribution to environmental variables and to project future giant panda habitat. Maxent (http://www.cs.princeton.edu/~schapire/maxent/) is a general-purpose species distribution model that can make predictions from incomplete information with a high prediction success, particularly in cases with presence-only data [41, 42]. Maxent estimates the species' distributions by finding the probability distribution of maximum entropy (i.e., the mean of each variable in the projected distribution is close to means of the observed data) subject to the constraints of where data is actually available [43]. For our study we created 1,500 random points inside the study area, which represents approximately 1 point for every $100 \, m^2$. We created 10,000 randomly sampled points in the study area and extracted the 19 BIOCLIM variables for both of the GCMs for the year 2080 to serve as background points for projecting future distributions. Digital elevation models (DEMs) at 90 m resolution from CGIAR-CSI SRTM [44] were used for elevation and to calculate slope and aspect; all three variables were added as model variables. Current climatic conditions of each point were interpolated and projected into the two future climate models. For both models we used 1 for the regularization multiplier, a convergence threshold of 10^{-5} and a maximum of 500 iterations based on the default recommendations.

Model performance was measured using the area under the receiver operating characteristic curve (AUC). AUC ranges from 0.5, which is no better than random, up to 1.0 which represents perfect discriminatory capacity; AUC values over 0.75 are considered useful [45]. Maxent estimates the importance of the variables with percent contribution and permutation importance values. Percent contribution represents how much the variable contributed to the model based on the path selected for a particular run. Permutation importance is determined by changing the predictors' values between presence and background points and observing how that affects the AUC. The permutation importance depends on the final model, not the path used in an individual run and therefore is better for evaluating the importance of a particular variable. Standard errors and confidence intervals for both of the models were calculated in R v2.13.0 [46] using ROCR [47], vcd [48], and boot [49] packages. We assessed how well Maxent could predict the known current giant panda using the same methods, except we used GCMs for the year 2000 instead of 2080. Maxent was 77% accurate in modeling current giant panda distribution.

We imported the Maxent probability distributions for each model into ArcGIS 9.3 (ESRI Inc., Redlands, CA) and converted them to presence/absence (0/1) based on the threshold value that maximizes training sensitivity and plus specificity [50].

2.3. Quantifying Suitable Habitat and Fragmentation. Areas with dense human populations and roads are not suitable for giant pandas and croplands do not provide suitable habitat, therefore we removed croplands, urban areas, and human disturbance buffers. Based on a framework developed by Liu et al. [51], we considered areas within 1410 m of cities and 210 m of roads (city and transportation network data from NIMA [52]) to be unsuitable. We used land cover data from Global Land Cover 2000 [53]. Human development excludes giant pandas from areas <1,200 m in elevation [30], so these areas were also removed from projected suitable habitat. After removing human disturbance and unsuitable land cover we calculated the size of all remaining patches. We then removed all patches <4 km^2 based on the average panda home range size [54]. For each of the three models (current giant panda distribution and two predictions) we calculated presence area within each mountain range and the amount and percent area inside protected areas.

We calculated the number of patches >4 km^2 and the number of these patches >200 km^2 remaining. An area this size would support typical giant panda home ranges (~4 km^2) for about 50 individuals, representing a population minimum needed to deter inbreeding depression [55, 56]. We used FRAGSTATS [57] to measure fragmentation indices within mountain ranges for each model, specifically the number of patches, mean patch size (MPS), and mean nearest neighbor distance (MNN). Nearest neighbor for each patch is the single shortest distance to another patch.

3. Results

3.1. Current Giant Panda Distribution. The current giant panda distribution includes at least 18 large patches (>200 km^2) still intact (Table 2; Figure 1). The Minshan Mountains, supporting more than 40% of the giant panda population, have the largest suitable habitat area and the Daxiangling Mountains have the smallest. Larger mountain ranges with more of the remaining giant panda distribution have higher proportions protected than the smaller mountain ranges. The Qinling, Minshan, and Qionglai Mountain

ranges combined account for nearly 90% (13,500 km^2) of the protected area within the giant panda distribution (Table 2). Percentages range from 25% in the Liangshan Mountains up to 72% protected in the Qinling Mountain range.

Fragmentation indices vary considerably across mountain ranges with an MPS ranging from 73 km^2 (Daxiangling) to 863 km^2 (Minshan) and MNN distances from 0.3 to 5.5 km (mean 1.5 km; Table 3). The Qinling and Minshan Mountains have lower levels of fragmentation, having the largest MPS (509 km^2 and 863 km^2, resp.) and below average MNN distances (0.3 km and 1.2 km). The three southern mountain ranges are highly fragmented with few large patches remaining.

3.2. Projected Impacts of Climate Change. Our models tested well with an AUC of 0.752, standard error of 0.010, and a 95% confidence interval of 0.992, 1.000. The most important variable based on permutation importance was mean temperature of the warmest quarter (19.6%), followed closely by seasonality of precipitation (18%, Table 1). The next variable on the list, dropping 10 percentage points, is annual temperature range (9.5%), followed by annual precipitation (8.3%) and temperature seasonality (8.0%). All other precipitation variables, such as precipitation of the driest quarter, driest month, warmest quarter, wettest month, coldest quarter, and wettest quarter had permutation importance of less than 3.0%. Average temperatures for the year, the wettest quarter, and the driest quarter were also less than 3.0%, along with slope and aspect.

Less than half of the current giant panda distribution is projected to be suitable by 2080 according to both climate models. Current distribution areas projected to be suitable in 2080 can be considered to be remaining habitat, as opposed to current distribution areas projected to be not suitable which can be considered lost habitat. Areas projected to be suitable in 2080 falling outside the current distribution represent potential new habitat areas. The amount and percentage of suitable habitat projected to be lost is nearly 60% for both climate scenarios (Table 2, Figure 2). Projected habitat lost varies between models and among mountain ranges. The far northern range in the Qinling Mountains is projected to fare best; the CGCM3 projects less than 1% loss while the HadCM3 projects a 17% loss in suitable habitat. For all other mountain ranges the models project higher losses of between 60 to 97%. Both models project the three southern ranges would retain little suitable habitat. Predicted losses are accompanied by declines in amount of suitable habitat that is protected. The Qinling Mountains fare the best with approximately 70% of the suitable habitat protected in 2080. Other mountain ranges show only 1–28% of suitable habitat area protected.

Despite the predicted losses within the current giant panda distribution, there is an overall increase in suitable habitat projected outside the current distribution (Tables 2 and 4, Figure 2). Both models project considerable amounts of potential new habitat outside the current distribution of giant pandas—an additional 34,200 km^2 (CGCM3) and 24,300 km^2 area (HadCM3) (Table 4). However, most potential new areas are not contiguous with the current

TABLE 2: Current and predicted suitable habitat for giant panda based on the Canadian Centre for Climate Modeling Analysis (CGCM3) and Hadley Center for Climate Modeling (HadCM3)[1].

Range	Remaining habitat in km²(%)			Lost habitat in km²(%)		Protected habitat in km²(%)			Number of patches (>200 km²)		
	Current	CGCM3	HadCM3	CGCM3	HadCM3	Current	CGCM3	HadCM3	Current	CGCM3	HadCM3
Qinling	4,068	4,063 (99)	3,387 (83)	5 (<1)	681 (17)	2,919 (72)	2,917 (72)	2,517 (62)	2	2	2
Minshan	12,076	2,694 (22)	4,866 (40)	9,383 (77)	7,210 (60)	8,144 (67)	1,989 (16)	3,395 (28)	5	4	7
Qionlai	6,551	4,089 (62)	2,218 (34)	2,462 (38)	4,333 (66)	2,461 (38)	1,869 (28)	1,265 (19)	4	4	2
Xiaoxiangling	1731	48 (3)	136 (8)	1,683 (97)	1,596 (92)	531 (31)	20 (<1)	27 (<1)	3	0	0
Daxiangling	365	37 (10)	39 (11)	328 (90)	326 (89)	123 (34)	22 (6)	20 (5)	1	0	0
Liangshan	3,481	590 (17)	1,369 (40)	2,891 (83)	2,112 (61)	885 (25)	136 (4)	292 (8)	3	1	3
Total	28,273	11,520 (41)	12,015 (43)	16,753 (59)	16,258 (58)	15,062	6,950 (25)	7,516 (27)	18	11	14

[1] Remaining habitat are the areas of the current range predicted to remain suitable and lost habitat are areas not predicted to remain suitable in 2080. Protected habitat estimates are based on the current protected area system and percentages were calculated against the current distribution.

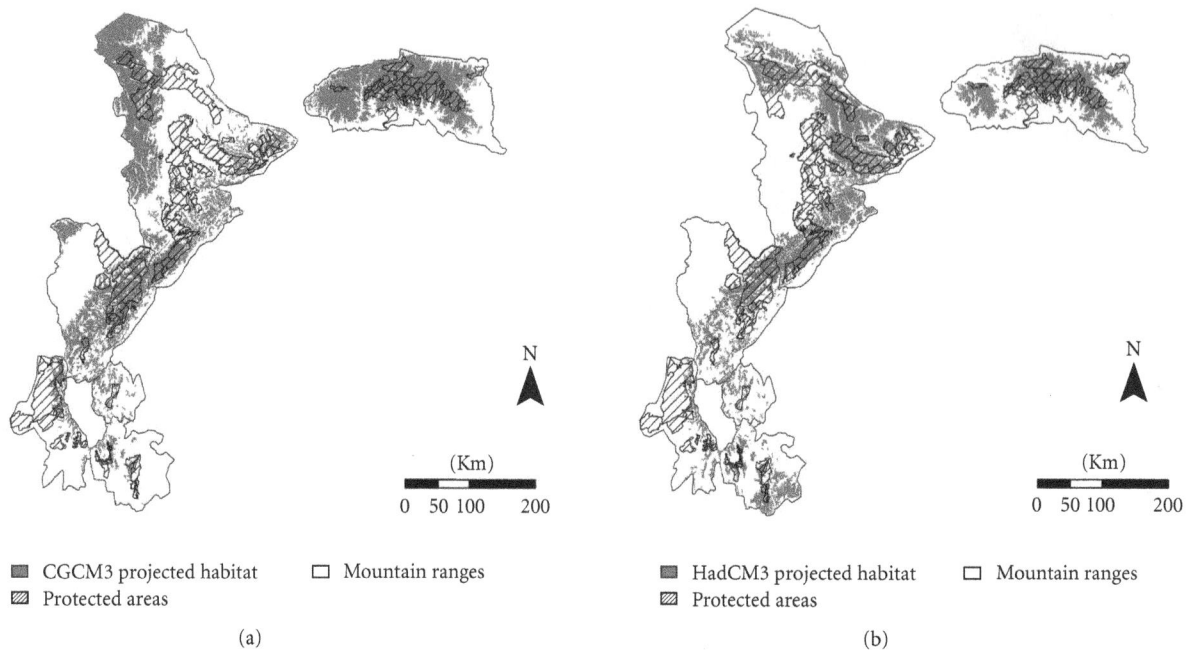

FIGURE 2: Predicted suitable habitat for the year 2080 based on (a) Canadian Centre for Climate Modeling and Analysis Coupled Model, version 3 (CGCM3), (b) Hadley Center for Climate Modeling Coupled Model, version 3 (HadCM3), and protected status based on the existing protected area system.

TABLE 3: Fragmentation metrics for the current and future suitable habitat for giant panda based on the Canadian Centre for Climate Modeling Analysis (CGCM3) and Hadley Center for Climate Modeling (HadCM3).

Range	Mean patch size (km²)			Mean nearest neighbor (km²)			Number of patches		
	Current	CGCM3	HadCM3	Current	CGCM3	HadCM3	Current	CGCM3	HadCM3
Qinling	509	508	338	0.3	0.3	0.6	8	8	10
Minshan	863	46	75	1.2	1.2	1.5	14	55	63
Qionglai	468	5	32	0.8	0.7	1.7	14	88	68
Xiaoxiangling	286	7	11	5.5	3.2	6.5	6	7	12
Daxiangling	73	11	10	2.5	3.8	1.0	5	3	4
Liangshan	387	65	47	0.6	4.2	1.6	9	9	29
Total	505	67	38	1.5	1.2	1.8	56	170	186

giant panda distribution and some patches are as far as 120 km away (average of 22 km for CGCM3 and 13 km for HadCM3). Potential new habitat includes some areas at elevations (>3,500 m) that are not often used by giant pandas today—approximately 10,100 km² (CGCM3) and 4,900 km² (HadCM3) are above 3,500 m (Table 4). Another concern is that much of this potential new habitat would not fall within the existing protected area system; only 12% of the CGCM3 and 14% of the HadCM3 potential new habitat area would be protected under the current protected area system (Table 4, Figure 2).

Substantial portions of both climate projection models were removed due to human disturbance (cropland, urban area, settlements, and roads), illustrating the highly fragmented nature of the remaining giant panda range. Approximately 21,840 km² and 11,690 km² were removed from the initial CGCM3 and HadCM2 models, respectively.

Comparing fragmentation metrics between current giant panda distribution and projected suitable habitat area for 2080 indicates major increases in fragmentation within all mountain ranges. The projected number of patches more than triples in both models (Tables 1 and 2). The MPS overall drops dramatically from 505 km² in the current distribution down to 67 km² (CGCM3) and 38 km² (HadCM3) in 2080 (Table 3). Within the mountain ranges both models project increased isolation between patches, with MNN increasing in all mountain ranges, with three exceptions; the CGCM3 projects a decrease in MNN within the Qionglai and Xiaoxiangling ranges and the HadCM3 projects a decrease in MNN in the Daxiangling range. Both models also show a reduction in large patches (>200 km²), from 18 to 14 or less (Table 2). However, both the Qinling and Minshan Mountains retain large patches, and fewer fragmentation effects are predicted.

TABLE 4: Predicted suitable habitat outside the current giant panda distribution based on the Canadian Centre for Climate Modeling Analysis (CGCM3) and Hadley Center for Climate Modeling (HadCM3).

Range	Potential new habitat in km² (%)		Protected new habitat in km² (%)		New habitat at elevations >3500 m		Spatial overlap in km² (%)	Protected overlap in km² (%)
	CGCM3	HadCM3	CGCM3	HadCM3	CGCM3	HadCM3		
Qinling	10,883 (32)	6,153 (25)	1,597 (29)	1,291 (26)	4 (<1)	4 (<1)	9,269 (41)	3,745 (45)
Minshan	16,907 (49)	12,320 (51)	2,299 (42)	2,500 (51)	6,909 (68)	2,209 (45)	7,109 (31)	2,345 (28)
Qionglai	4,359 (13)	3,003 (12)	1,005 (18)	893 (18)	2,897 (29)	2,106 (43)	4,466 (20)	1,827 (22)
Xiaoxiangling	1,102 (3)	883 (4)	539 (10)	186 (4)	254 (3)	267 (5)	561 (2)	170 (2)
Daxiangling	702 (2)	615 (3)	38 (1)	38 (1)	0 (0)	0 (0)	531 (2)	44 (1)
Liangshan	246 (1)	1,303 (5)	1 (<1)	2 (<1)	73 (1)	288 (6)	634 (3)	112 (1)
Total	34,199	24,278	5,477	4,911	10,136	4,874	22,570	8,245

FIGURE 3: Spatial agreement for predictions of suitable habitat in 2080 based on the Canadian Centre for Climate Modeling and Analysis Coupled Model version 3 and the Hadley Center for Climate Modeling Coupled Model version 3, the current giant panda distribution, and protected status based on the existing protected area system.

The projected suitable habitat for the two models has only 38% spatial agreement (Table 4, Figure 3). The most similar projections are found in the Qinling Mountains (61%) while the lowest are found in the Liangshan Mountains (22%).

4. Discussion

4.1. Climate Change Effects on Geographic Distribution and Extent of Giant Panda Habitat. Our results project major losses within the giant panda's present distribution by 2080, supporting a major general prediction of climate change–the shift of vegetation regimes towards higher altitudes and latitudes. About 90% of the future suitable habitat occurs in the three northern mountain ranges while the southern mountain ranges show a >80% decrease in suitable habitat. The far northern Qinling Mountains fare best, retaining the largest proportion of suitable habitat, showing a significant increase in new suitable habitat, maintaining large MPSs and small MNNs. Generally giant pandas are projected to follow climate effect patterns (i.e., altitudinal and latitudinal shifts) previously observed for many bird and butterfly species [58–60], mammals [61, 62], and a range of other taxa [18]. Many of the previous studies give direct evidence that northward shifts are already happening.

Our results project upward elevational shifts, with more than a quarter of the new presence areas predicted to be above 3,500 m. Evidence of an overall upward shift in optimum elevation due to climate change exists for many plant species in temperate forests of western Europe between 1905

and 2005 [63]. Similar patterns have emerged for some butterflies in the Czech Republic [64] and Spain [65], vegetation in the Alps [66], and some birds in Southeast Asia [67].

Giant pandas may be particularly at risk from shifting vegetation regimes. Not only do they specialize on bamboo, but they may eat only two or three bamboo species, depending on their region, despite the presence of other bamboo species (e.g., [2, 68, 69]). We currently cannot predict impacts of climate change on various bamboo species or whether they will remain at lower elevations, shift upslope, or do both. Here we make the assumption that all future suitable habitat will be usable; however, it is possible that some of the higher elevation areas are unsuitable for vegetation due to rocky cover.

Our results may be a conservative estimate of climate change effects on giant pandas. First our models are based on the assumption that agriculture will not advance above ~1,200 m. This assumption may not hold if warming results in higher elevations becoming suitable for crops. If agricultural use shifts up to higher elevations it will exclude giant pandas from those areas.

Finally, we may be underestimating future forest cover. During the past decade the Chinese government has enacted two major initiatives to protect forest and restore giant panda habitat. These include the Grain-to-Green Program, which compensates farmers for returning steep cropland to forest, and the Natural Forest Conservation Program, which bans harvest of natural forests and provides economic incentives to locals for enforcement against tree harvest [7, 70]. Research in Wolong Nature Reserve suggests that these programs have helped restore habitat [71]. Expansion of suitable areas into higher elevations could also increase connectivity in some places.

4.2. *Adaptation Strategies.* The momentum of climate change already underway will likely result in changes to natural systems, regardless of any current or future efforts to mitigate emissions [72, 73]. Results of our work could help inform adaptation strategies to help ameliorate predicted impacts of climate change on giant pandas and their habitat by identifying areas that are likely to have drastic changes, those that are expected to fare better and are already well protected, and those that are expected to fare well but are not currently protected.

China's government has already made key steps that would top the list for any adaptation strategies developed for giant pandas. In addition to the restoration programs currently underway they have greatly expanded the amount of protected area in recent decades. The number of giant panda protected areas has increased from only 12 in 1980 to 33 by the year 2000 [2, 68], and currently there are more than 59 protected areas [74]. Still, just 60% of giant pandas are currently found within protected areas [29]. Our results show that there is great potential for expanding protected status to areas identified as suitable, as well as for increasing connectivity within ranges to help facilitate movements between small populations.

Both our models predict the Qinling Mountains will maintain most of its habitat and that fragmentation will be minimal even under various climate change scenarios. Suitable habitat is estimated to double or triple in the Qinling Mountains by 2080, though only 15% would be protected and extending protection to new areas could help enhance the Qinling population. In contrast, the Minshan and the Qionglai are expected to lose more than half their current suitable habitat. Both mountain ranges are predicted to have a considerable amount of new areas becoming suitable, but with low levels of protection. Both our models project suitable habitat buffering existing protected areas in the far south and the northwest corner of the Minshan Mountains. In the Qionglai Mountains both our models project a substantial amount of suitable habitat between the large group of protected areas and the lone protected area to in the south to remain suitable in 2080. These areas could be especially important because they already support giant panda populations but are not yet protected, and they could enhance connectivity within the Qionglai range. The Minshan and the Qionglai mountain ranges have the largest and second largest giant panda populations and should be prioritized for increased protection to help stabilize them against the impacts of climate change.

Our results predict a particularly dire situation in the small southern mountain ranges. These three mountain ranges are predicted to be the most fragmented as a result of climate change—no large patches are predicted for the Xiaoxingling and Daxiangling and only a few for the Liangshan. Though habitat is expected to decline in the three larger ranges, a similar number of large patches is predicted to remain. According to the 2000-2001 Third National Survey, the 3 smaller ranges currently support 115 (Liangshan), 32 (Xiaoxiangling), and 29 (Daxiangling) giant pandas. The drastic decline predicted for the southern ranges raises difficult questions which should be considered when developing adaptation strategies. Small populations currently reported from the south are likely the result of the isolation and fragmentation of subpopulations, which may already be causing reduced genetic diversity. While it is important to maintain these small populations for maintenance of the overall genetic diversity of the metapopulation, it may be more effective to prioritize protection of areas that are most likely to remain suitable even with expected changes in climate. Translocation is a tool that may be useful in adaptation strategies if habitat losses in the south become severe enough to warrant moving small populations to areas where they have a better chance for survival. This could help bolster other populations against risk of extinction.

Though our climate models only have a 40% agreement, there is still a large amount of area (22,570 km^2) that both models predict as suitable in 2080 and less than a quarter of this area is currently protected. These areas, particularly those close to current giant panda populations, should be explored via field surveys to assess suitability and should potentially become high priorities for protection. Once key areas are identified they should also be considered for future protection and for new conservation initiatives, such as planning for reintroduction of captive pandas into the wild.

Corridors and increased connectivity also must be prioritized in any adaptation strategy designed to mitigate impacts of climate change on giant pandas. Our model estimates of forest fragmentation, one of the greatest threats to giant pandas, increase dramatically by the year 2080, with reduced patch size, increased isolation, and fewer large patches that can support populations in the long term. Much of the habitat in 2080 is not connected to the current giant panda distribution. Based on radiotelemetry data giant pandas typically move <500 m/day [54] and thus are unlikely to be able to adjust to increased fragmentation at the necessary scale predicted by our models. This is especially critical to giant panda survival since previous research demonstrated that increased connectivity is needed to improve gene flow and maintain genetic diversity. A majority of the remaining populations have 20 pandas or fewer [3, 5, 75], putting them at risk of inbreeding and extirpation [4, 76]. Enhancing their chance for survival will depend on improving connectivity by increasing protection of suitable habitat and establishing corridors to connect isolated populations [4, 68, 76].

4.2.1. *Measuring Influence of Climate Variables.* Based on Maxent's permutation importance values, temperature during the warmest quarter was the most important variable and annual amount of precipitation was high—both have also been shown to influence growth rates of bamboo [77]. Permutation importance values also illustrate the influence of seasonality on the model, with both precipitation and temperature seasonality along with annual temperature range reported in the top five most important variables. Extensive analysis of satellite imagery and the CO_2 records between 1981–1991 has shown that changes in seasonality are associated with changes in air temperatures and provide evidence that the warmer temperatures are promoting increased vegetative growth [78]. Seasonality is a key factor in the growth and distribution of bamboo and will be a key factor in future habitat suitability.

4.2.2. Model Limitations. Though climate envelope models are widely used in climate change research to predict future species distributions [26, 27], there are limitations. Ibáñez et al. [79] point out that this approach effectively treats the realized niche as though it were the fundamental niche. Changes in climate may also affect how a species can disperse, influence reproductive capacity, and disrupt ecosystem functioning as key species move in or out of an area at varying rates [80], while the landscape mosaic and human activities may impede or facilitate migration to different degrees for different species [81]. We expanded the current giant panda distribution layer to include lower elevations to better reflect the fundamental niche of giant pandas, and have incorporated human use as much as possible by removing incompatible human disturbance based on current data. Though we cannot include all potential influences in the model, our results represent the current state-of-the-art in analyzing the impacts of climate change on species distributions and are based on the best data available for giant pandas. We believe that for landscape purposes it is likely a good representation of how climate change will affect the patterns and distribution of giant panda habitat.

4.3. Conclusions. Giant panda habitat and the effectiveness of protecting this habitat will be severely affected by climate change. Using well-established modeling procedures we provide essential guidance for developing adaptation strategies, designing future surveys, and prioritizing protection of giant panda habitat. Our results are consistent with previous studies on climate change effects on montane species. Our research provides compelling evidence to increase protected area development in the northern and central ranges of the current giant panda distribution and for ensuring increased connectivity between currently existing and potential future suitable areas.

Acknowledgments

The authors would like to thank Peter Leimgruber for reviewing and editing many versions of this paper, and for providing valuable comments and suggestions. This paper was funded by the Friends of the National Zoo.

References

[1] J. Schipper, J. S. Chanson, F. Chiozza et al., "The status of the world's land and marine mammals: diversity, threat, and knowledge," *Science*, vol. 322, no. 5899, pp. 225–230, 2008.

[2] G. B. Schaller, J. Hu, W. Wenshi, and J. Zhu, *The Giant Pandas of Wolong*, University of Chicago Press, Chicago, Ill, USA, 1985.

[3] Ministry of Forestry and World Wide Fund for Nature, *National Conservation Management Plan for the Giant Panda and Its Habitat: Sichuan, Shaanxi and Gansu Provinces, the People's Republic of China*, Ministry of Forestry, Beijing and WWF International, Gland, Switzerland, 2006.

[4] S. J. O'Brien, W. Pan, and Z. Lu, "Pandas, people and policy," *Nature*, vol. 369, no. 6477, pp. 179–180, 1994.

[5] J. MacKinnon and R. De Wulf, "Designing protected areas for giant pandas in China," in *Mapping the Diversity of Nature*, R. I. Miller, Ed., Chapman and Hall, London, UK, 1994.

[6] J. Hu, *Research on the Giant Panda*, Shanghai Publishing House of Science and Technology, Shanghai, China, 2001.

[7] W. Xu, Z. Ouyang, A. Viña, H. Zheng, J. Liu, and Y. Xiao, "Designing a conservation plan for protecting the habitat for giant pandas in the Qionglai mountain range, China," *Diversity and Distributions*, vol. 12, no. 5, pp. 610–619, 2006.

[8] R. Lande, "Genetics and demography in biological conservation," *Science*, vol. 241, no. 4872, pp. 1455–1460, 1988.

[9] K. Ralls, K. Brugger, and J. Ballou, "Inbreeding and juvenile mortality in small populations of ungulates," *Science*, vol. 206, no. 4422, pp. 1101–1103, 1979.

[10] S. H. Schneider and T. L. Root, "Climate change," in *Status and Trends of the Nation's Biological Resources*, M. J. Mac, P. A. Ople, C. E. Puckett Haecke, and P. D. Doran, Eds., United States Geological Survey, Reston, Va, USA, 1998.

[11] IPCC, "Climate Change 2001: The Scientific Basis in Contribution of Working Group I to the 3rd Assessment Report of the Intergovernmental Panel on Climate Change," Cambridge University Press, Cambridge, UK, 2001.

[12] J. Reilly, P. H. Stone, C. E. Forest, M. D. Webster, H. D. Jacoby, and R. G. Prinn, "Climate change: uncertainty and climate change assessments," *Science*, vol. 293, no. 5529, pp. 430–433, 2001.

[13] T. M. L. Wigley and S. C. B. Raper, "Interpretation of high projections for global-mean warming," *Science*, vol. 293, no. 5529, pp. 451–454, 2001.

[14] K. A. McDonald and J. H. Brown, "Using montane mammals to model extinctions due to global change," *Conservation Biology*, vol. 6, no. 3, pp. 409–415, 1992.

[15] J. P. McCarty, "Ecological consequences of recent climate change," *Conservation Biology*, vol. 15, no. 2, pp. 320–331, 2001.

[16] A. D. Barnosky, E. A. Hadly, and C. J. Bell, "Mammalian response to global warming on varied temporal scales," *Journal of Mammalogy*, vol. 84, no. 2, pp. 354–368, 2003.

[17] E. P. Lessa, J. A. Cook, and J. L. Patton, "Genetic footprints of demographic expansion in North America, but not Amazonia, during the Late Quaternary," *Proceedings of the National Academy of Sciences of the United States of America*, vol. 100, no. 18, pp. 10331–10334, 2003.

[18] C. Parmesan and G. Yohe, "A globally coherent fingerprint of climate change impacts across natural systems," *Nature*, vol. 421, no. 6918, pp. 37–42, 2003.

[19] T. L. Root, J. T. Price, K. R. Hall, S. H. Schneider, C. Rosenzweig, and J. A. Pounds, "Fingerprints of global warming on wild animals and plants," *Nature*, vol. 421, no. 6918, pp. 57–60, 2003.

[20] R. Guralnick, "The legacy of past climate and landscape change on species' current experienced climate and elevation ranges across latitude: a multispecies study utilizing mammals in western North America," *Global Ecology and Biogeography*, vol. 15, no. 5, pp. 505–518, 2006.

[21] J. J. Lawler, D. White, R. P. Neilson, and A. R. Blaustein, "Predicting climate-induced range shifts: model differences and model reliability," *Global Change Biology*, vol. 12, no. 8, pp. 1568–1584, 2006.

[22] R. L. Peters and J. D. S. Darling, "The greenhouse effect and nature reserves. Global warming would diminish biological diversity by causing extinctions among reserve species," *BioScience*, vol. 35, no. 11, pp. 707–717, 1985.

[23] J. L. Isaac, J. Vanderwal, C. N. Johnson, and S. E. Williams, "Resistance and resilience: quantifying relative extinction risk in a diverse assemblage of Australian tropical rainforest vertebrates," *Diversity and Distributions*, vol. 15, no. 2, pp. 280–288, 2009.

[24] IUCN, *Species Susceptibility to Climate Change Impacts*, IUCN, World Conservation Union, Gland, Switzerland, 2008.

[25] D. G. Reid, A. H. Taylor, Hu Jinchu, and Qin Zisheng, "Environmental influences on bamboo Bashania fangiana growth and implications for giant panada conservation," *Journal of Applied Ecology*, vol. 28, no. 3, pp. 855–868, 1991.

[26] B. Huntley, R. E. Green, Y. C. Collingham et al., "The performance of models relating species geographical distributions to climate is independent of trophic level," *Ecology Letters*, vol. 7, no. 5, pp. 417–426, 2004.

[27] R. J. Hijmans and C. H. Graham, "The ability of climate envelope models to predict the effect of climate change on species distributions," *Global Change Biology*, vol. 12, no. 12, pp. 2272–2281, 2006.

[28] C. Wang, *The Forests of China*, Harvard University Press, Cambridge, Mass, USA, 1961.

[29] State Forestry Administration of China, *The Third National Survey Report on Giant Panda in China*, Science Press, Beijing, China, 2006.

[30] W. Pan, Z. Gao, and Z. Lü, *The Giant Panda's Natural Refuge in the Qinling Mountains*, Peking University Press, Beijing, China, 1988.

[31] R. J. Hijmans, S. E. Cameron, J. L. Parra, P. G. Jones, and A. Jarvis, "Very high resolution interpolated climate surfaces for global land areas," *International Journal of Climatology*, vol. 25, no. 15, pp. 1965–1978, 2005.

[32] G. M. Flato, G. J. Boer, W. G. Lee et al., "The Canadian centre for climate modelling and analysis global coupled model and its climate," *Climate Dynamics*, vol. 16, no. 6, pp. 451–467, 2000.

[33] C. Gordon, C. Cooper, C. A. Senior et al., "The simulation of SST, sea ice extents and ocean heat transports in a version of the Hadley Centre coupled model without flux adjustments," *Climate Dynamics*, vol. 16, no. 2-3, pp. 147–168, 2000.

[34] Intergovernmental Panel on Climate Change (IPCC), *Special Report on Emissions Scenarios*, Edited by N. Narkicenovic and R. Swart, Cambridge University Press, Cambridge, UK, 2000.

[35] IPCC, "Summary for policymakers," in *Climate Change 2007: The Physical Science Basis. Contribution of Working Group I to the Fourth Assessment Report of the Intergovernmental Panel on Climate Change*, S. Solomon, D. Qin, M. Manning et al., Eds., Cambridge University Press, Cambridge, UK, 2007.

[36] T. A. Boden, G. Marland, and R. J. Andres, *Global, Regional, and National Fossil-Fuel CO2 Emissions*, Carbon Dioxide Information Analysis Center, Oak Ridge National Laboratory, U.S. Department of Energy, Oak Ridge, Tenn, USA, 2011.

[37] United Nations, Department of Economic and Social Affairs, Population Division, *World Fertility Report 2009, New York, NY, USA, 2011*.

[38] The World Bank, *Global Purchasing Power parities and Real Expenditures: 2005 International Comparison Program*, World Bank, Washington, DC, USA, 2008.

[39] Yale Center for Environmental Policy (YCELP) and Center for International Earth Science Information Network, *2010 Environmental Performance Index*, Socioeconomic Data and Applications Center (SEDAC), Columbia University, New York, NY, USA, http://sedac.ciesin.columbia.edu/es/epi/downloads.html.

[40] R. Tol, "Memorandum by Defra/HM treasury," in *The Economics of Climate Change*, Second Report of the 2005-2006 session, UK Parliament House of Lords Economics Affairs Select Committee, 2006.

[41] H. Nix, "A biogeographic analysis of Australian elaphid snakes," in *Atlas of Elapid Snakes of Australia*, Australian Government Publishing Service, 1986.

[42] J. Elith, C. H. Graham, R. P. Anderson et al., "Novel methods improve prediction of species' distributions from occurrence data," *Ecography*, vol. 29, no. 2, pp. 129–151, 2006.

[43] S. J. Phillips, R. P. Anderson, and R. E. Schapire, "Maximum entropy modeling of species geographic distributions," *Ecological Modelling*, vol. 190, no. 3-4, pp. 231–259, 2006.

[44] A. Jarvis, H. Reuter, E. Nelson, and E. Guevara, *Hole-Filled Seamless SRTM Data V3*, International Centre for Tropical Agriculture (CIAT), 2006.

[45] J. Elith, "Quantitative methods for modeling species habitat: comparative performance and an application to Australian plants," in *Quantitative Methods for Conservation Biology*, S. Ferson and M. Burgman, Eds., pp. 39–58, Springer, 2002.

[46] R Development Core Team, *R: A Language and Environment for Statistical Computing*, R Foundation for Statistical Computing, Vienna, Austria, 2010.

[47] T. Sing, O. Sander, N. Beerenwinkel, and T. Lengauer, *ROCR: Visualizing the performance of scoring classifiers*, R package version 1.0-4, 2009, http://CRAN.R-project.org/package=ROCR.

[48] D. Meyer, A. Zeileis, and K. Hornik, *vcd: Visualizing Categorical Data*, R package version 1.2-9, 2010.

[49] A. Canty and B. Ripley, *Boot: Bootstrap R (S-Plus) Functions*, R package version 1.2-42, 2010.

[50] C. Liu, P. M. Berry, T. P. Dawson, and R. G. Pearson, "Selecting thresholds of occurrence in the prediction of species distributions," *Ecography*, vol. 28, no. 3, pp. 385–393, 2005.

[51] J. Liu, Z. Ouyang, W. W. Taylor, R. Groop, Y. Tan, and H. Zhang, "A framework for evaluating the effects of human factors on wildlife habitat: the case of giant pandas," *Conservation Biology*, vol. 13, no. 6, pp. 1360–1370, 1999.

[52] National Imagery and Mapping Agency (NIMA), Military Specification MIL-V-89033 Vector Smart Map (VMap) Level 1, 1995.

[53] European Commission Joint Research Centre, *Global Land Cover 2000 database*, 2003.

[54] X. Liu, A. K. Skidmore, T. Wang, Y. Yong, and H. H. T. Prins, "Giant panda movements in Foping Nature Reserve, China," *Journal of Wildlife Management*, vol. 66, no. 4, pp. 1179–1188, 2002.

[55] I. R. Franklin, "Evolutionary change in small populations," in *Conservation Biology: An Evolutionary-Ecological Perspective*, M. E. Soule and B. A. Wilcox, Eds., Sinauer, Sunderland, Mass, USA, 1980.

[56] M. E. Soule, "Thresholds for survival: maintaining fitness and evolutionary potential," in *Conservation Biology: An Evolutionary-Ecological Perspective*, M. E. Soule and B. A. Wilcox, Eds., Sinauer, Sunderland, Mass, USA, 1980.

[57] K. McGarigal, S. A. Cushman, M. C. Neel, and E. Ene, *FRAGSTATS: Spatial Pattern Analysis Program for Categorical Maps*, Computer Software Program, University of Massachusetts, Amherst, Mass, USA, 2002.

[58] C. Parmesan, "Climate and species' range," *Nature*, vol. 382, no. 6594, pp. 765–766, 1996.

[59] C. D. Thomas, A. Cameron, R. E. Green et al., "Extinction risk from climate change," *Nature*, vol. 427, no. 6970, pp. 145–148, 2004.

[60] G. Grabherr, M. Gottfried, and H. Paull, "Climate effects on mountain plants," *Nature*, vol. 369, no. 6480, p. 448, 1994.

[61] P. Hersteinsson and D. W. Macdonald, "Interspecific competition and the geographical distribution of red and Arctic foxes Vulpes vulpes and Alopex lagopus," *Oikos*, vol. 64, no. 3, pp. 505–515, 1992.

[62] S. Payette, "Recent porcupine expansion at tree line: a dendro-ecological analysis," *Canadian Journal of Zoology*, vol. 65, no. 3, pp. 551–557, 1987.

[63] J. Lenoir, J. C. Gégout, P. A. Marquet, P. De Ruffray, and H. Brisse, "A significant upward shift in plant species optimum elevation during the 20th century," *Science*, vol. 320, no. 5884, pp. 1768–1771, 2008.

[64] M. Konvicka, M. Maradova, J. Benes, Z. Fric, and P. Kepka, "Uphill shifts in distribution of butterflies in the Czech Republic: effects of changing climate detected on a regional scale," *Global Ecology and Biogeography*, vol. 12, no. 5, pp. 403–410, 2003.

[65] R. J. Wilson, D. Gutiérrez, J. Gutiérrez, D. Martínez, R. Agudo, and V. J. Monserrat, "Changes to the elevational limits and extent of species ranges associated with climate change," *Ecology Letters*, vol. 8, no. 11, pp. 1138–1146, 2005.

[66] H. M. Pauli, M. Gottfried, and G. Grabherr, "Effects of climate change on the alpine and nival vegetation of the Alps," *Journal of Mountain Ecology*, vol. 7, pp. 9–12, 2003.

[67] K. S. H. Peh, "Potential effects of climate change on elevational distributions of tropical birds in Southeast Asia," *Condor*, vol. 109, no. 2, pp. 437–441, 2007.

[68] C. J. Loucks, Z. Lu, E. Dinerstein et al., "Giant pandas in a changing landscape," *Science*, vol. 294, no. 5546, p. 1465, 2001.

[69] W. Zhang, Y. Hu, B. Chen et al., "Evaluation of habitat fragmentation of giant panda (*Ailuropoda melanoleuca*) on the north slopes of Daxiangling Mountains, Sichuan province, China," *Animal Biology*, vol. 57, no. 4, pp. 485–500, 2007.

[70] P. Zhang, G. Shao, G. Zhao et al., "China's forest policy for the 21st century," *Science*, vol. 288, no. 5474, pp. 2135–2136, 2000.

[71] A. Viña, S. Bearer, X. Chen et al., "Temporal changes in giant panda habitat connectivity across boundaries of Wolong Nature Reserve, China," *Ecological Applications*, vol. 17, no. 4, pp. 1019–1030, 2007.

[72] J. R. Mawdsley, R. O'Malley, and D. S. Ojima, "A review of climate-change adaptation strategies for wildlife management and biodiversity conservation," *Conservation Biology*, vol. 23, no. 5, pp. 1080–1089, 2009.

[73] N. E. Heller and E. S. Zavaleta, "Biodiversity management in the face of climate change: a review of 22 years of recommendations," *Biological Conservation*, vol. 142, no. 1, pp. 14–32, 2009.

[74] J. Ran, B. Du, and B. Yue, "Conservation of the endangered giant panda ailuropoda melanoleuca in China: successes and challenges," *Oryx*, vol. 43, no. 2, pp. 176–178, 2009.

[75] G. B. Schaller, *The Last Panda*, University Of Chicago Press, Chicago, Ill, USA, 1993.

[76] S. J. O'brien and J. A. Knight, "The future of the giant panda," *Nature*, vol. 325, no. 6107, pp. 758–759, 1987.

[77] N. Rao, X. P. Zhang, and S. L. Zhu, *Selected Papers on Recent Bamboo Research in China*, Bamboo Information Centre, Chinese Academy of Forestry, 1991.

[78] R. B. Myneni, C. D. Keeling, C. J. Tucker, G. Asrar, and R. R. Nemani, "Increased plant growth in the northern high latitudes from 1981 to 1991," *Nature*, vol. 386, no. 6626, pp. 698–702, 1997.

[79] I. Ibáñez, J. S. Clark, M. C. Dietze et al., "Predicting biodiversity change: outside the climate envelope, beyond the species-area curve," *Ecology*, vol. 87, no. 8, pp. 1896–1906, 2006.

[80] M. Loreau, S. Naeem, P. Inchausti et al., "Ecology: biodiversity and ecosystem functioning: current knowledge and future challenges," *Science*, vol. 294, no. 5543, pp. 804–808, 2001.

[81] Y. C. Collingham and B. Huntley, "Impacts of habitat fragmentation and patch size upon migration rates," *Ecological Applications*, vol. 10, no. 1, pp. 131–144, 2000.

3

A GIS Framework for Fish Habitat Prediction at the River Basin Scale

Marcia S. Meixler[1] and Mark B. Bain[2]

[1] Department of Ecology, Evolution and Natural Resources, Rutgers University, 14 College Farm Road, New Brunswick,
 NJ 08901, USA
[2] Department of Natural Resources, Cornell University, Ithaca, NY 14850, USA

Correspondence should be addressed to Marcia S. Meixler, meixler@aesop.rutgers.edu

Academic Editor: Bruce Leopold

We present a geographic information system (GIS) framework to classify stream habitats and provide fish distribution predictions comprehensively at the landscape scale. Stream segments were classified into one of eighteen habitat types using three landscape attributes: stream size (three categories), stream quality (three categories), and water quality (two categories). An extensive literature search was undertaken to classify fish species into the same eighteen habitat types based on preferences for the three landscape attributes. We tested our framework in 39 sites throughout the upper Allegheny River basin in western New York. No difference was detected between observed and predicted numbers of fish species among stream habitats. Further, field collected bankfull width measurements, stream quality ratings, and water quality sampling results were largely consistent with predicted values. The habitat type expected to have the greatest fish species richness was large streams or small rivers with intact stream quality and suitable water quality. Our framework is rapidly applied, comprehensive, inexpensive, and built on widely available data thereby offering an efficient alternative to traditional field-based efforts for regional habitat classification and fish distribution prediction.

1. Introduction

Declines in biodiversity [1, 2] driven by climate change [3], overexploitation, water pollution, flow modification, habitat degradation, and invasion by exotic species [1] have prompted efforts aimed at development of laws and policies for sound ecosystem planning and management [4, 5]. Conservation of areas high in species richness is often in conflict with resource extraction and land development [6]. It is critical that species-rich areas be protected because such areas tend to support a high number of rare species [7], and proper management of these regions optimizes resources for conservation [8]. In addition, ecosystems high in diversity have been found to promote increased productivity [9], resource utilization [10], and resistance to disturbance [11].

Much effort has thus been dedicated to develop methods to manage regions for conservation of biodiversity [12–14]. It is clear that basic habitat information is needed to make informed conservation decisions; however, comprehensive field sampling over large study areas can be too costly in time and labor [15]. Thus, geographic information system (GIS) models that can synthesize multiple landscape dimensions have become particularly valuable in regions where biological surveys have not been completed or are difficult to perform (e.g., [5, 16]).

GIS models have become quite common for habitat prediction in landscape scale conservation planning [17–20], biodiversity conservation planning [21], and spatial pattern evaluation over large regions [15, 16, 22]. However, advancements in the use of GIS models to remotely predict biotic communities have primarily been confined to terrestrial environments (e.g., [8, 23]). Far less attention has been paid to the development of landscape models to predict aquatic communities [24]. Most existing aquatic habitat classifications are hierarchical (e.g., [25–29]), have extensive data requirements (e.g., [30, 31]), or are based on only a single landscape attribute (e.g., [32–34]). Thus, although many landscape attributes can be combined in

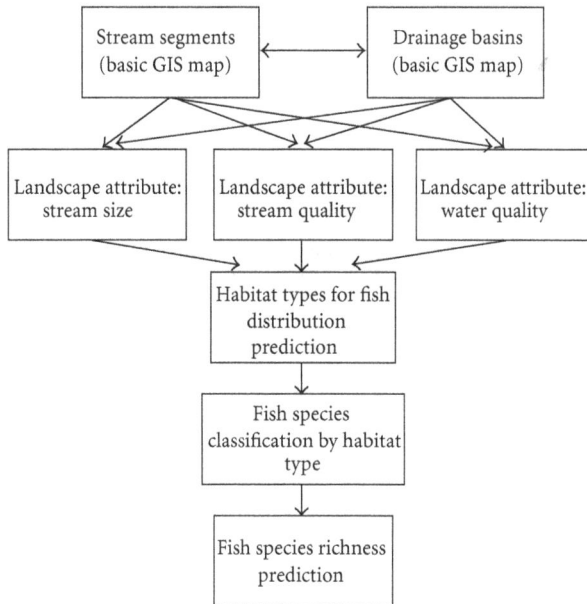

FIGURE 1: GIS framework for stream habitat classification and fish distribution prediction.

a GIS for conservation planning, few complete frameworks are available to classify aquatic habitats and predict fish species presence.

We hypothesize and test the assertion that a framework can be built from fundamental principles to classify stream habitats and provide fish distribution predictions comprehensively at the basin scale for regional application. Our framework is rapidly applied, regional in scale, inexpensive, and built on widely available data thereby offering an efficient alternative to traditional field-based efforts for regional habitat classification and fish distribution prediction.

2. Methods

GIS AML (Arc Macro Language) scripts were used to assess landscape attributes from digital maps and classify stream segments into habitat types. The automated nature of this GIS classification system allowed efficient assessment of stream habitats across large regions. The first step in our habitat classification procedure was to define stream segments as the section of stream or river from tributary confluence to tributary confluence on United States Environmental Protection Agency (US EPA) River Reach File Version 3.0 maps at the 1 : 100,000 scale. Next we classified each stream segment as one of eighteen habitat types using three landscape attributes: stream size, stream quality, and water quality (Figure 1). These attributes were chosen for their influence on fish species composition, availability of data, and ability to link to species biology.

Our methods for classifying streams by size were not meant to be an analysis of biological associations with stream size, but a quantification of the judgments used by ichthyologists when they qualitatively describe habitat for fish species (e.g., [35]). Stream segments with drainage areas less than $100 \, km^2$ were defined as "small streams" (Table 1). This class of largely wadable waters includes all first and second order streams and the lower 68% (<1 standard deviation above mean drainage area) of thirrd order streams. Small streams were predicted to have channels that average no more than 20 m wide and 1 m deep in moderate flow periods. We judged that streams of this size would be classified by ichthyologists as small streams in species biology descriptions. Stream segments with drainage areas from $100 \, km^2$ to less than $3,000 \, km^2$ were defined as "large streams or small rivers" or mid-sized flow waters. These waters are commonly shallow enough for light to reach most of the substrate. Large streams or small rivers were predicted to have channels that average 20–60 m wide with water depths averaging 1–3 m in moderate flow periods. We judged that streams of this nature would be classified by field biologists as large streams, small rivers, or mid-sized flowing waters in species biology descriptions. The final stream size category was "large rivers," which includes stream segments with drainage areas greater than $3,000 \, km^2$ and describes waters that are mostly navigable by motor boats. This size criterion was selected after reviewing drainage areas at USGS gauge sites on streams regarded as large rivers in New York, and the drainage areas for a wide range of stream and river sites described by Barnes [36]. Equations in Dunne and Leopold [37] were used to relate channel size to drainage area.

Fish species vary greatly in their need for microhabitat conditions in streams and rivers. The level of habitat specificity is described for most fish species in ichthyological reference books (e.g., [35]) and microhabitat studies. However, it was not possible to determine microhabitats available for all stream segments in a watershed. As a surrogate, we assumed that the natural range of microhabitat diversity would occur in segments that experience natural channel erosion and deposition processes. Where human land uses impinged on the stream, we assumed that channel control measures and modifications were likely and that these hydromorphological pressures disrupted normal fluvial geomorphology processes. Thus, quality for each stream segment was classified using a GIS script in AML to generate areas and percentages of landcover (EROS Data Center 1991–1993), and total length of roads (New York Department of Environmental Conservation 1993) and railroads (New York Department of Environmental Conservation 1993) in a 30 m buffer [38, 39] surrounding each stream segment. Stream segments were then classified into stream quality categories of "intact," "modified," or "highly altered" based on the composition of landcover and total length of roads and railroads in the 30 m buffer. Stream segments with no urban or agricultural lands, no railroad tracks, and no roads were classified as "intact." Stream segments with 0–5% urban or 0–40% agricultural lands [40, 41] or summed road or railroad lengths less than half of the stream length [42] were classified as "modified." Stream segments with greater than 5% urban lands, greater than 40% agricultural lands, or summed road or railroad lengths greater than or equal to half of the stream length were classified as "highly altered" (Table 1; [40–43]).

TABLE 1: Landscape attribute classification criteria.

Landscape attribute	Classification metrics	Criteria		
Stream size	Drainage area (km^2)	<100	100–3000	>3000
		Small streams	Large streams/small rivers	Large rivers
Stream quality	Anthropogenic pressure from land use, roads, and railroads (%)	No urban areas, no agricultural areas, no roads, and no railroads within 30 m buffer	0–5% urban areas or 0–40% agricultural areas or summed roads or railroads less than half the length of stream within 30 m buffer	>5% urban areas or >40% agricultural areas or summed roads or railroads ≥ half the length of stream within 30 m buffer
		Intact	Modified	Highly altered
Water quality	Total N, total P, suspended sediment	Stream segment is above EPA criteria thresholds for all three pollutants	Stream segment is below EPA criteria threshold for any of the three pollutants	
		Suitable water quality	Degraded water quality	

Nonpoint source (NPS) pollution, caused by agricultural and urban land use, is the primary source of stream impairment in the United States, and elevated sedimentation is the principal pollutant causing stream degradation [44]. Each of these effects, in turn, can threaten fish populations in aquatic systems [45–47]. In our model, water quality was classified using an adaptation of a GIS nonpoint source runoff model originally developed by Adamus and Bergman [48]. We used inputs of landcover (EROS Data Center 1991–1993), soils (STATSGO 1994), average annual rainfall (Northeast Regional Climate Center 1961–1990), runoff coefficients, and pollutant concentrations to determine annual pollutant loading of total phosphorous (TP), total nitrogen (TN), and suspended sediment (SS) to each stream segment from its drainage basin. We adapted the model [49] to compare predicted concentrations to the allowable US EPA pollutant criteria thresholds for the study area (ecoregion 11) for TP and TN, which are 0.01 and 0.31 mg/L, respectively [50], and the strictest SS 30-day average in warm water streams (90 mg/L; 51). A stream segment was classified as having acceptable water quality for each pollutant if its estimate was below the pollution criteria threshold; otherwise, the stream segment was considered substandard. If all three pollutants were considered within acceptable levels for a single stream segment, the reach was classified as having "suitable water quality." Otherwise, the stream segment was classified as having "degraded water quality" (Table 1).

The eighteen habitat types were determined based on combinations of stream size (three categories), stream quality (three categories), and water quality (two categories). Fish species were then classified into one of these eighteen habitat types based on their associations with the same three landscape attributes (Figure 1). An extensive literature search was undertaken to classify each of the 114 fish species in the upper Allegheny River basin into the eighteen habitat types (Table 2). Classifications were achieved by researching preferences for stream size and tolerances for stream quality degradation and water quality degradation for each species [35, 51–56]. A single fish species can be classified into several habitat types. Once species were classified, totals for each habitat type were tallied and the number of fish species was predicted for each stream segment in the study area (Figure 2).

Our model was developed and tested in the upper Allegheny River basin in western New York State, USA. The upper Allegheny River basin comprises approximately 4,870 km^2 (488 stream segments) north of the Pennsylvania-New York state line (Figure 3) in Chautauqua, Cattaraugus, and Allegany counties. A variety of land uses are present in the region including agricultural farming (crop and dairy 28%) and residential/urban development (1.5%). Primary and secondary growth forest (67%) and wetlands/lakes (3.5%) comprise the remainder of the land in the region.

A survey of 39 sites in the upper Allegheny River basin was completed between late May and mid-August 1998. Sites were originally chosen using stratified random sampling to maintain an equal number of sites in each habitat type. However, several sites chosen randomly were inaccessible, located in dense wetlands or dry. Such sites were replaced with suitable locations elsewhere but of the same habitat type.

In sites accessible with heavy equipment, a distance ten times the average wetted width was electro-fished once using a Honda EX1000 generator and 15 Amp Coffelt VVP-2C transformer at approximately 300 volts. Fish were identified, enumerated, and a representative proportion were measured before release. Streams measuring in excess of 10 m in width were electro-fished in intervals of ten minutes until no new species of fish were collected over a period of three intervals. Fish collections from streams inaccessible from the road were dropped from the analysis because differences in gear proved too great to allow aggregation with the rest of the data.

Bankfull width measurements were taken in riffle, pool, and run sections, where possible, at each of the sites using

(a)

■ Intact
■ Modified
■ Highly altered

(b)

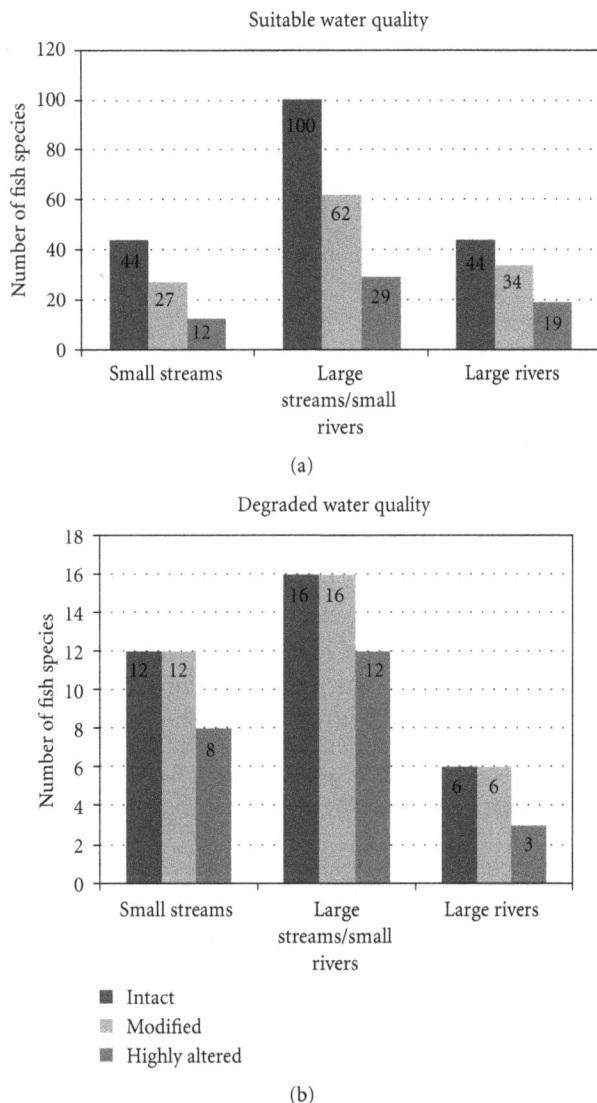

FIGURE 2: Predicted number of fish species by habitat type.

a measuring tape. To compare with stream size predictions from the model, we converted observed bankfull width measurements to drainage area estimates using relations in Dunne and Leopold [37] for the eastern United States.

Physical quality of the stream channel was assessed using a rapid bioassessment protocol [57] adapted for the upper Allegheny River basin. Questions characterizing stream quality addressed existence of retention devices, channel structure, channel sediments, stream-bank structure, bank undercutting, stony substrate, stream bottom, and riffle/pool spacing. Each indicator was given a numerical score and scores were summed to provide an overall rating of stream quality for each site. Numerical ratings were then converted to the categories "intact," "modified," and "highly altered" using cutoffs provided by Petersen [57] to match the categories used by our model. All questions were answered by the same observer throughout the study to maintain uniformity of responses.

TP, TN, and SS measurements were taken at the downstream end of each site in riffle, pool, and run areas, where applicable, between 27–30 July 1998, except for one site which was dry. This time period was chosen to take advantage of conditions when the nitrogen content was at its lowest point and water was the clearest. Three water samples of 250 mL were obtained for suspended sediment measurements at each of the sites and stored in a cooler with ice. After the field day was completed, samples were pumped through preweighed filters (cellulose nitrate filter membranes; 45 microns), dried in an oven at 103–105°C for one hour then in a dessicator for 24 hours, after which the filters were weighed again. Three additional water samples of 100 mL were taken from each of the sites for total dissolved nitrogen and total dissolved phosphorous measurements. These were stored in a freezer until processing could be completed at a lab.

Observed field data were tested against model predictions for number of fish species and all landscape attributes. Observed and predicted number of fish species were assessed using correlation and the paired t-test. Drainage areas, predicted based on spatial data and calculated based on measured bankfull widths, were compared using correlation. A sign test was used to compute the probability of obtaining stream quality and water quality results by chance, and these results were used to judge confidence in our findings. Data were analyzed using Minitab statistical software. Statistical significance was tested at $\alpha = 0.05$.

3. Results

No difference was detected between observed and predicted number of fish species in a paired t-test ($T = -1.085$, $P = 0.285$), and the two ranked datasets were weakly correlated ($r = 0.39$, $P = 0.018$) but significant. The habitat type expected to have the greatest fish species richness was clearly large streams or small rivers with intact stream quality and water quality suitable for life support. These rare (<1%) stream segments averaged 0.6 km long, considerably shorter than stream segments experiencing some form of degradation (2.7 km).

Based on the number of fish species predicted in each habitat type (Figure 2), we would expect the most significant decline in number of fish species to follow degradation in water and physical stream quality. On average, 56% of species were expected to disappear with degradations in water quality whereas an average loss of 37% of the species was related to radical changes in stream quality from intact to highly altered channels. No loss in species was expected with moderate physical stream degradation (intact to modified) once water quality has been degraded. An increase in species numbers by 28% was expected from small streams to large streams or small rivers, whereas further increases in stream size to large rivers resulted in a predicted reduction of 40% of the species.

Observed drainage areas, converted from field collected bankfull width measurements using Dunne and Leopold [37], were compared to predicted drainage areas from the model and were found to be strongly correlated

Predicted fish species

No data
1–19
20–39
40–59
60–79
80–99
100+
Field data collection sites

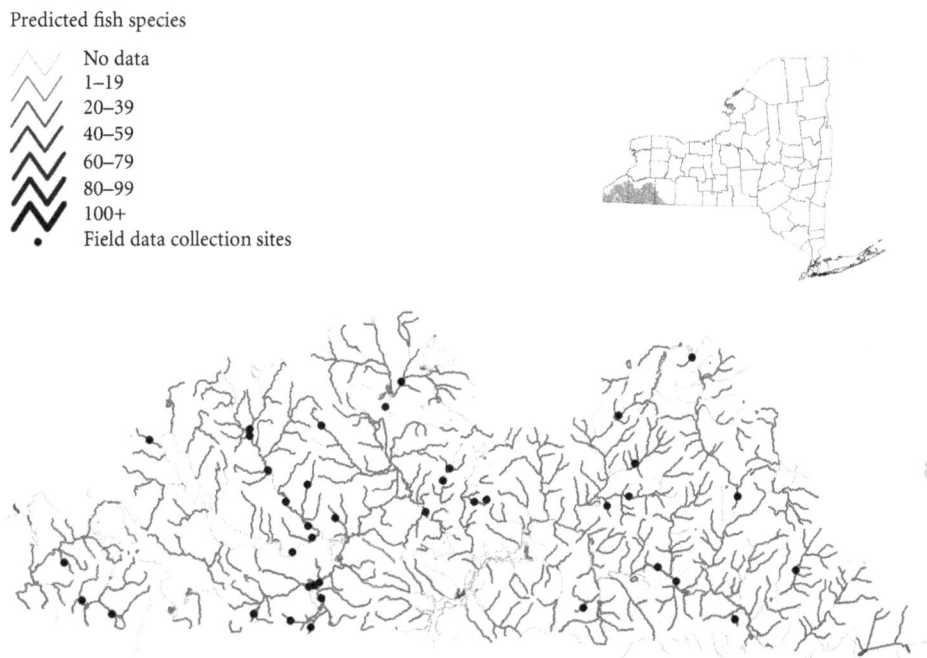

FIGURE 3: Predicted number of fish species and field data collection sites (black dots) in the upper Allegheny River basin in western New York.

TABLE 2: Stream size, stream quality, and water quality preferences and tolerances for fish species in the upper Allegheny River basin. Size: S: small, M: midsize, L: large. Stream quality and water quality tolerances: I: intolerant, M: moderately tolerant, T: tolerant.

Common name	Species name	Size	Stream quality tolerance	Water quality tolerance
American eel	*Anguilla rostrata*	SML	T	I
American brook lamprey	*Lampetra appendix*	SM	M	I
Banded darter	*Etheostoma zonale*	M	I	I
Banded killifish	*Fundulus diaphanus*	ML	M	T
Bigeye chub	*Hybopsis amblops*	ML	I	I
Bigmouth shiner	*Hybopsis dorsalis*	SM	T	I
Black crappie	*Pomoxis nigromaculatus*	LM	M	I
Black red horse	*Moxostoma duquesnii*	M	I	I
Blackchin shiner	*Notropis heterodon*	SM	M	I
Blacknose dace	*Rhinichthys atratulus*	S	M	T
Blacknose shiner	*Notropis heterolepis*	S	I	I
Blackside darter	*Percina maculata*	M	T	I
Bluebreast darter	*Etheostoma camurum*	M	M	I
Bluegill	*Lepomis macrochirus*	M	M	I
Bluntnose minnow	*Pimephales notatus*	M	T	T
Brindled madtom	*Noturus miurus*	LM	M	I
Brook silverside	*Labidesthes sicculus*	ML	I	I
Brook stickleback	*Culaea inconstans*	S	M	I
Brook trout	*Salvelinus fontinalis*	S	I	I
Brown bullhead	*Ameiurus nebulosus*	ML	T	T
Brown trout	*Salmo trutta*	SM	I	I

TABLE 2: Continued.

Common name	Species name	Size	Stream quality tolerance	Water quality tolerance
Burbot	*Lota Iota*	LM	T	I
Central mudminnow	*Umbra limi*	SM	T	T
Central stoneroller	*Campostoma anomalum*	SM	M	I
Chain pickerel	*Esox niger*	M	M	I
Channel catfish	*Ictalurus punctatus*	LM	T	I
Channel darter	*Percina copelandi*	LM	I	I
Common carp	*Cyprinus carpio*	LM	T	T
Common shiner	*Luxilus cornutus*	SM	I	I
Creek chub	*Semotilus atromaculatus*	SM	T	T
Creek chubsucker	*Erimyzon oblongus*	SM	T	I
Cutlips minnow	*Exoglossum maxillingua*	SM	I	I
Eastern mudminnow	*Umbra pygmaea*	SM	T	T
Eastern sand darter	*Ammocrypta pellucida*	M	I	I
Eastern silvery minnow	*Hybognathus regius*	ML	T	I
Emerald shiner	*Notropis atherinoides*	LM	T	I
Fantail darter	*Etheostoma flabellare*	SM	M	I
Fathead minnow	*Pimephales promelas*	SM	T	T
Flathead catfish	*Pylodictis olivaris*	LM	T	I
Freshwater drum	*Aplodinotus grunniens*	L	T	I
Gilt darter	*Percina evides*	M	I	I
Gizzard shad	*Dorosoma cepedianum*	ML	T	I
Golden red horse	*Moxostoma erythrurum*	ML	M	I
Golden shiner	*Notemigonus crysoleucas*	LM	T	T
Grass pickerel	*Esox americanus*	S	M	I
Gravel chub	*Erimystax x-punctatus*	ML	M	I
Green sunfish	*Lepomis cyanellus*	MS	T	T
Greenside darter	*Etheostoma blennioides*	M	I	I
Highfin carpsucker	*Carpiodes velifer*	LM	I	I
Horneyhead chub	*Nocomis biguttatus*	MS	I	I
Iowa darter	*Etheostoma exile*	SM	I	I
Johnny darter	*Etheostoma nigrum*	M	M	I
Largemouth bass	*Micropterus salmoides*	L	T	I
Logperch	*Percina caprodes*	M	I	I
Longear sunfish	*Lepomis megalotis*	M	I	I
Longhead darter	*Percina macrocephala*	M	M	I
Longnose dace	*Rhinichthys cataractae*	SM	M	T
Longnose gar	*Lepisosteus osseus*	LM	M	I
Margined madtom	*Noturus insignis*	M	I	I
Mimic shiner	*Notropis volucellus*	ML	M	I
Mottled sculpin	*Cottus bairdii*	SM	I	I
Mountain brook lamprey	*Ichthyomyzon greeleyi*	SM	I	I
Mountain madtom	*Noturus eleutherus*	ML	I	I
Muskellunge	*Esox masquinongy*	LM	M	I
Northern hog sucker	*Hypentelium nigricans*	MS	I	I
Northern madtom	*Noturus stigmosus*	SM	M	I
Northern pike	*Esox lucius*	ML	M	I
Northern redbelly dace	*Phoxinus eos*	S	M	I
Ohio lamprey	*Ichthyomyzon bdellium*	SML	I	

TABLE 2: Continued.

Common name	Species name	Size	Stream quality tolerance	Water quality tolerance
Pearl dace	*Margariscus margarita*	S	T	I
Popeye shiner	*Notropis ariommus*	M	I	I
Pumpkinseed	*Lepomis gibbosus*	M	M	I
Quillback	*Carpiodes cyprinus*	LM	M	I
Rainbow darter	*Etheostoma caeruleum*	SM	I	I
Rainbow smelt	*Osmerus mordax*	M	T	I
Rainbow trout	*Oncorhynchus mykiss*	SM	I	I
Redear sunfish	*Lepomis microlophus*	M	M	I
Redfin shiner	*Lythrurus umbratilis*	ML	T	I
Redside dace	*Clinostomus elongatus*	S	I	I
River chub	*Nocomis micropogon*	MS	T	T
River redhorse	*Moxostoma carina tum*	M	I	I
Rock bass	*Ambloplites rupestris*	M	I	I
Rosyface shiner	*Notropis rubellus*	M	I	I
Rosyside dace	*Clinostomus funduloides*	S	I	I
Sand shiner	*Notropis stramineus*	M	M	I
Sauger	*Sander canadensis*	LM	T	I
Shorthead red horse	*Moxostoma macrolepidotum*	ML	I	I
Silver redhorse	*Moxostoma anisurum*	ML	I	I
Silver shiner	*Notropis photogenis*	M	I	I
Silverjaw minnow	*Notropis buccatus*	MS	M	I
Smallmouth bass	*Micropterus dolomieu*	M	M	I
Snubnose darter	*Etheostoma simoterum*	SM	M	I
Southern red belly dace	*Phoxinus erythrogaster*	S	I	I
Spotfin shiner	*Cyprinella spiloptera*	ML	M	I
Spottail shiner	*Notropis hudsonius*	SML	I	I
Spotted darter	*Etheostoma macula tum*	M	I	I
Steelcolor shiner	*Cyprinella whipplei*	SM	M	I
Stonecat	*Noturus flavus*	M	I	I
Streamline chub	*Erimystax dissimilis*	ML	I	I
Striped shiner	*Luxilus chrysocephalus*	M	M	I
Tadpole madtom	*Noturus gyrinus*	SML	M	T
Tessellated darter	*Etheostoma olmstedi*	SML	M	T
Tippecanoe darter	*Etheostoma tippecanoe*	M	I	I
Tonguetied minnow	*Exoglossum laurae*	M	I	I
Trout-perch	*Percopsis omiscomaycus*	LM	T	I
Variegate darter	*Etheostoma varia tum*	M	I	I
Walleye	*Sander vitreus*	LM	T	I
Warmouth	*Lepomis gulosus*	L	T	I
White bass	*Morone chrysops*	L	M	I
White catfish	*Ameiurus catus*	LM	T	I
White crappie	*Pomoxis annularis*	LM	T	I
White sucker	*Catostomus commersonii*	MS	T	T
Yellow bullhead	*Ameiurus natalis*	SM	T	T
Yellow perch	*Perca flavescens*	ML	M	I

($r = 0.789$, $P < 0.001$). We also evaluated observed drainage areas to determine if the classification criterion used to differentiate small streams from large streams and small rivers was appropriately placed. Both the average cumulative drainage area for observed small streams ($47\,km^2$) and for large streams and small rivers ($440\,km^2$) were well within the appropriate ranges ($0–100\,km^2$ and $100–3,000\,km^2$, resp.) for their category.

Approximately 40% of the stream segments in the study area were predicted to have highly altered stream quality with most of the remaining stream segments classified as modified (57%). Intact stream segments were predicted to be present primarily in small streams (70%) and large streams and small rivers (26%). Modified and highly altered stream segments were widely dispersed throughout the watershed. Predicted stream quality classifications matched observed for all but nine stream segments (77%). The probability of obtaining 30 matching classifications out of 39 comparisons was <0.001 indicating that the high rate of matches is a significant and highly confident result.

Approximately 85% of the stream segments in the study area were classified as having degraded water quality. Most stream segments with degraded water quality were located in the western side of the watershed where agricultural and urban land uses were concentrated and the far eastern side of the watershed. The high-quality stream segments were largely located south of the Allegheny River in an area protected by the New York State Park system. Thus, water quality degradation appeared more clustered, regional, and prevalent than stream quality degradation.

Predicted TN classifications matched observed for all but five stream segments (90%). Predicted and observed TP classifications matched for 26 of the stream segments (67%), and SS classifications matched for all stream segments (100%). The probability of obtaining 35 and 39 matching classifications out of 39 comparisons was <0.001 indicating that the high rates of TN and SS matches were significant and highly confident results. The probability of obtaining 26 matching classifications out of 39 stream segment comparisons was 0.027 indicating that rate of TP matches was slightly lower but still a significant result.

4. Discussion

This study proposed a framework that used standard GIS methods and data to classify fish habitats at the river basin scale. The framework is composed of the landscape attributes stream size, stream quality, and water quality and was tested with a survey of fishes, stream size quantification, stream quality ratings, and water quality sampling. Ranked predicted and observed fish species by habitat were correlated but there was considerable variability. Large streams and small rivers with intact stream quality and good water quality were predicted to have the greatest number of fish species. We found no significant differences between predicted and observed number of fish species for this class. While this test was one point in time, results suggest that fish species were associated with the correct habitat class.

Sources of variability in our observed fish species data stem from misidentification of uncommon fish, gear and field technician inefficiency in fish capture, and escaped fish. Sources of variability in our predicted fish species data stem from inexact model parameterization or model structure. In six sites, observed and predicted fish species values were equal, and in seventeen sites predicted values were greater than observed values. Many of the sources of variability in our observed data would result in lower species richness estimates (i.e., misidentification of uncommon fish and inefficiency of fish capture). Thus, with improvements in observed data sampling techniques we would expect our observed values to increase and our observed and predicted fish species correlation to become stronger.

Prediction accuracy from our model was promising across all three landscape attributes. We found a strong significant correlation between predicted and observed drainage areas. The model succeeded in using predicted cumulative drainage area to characterize bankfull width with significant accuracy. This lends further support that drainage area at any point is correlated closely with many size characteristics and can be used as a general measure of stream size with confident accuracy. Examination of cumulative drainage area criteria to differentiate small streams from large streams and small rivers indicated appropriate placement for the upper Allegheny River basin.

Predicted stream quality classifications matched field observations significantly more often than chance alone would indicate. Scarcity of intact stream channels was likely a consequence of agricultural activity, presence of primary and secondary roads, or human-related alteration to the stream channel or banks. This scarcity, combined with a shortage of large streams and small rivers (10% of the stream segments) and high water quality streams (15%), caused few stream segments to have the necessary qualifications for greatest predicted number of fish species.

Our adapted nonpoint source pollution load screening model was designed to predict if TP, TN, and SS concentrations were greater than US EPA criteria for acceptable water quality. The coarse, simple GIS approach of the model was intended for annual prediction of parameters and could not accurately reflect subtle changes in pollutant concentrations. Field samples, collected over one brief time interval at baseflow conditions, were not a thorough test of annual water quality averages. Additional error may have been caused by misclassification errors in creation of the digital data and inappropriate runoff coefficients and pollutant concentration values for western New York State. Despite all these sources of possible error, the model was still able to accurately match observed water quality classifications for all parameters. We reported a much more thorough test of the water quality model and details about GIS operations in Meixler and Bain [49].

Our findings indicate that our GIS framework can provide coarse, landscape scale fish habitat classifications that can be related to expected fish distributions. Effective conservation of biodiversity in aquatic communities requires the identification and protection of key locations within river basins and regionally. Our fish habitat GIS framework meets this need by providing rapid, comprehensive, inexpensive, landscape scale habitat patterns from widely available data given reasonable time and resources. Further, GIS modeling can be readily updated to reflect changes in land use patterns [15]. Note that use of our model in other ecoregions may require adjustment and verification of model parameters. Thus, some field verification may be required for early adopters of our model. However, it is clear that the GIS framework presented here has considerable potential to classify fish habitat classes and predict fish distributions at the landscape scale to better inform management decisions.

Acknowledgments

This paper was supported by the United States Geological Survey under cooperative agreements no. 1434-HQ-97-RU-01553 and RWO no. 40 and benefited from research performed in the study area under a grant from the Nature Conservancy. Special thanks go to Greg Galbreath for compiling the fish data in Table 2. The authors also wish to thank Jordan Gass and Andrew Koo for fieldwork assistance and Magdeline Laba and Steve Smith for GIS aid in support of this project.

References

[1] D. Dudgeon, A. H. Arthington, M. O. Gessner et al., "Freshwater biodiversity: importance, threats, status and conservation challenges," *Biological Reviews of the Cambridge Philosophical Society*, vol. 81, no. 2, pp. 163–182, 2006.

[2] J. Speth, *The Bridge at the Edge of the World: Capitalism, the Environment and Crossing from Crisis to Stability*, Yale University Press, London, UK, 2008.

[3] M. A. Palmer, C. A. Reidy Liermann, C. Nilsson et al., "Climate change and the world's river basins: anticipating management options," *Frontiers in Ecology and the Environment*, vol. 6, no. 2, pp. 81–89, 2008.

[4] Z. Naveh and A. Lieberman, *Landscape Ecology: Theory and Application*, Springer, New York, NY, USA, 1994.

[5] L. A. Bojorquez-Tapia, I. Azuara, E. Ezcurra, and O. Flores-Villela, "Identifying conservation priorities in Mexico through geographic information systems and modeling," *Ecological Applications*, vol. 5, no. 1, pp. 215–231, 1995.

[6] K. Henle, D. Alard, J. Clitherow et al., "Identifying and managing the conflicts between agriculture and biodiversity conservation in Europe-A review," *Agriculture, Ecosystems and Environment*, vol. 124, no. 1-2, pp. 60–71, 2008.

[7] D. H. Wright and J. H. Reeves, "On the meaning and measurement of nestedness of species assemblages," *Oecologia*, vol. 92, no. 3, pp. 416–428, 1992.

[8] J. M. Scott, F. Davis, and B. Csuti, "Gap Analysis: a geographic approach to protection of biological diversity," *Wildlife Monographs*, vol. 123, pp. 1–41, 1993.

[9] D. Tilman, D. Wedin, and J. Knops, "Productivity and sustainability influenced by biodiversity in grassland ecosystems," *Nature*, vol. 379, no. 6567, pp. 718–720, 1996.

[10] D. Tilman, "Community invasibility, recruitment limitation, and grassland biodiversity," *Ecology*, vol. 78, no. 1, pp. 81–92, 1997.

[11] C. Folke, S. Carpenter, B. Walker et al., "Regime shifts, resilience, and biodiversity in ecosystem management," *Annual Review of Ecology, Evolution, and Systematics*, vol. 35, pp. 557–581, 2004.

[12] R. L. Kitching, "Biodiversity—political responsibilities and agendas for research and conservation," *Pacific Conservation Biology*, vol. 1, pp. 279–283, 1994.

[13] K. D. Smythe, J. C. Bernabo, T. B. Carter, and P. R. Jutro, "Focusing biodiversity research on the needs of decision makers," *Environmental Management*, vol. 20, no. 6, pp. 865–872, 1996.

[14] T. Tscharntke, A. M. Klein, A. Kruess, I. Steffan-Dewenter, and C. Thies, "Landscape perspectives on agricultural intensification and biodiversity—ecosystem service management," *Ecology Letters*, vol. 8, no. 8, pp. 857–874, 2005.

[15] K. Tucker, S. P. Rushton, R. A. Sanderson, E. B. Martin, and J. Blaiklock, "Modelling bird distributions—a combined GIS and Bayesian rule-based approach," *Landscape Ecology*, vol. 12, no. 2, pp. 77–93, 1997.

[16] M. S. Meixler, M. B. Bain, and M. Todd Walter, "Predicting barrier passage and habitat suitability for migratory fish species," *Ecological Modelling*, vol. 220, no. 20, pp. 2782–2791, 2009.

[17] P. Angelstam, J.-M. Roberge, and A. Lõhmus, "Habitat modelling as a tool for landscape-scale conservation—a review of parameters for focal forest birds," *Ecological Bulletins*, vol. 51, pp. 427–453, 2004.

[18] M. Powell, A. Accad, and A. Shapcott, "Geographic information system (GIS) predictions of past, present habitat distribution and areas for re-introduction of the endangered subtropical rainforest shrub Triunia robusta (Proteaceae) from south-east Queensland Australia," *Biological Conservation*, vol. 123, no. 2, pp. 165–175, 2005.

[19] D. Scott, H. Moller, D. Fletcher et al., "Predictive habitat modelling to estimate petrel breeding colony sizes: sooty shearwaters (*Puffinus griseus*) and mottled petrels (*Pterodroma inexpectata*) on Whenua Hou Island," *New Zealand Journal of Zoology*, vol. 36, no. 3, pp. 291–306, 2009.

[20] K. Stachura-Skierczyńska, T. Tumiel, and M. Skierczyński, "Habitat prediction model for three-toed woodpecker and its implications for the conservation of biologically valuable forests," *Forest Ecology and Management*, vol. 258, no. 5, pp. 697–703, 2009.

[21] T. W. Norton and J. E. Williams, "Habitat modelling and simulation for nature conservation: a need to deal systematically with uncertainty," *Mathematics and Computers in Simulation*, vol. 33, no. 5-6, pp. 379–384, 1992.

[22] W. R. Gordon, "A role for comprehensive planning, geographical information system (GIS) technologies and program evaluation in aquatic habitat development," *Bulletin of Marine Science*, vol. 55, no. 2-3, pp. 995–1013, 1994.

[23] A. S. L. Rodrigues, H. R. Akçakaya, S. J. Andelman et al., "Global gap analysis: priority regions for expanding the global protected-area network," *BioScience*, vol. 54, no. 12, pp. 1092–1100, 2004.

[24] P. L. Angermeier and M. R. Winston, "Characterizing fish community diversity across Virginia landscapes: prerequisite for conservation," *Ecological Applications*, vol. 9, no. 1, pp. 335–349, 1999.

[25] C. A. Frissell, W. J. Liss, C. E. Warren, and M. D. Hurley, "A hierarchical framework for stream habitat classification: viewing streams in a watershed context," *Environmental Management*, vol. 10, no. 2, pp. 199–214, 1986.

[26] P. B. Moyle and J. P. Ellison, "A conservation-oriented classification system for the inland waters of California," *California Fish and Game*, vol. 77, pp. 161–180, 1991.

[27] D. L. Rosgen, "A classification of natural rivers," *Catena*, vol. 22, no. 3, pp. 169–199, 1994.

[28] P. L. Angermeier and I. J. Schlosser, "Conserving aquatic biodiversity: beyond species and populations," *American Fisheries Society Symposium*, vol. 17, pp. 402–414, 1995.

[29] J. V. Higgins, M. T. Bryer, M. L. Khoury, and T. W. Fitzhugh, "A freshwater classification approach for biodiversity conservation planning," *Conservation Biology*, vol. 19, no. 2, pp. 432–445, 2005.

[30] S. S. Wall, C. R. Berry, C. M. Blausey, J. A. Jenks, and C. J. Kopplin, "Fish-habitat modeling for gap analysis to conserve the endangered Topeka shiner (*Notropis topeka*)," *Canadian Journal of Fisheries and Aquatic Sciences*, vol. 61, no. 6, pp. 954–973, 2004.

[31] S. P. Sowa, G. Annis, M. E. Morey, and D. D. Diamond, "A gap analysis and comprehensive conservation strategy for riverine ecosystems of Missouri," *Ecological Monographs*, vol. 77, no. 3, pp. 301–334, 2007.

[32] F. B. Lotspeich, "Watersheds as the basic ecosystem: this conceptual framework provides a basis for a natural classification system," *Water Resources Bulletin*, vol. 16, pp. 581–586, 1980.

[33] R. J. Naiman, D. G. Lonzarich, T. J. Beechie, and S. C. Ralph, "General principles of classification and the assessment of conservation potential in rivers," in *River Conservation and Management*, P. J. Boon, P. Calow, and G. E. Petts, Eds., pp. 93–124, John Wiley & Sons, Chichester, UK, 1992.

[34] L. P. Aadland, "Stream habitat types: their fish assemblages and relationship to flow," *North American Journal of Fisheries Management*, vol. 13, no. 4, pp. 790–806, 1993.

[35] C. L. Smith, *The Inland fishes of New York State*,, The New York State Department of Environmental Conservation, Albany, NY, USA, 1985.

[36] H. H. Barnes Jr., "Roughness characteristics of natural channels," Water Supply Paper 1849, U. S. Geological Survey, Washington, DC, USA, 1967.

[37] T. Dunne and L. B. Leopold, *Water in Environmental Planning*, W. H. Freeman, New York, NY, USA, 1978.

[38] A. J. Castelle, A. W. Johnson, and C. Conolly, "Wetland and stream buffer size requirements—a review," *Journal of Environmental Quality*, vol. 23, no. 5, pp. 878–882, 1994.

[39] N. E. Roth, J. David Allan, and D. L. Erickson, "Landscape influences on stream biotic integrity assessed at multiple spatial scales," *Landscape Ecology*, vol. 11, no. 3, pp. 141–156, 1996.

[40] J. D. Allan, "Landscapes and riverscapes: the influence of land use on stream ecosystems," *Annual Review of Ecology, Evolution, and Systematics*, vol. 35, pp. 257–284, 2004.

[41] K. M. Mattson and P. L. Angermeier, "Integrating human impacts and ecological integrity into a risk-based protocol for conservation planning," *Environmental Management*, vol. 39, no. 1, pp. 125–138, 2007.

[42] R. T. T. Forman and L. E. Alexander, "Roads and their major ecological effects," *Annual Review of Ecology and Systematics*, vol. 29, pp. 207–231, 1998.

[43] M. S. Meixler and M. B. Bain, "Landscape scale assessment of stream channel and riparian habitat restoration needs," *Landscape and Ecological Engineering*, vol. 6, no. 2, pp. 235–245, 2010.

[44] United States Environmental Protection Agency (US EPA), National water quality inventory: report to congress, 2002 Reporting Cycle, EPA 841-R-07-001, Office of Water, US Environmental Protection Agency, Washington, DC, USA, 2007.

[45] D. E. Blockstein, "An aquatic perspective on U.S. biodiversity policy," *Fisheries*, vol. 17, pp. 26–30, 1992.

[46] M. Cairns and R. T. Lackey, "Biodiversity and management of natural resources: the issues," *Fisheries*, vol. 17, pp. 6–10, 1992.

[47] R. M. Hughes and R. F. Noss, "Biological diversity and biological integrity: current concerns for lakes and streams," *Fisheries*, vol. 17, pp. 11–19, 1992.

[48] C. L. Adamus and M. J. Bergman, "Estimating nonpoint source pollution loads with GIS screening model," *Water Resources Bulletin*, vol. 31, no. 4, pp. 647–655, 1995.

[49] M. S. Meixler and M. B. Bain, "A water quality model for regional stream assessment and conservation strategy development," *Environmental Management*, vol. 45, no. 4, pp. 868–880, 2010.

[50] United States Environmental Protection Agency (US EPA), Ambient water quality criteria recommendation: rivers and streams, EPA/822/b-00/020, US Environmental Protection Agency, Washington, DC, USA, 2000.

[51] W. Swietlik, *Developing Water Quality Criteria for Suspended and Bedded Sediments (SABS): Potential Approaches*, United States Environmental Protection Agency, Office of Water, Washington, DC, USA, 2003.

[52] I. J. Schlosser, "Trophic structure, reproductive success, and growth rate of fishes in a natural and modified headwater stream," *Canadian Journal of Fisheries and Aquatic Sciences*, vol. 39, no. 7, pp. 968–978, 1982.

[53] J. R. Karr, K. D. Fausch, P. L. Angermeier, P. R. Yant, and I. J. Schlosser, *Assessing Biological Integrity in Running Waters: A Method and Its Rationale*, Illinois Natural History Survey, Sullivan , Ill, USA, 1986.

[54] Ohio Environmental Protection Agency, *Biological Criteria for the Protection of Aquatic Life: Vol. III. Standardized Biological Field Sampling and Laboratory Methods for Assessing Fish and Macroinvertebrate Communities*, Ohio Environmental Protection Agency, Division of Water Quality Monitoring and Assessment, Columbus, Ohio, USA, 1989.

[55] W. D. Crumby, M. A. Webb, F. J. Bulow, and H. J. Cathey, "Changes in biotic integrity of a river in north-central Tennessee," *Transactions of the American Fisheries Society*, vol. 119, pp. 885–893, 1990.

[56] R. G. Bramblett and K. D. Fausch, "Variable fish communities and the index of biotic integrity in a western Great Plains river," *Transactions of the American Fisheries Society*, vol. 120, pp. 752–769, 1991.

[57] R. C. Petersen, "The RCE: a riparian, channel, and environmental inventory for small streams in the agricultural landscape," *Freshwater Biology*, vol. 27, no. 2, pp. 295–306, 1992.

4

Ant-Related Oviposition and Larval Performance in a Myrmecophilous Lycaenid

Matthew D. Trager,[1,2] **Matthew D. Thom,**[2,3] **and Jaret C. Daniels**[2,3]

[1] *USDA Forest Service, National Forests of Florida, 325 John Knox Road, Suite F-100, Tallahassee, FL 32303, USA*
[2] *Florida Museum of Natural History, 3215 Hull Road, Gainesville, FL 32611, USA*
[3] *Department of Entomology and Nematology, University of Florida, P.O. Box 110620, Gainesville, FL 32611, USA*

Correspondence should be addressed to Matthew D. Trager; mdtrager@fs.fed.us

Academic Editor: Mats Olsson

We experimentally assessed ant-related oviposition and larval performance in the Miami blue butterfly (*Cyclargus thomasi bethunebakeri*). Ant tending had sex-dependent effects on most measures of larval growth: female larvae generally benefitted from increased tending frequency whereas male larvae were usually unaffected. The larger size of female larvae tended by ants resulted in a substantial predicted increase in lifetime egg production. Oviposition by adult females that were tended by *C. floridanus* ants as larvae was similar between host plants with or without ants. However, they laid relatively more eggs on plants with ants than did females raised without ants, which laid less than a third of their eggs on plants with ants present. In summary, we found conditional benefits for larvae tended by ants that were not accompanied by oviposition preference for plants with ants present, which is a reasonable result for a system in which ant presence at the time of oviposition is not a reliable indicator of future ant presence. More broadly, our results emphasize the importance of considering the consequences of variation in interspecific interactions, life history traits, and multiple measures of performance when evaluating the costs and benefits of mutualistic relationships.

1. Introduction

For egg-laying animals that do not provide parental care, oviposition location is among the most important maternal decisions affecting subsequent offspring performance [1, 2]. Consequently, when growth and survival of immature stages are strongly influenced by other species—either positively or negatively—spatial patterns of oviposition often reflect these interactions [3–7]. In mutualistic relationships, we might expect preferential oviposition and improved performance when and where mutualists are present, but tradeoffs and constraints in both oviposition choices and immature growth strategies may make such simple correlations rare.

Most lycaenid butterfly species interact with ants, and many of these relationships include substantial benefits for lycaenid larvae [8]. However, the costs and benefits of ant tending for larval growth of lycaenids vary substantially among systems. For example, larvae of *Jalmenus evagoras* and *Glaucopsyche lygdamus* pupate at a greater mass when

untended by ants [9–11] but are dependent on ants for protection under natural conditions. By contrast, larvae of *Paralucia aurifera* and *Hemiargus isola* pupate at a greater mass and developed faster when reared with ants [12, 13]. Other studies have found that the consequences of ant tending on larval performance are conditional upon the sex of the larvae [14] or the identity of the tending ant species [13, 15]. In a comparative study of five lycaenid species, Fiedler and Saam [16] suggested that compensatory or slightly overcompensatory growth in response to the costs of nectar production for ants may be common among facultatively myrmecophilous lycaenids.

Although interactions with ants influence lycaenid growth decisions, they are also subject to many of the same environmental pressures and life history tradeoffs experienced by other immature insects. In particular, the size and age at which metamorphosis occurs are important life history traits for organisms with complex life cycles, particularly for taxa in which most or all growth occurs during immature development [17, 18]. Large phenotypes often confer substantial advantages, but theoretical and empirical studies suggest that physiological limitations, tradeoffs in resource allocation, and variation in environmental conditions may constrain both large size per se and the growth rates required to attain larger sizes [19, 20]. For many insects, the primary benefit of larger size is higher potential reproductive output and the primary risk of prolonging the immature growth period to accommodate the larger size at maturity is increased exposure to predators and parasitoids [17, 21, 22]. Behavioral and physiological adaptations to deal with this tradeoff include altering the location and timing of feeding to minimize predation and the evolution of phenotypically plastic growth rates that allow rapid growth when resources are abundant or risks associated with growth are low [23–26].

Ant-related oviposition by lycaenids is somewhat less studied but appears to be at least as variable as ant-related larval growth. Atsatt [27] proposed that ovipositing females should evolve searching behaviors for ants when ant abundance and predictability are high and should be accompanied by obligate and usually species-specific larval relationships with ants. Pierce and Elgar [28] suggested a more continuous relationship in which the degree of preference demonstrated by ovipositing lycaenids for host plants where ants are present should correspond with the importance of the ant interaction for larval survival and growth. Several individual studies and comparative empirical data support this hypothesis [29–31], but relatively few lycaenid species have been shown to preferentially oviposit on host plants with ants present in experimental settings.

Oviposition decisions are complex and may not be correlated with offspring performance for a variety of reasons. For example, in stochastic environments ovipositing females may not discriminate among potential oviposition sites because there are no reliable indicators of future offspring performance at the time of oviposition [32]. Alternatively, variation in natal experiences (e.g., habitat or diet) may affect oviposition decisions later in life [33–35]. Additionally, for short-lived species or older females, optimal oviposition preferences may be superseded by the evolutionary imperative to distribute eggs prior to death [36]. Therefore, we would only expect strong oviposition preference for mutualists in systems that meet the following criteria: (1) the interaction is very important for immature performance, (2) the conditions at the time of oviposition accurately indicate mutualist presence, or (3) the immature organisms have low likelihood of encountering mutualists if they are initially absent [1].

We experimentally tested effects of ant interactions on larval performance and adult oviposition for the Miami blue butterfly (*Cyclargus thomasi bethunebakeri*). This lycaenid engages in facultative mutualistic interactions with several ant species [37], but preliminary laboratory studies suggested that ant tending had no effects on the timing of larval development or size at pupation [38]. However, based on sex-dependent effects of size on reproductive performance [39], production of nutritional rewards for ant defenders, and nearly constant tending by ants [37], we expected frequency of ant tending to influence larval development in a more detailed study. We then tested the effects of interaction with

FIGURE 1: *Camponotus floridanus* workers tending two fourth instar *C. thomasi bethunebakeri* larvae. Note the everted tentacular organs in the upper larva and the ant consuming secretions from the dorsal nectary organ of the lower larva.

ants during the larval stage and female age on oviposition preference for host plants with and without ants present. We interpret the results of these studies in the context of oviposition and larval growth theories, with particular attention to how and why this system deviates from a simple positive correlation between oviposition preference and larval performance.

2. Materials and Methods

2.1. Study System. The Miami blue butterfly is a small lycaenid native to southern coastal Florida and the Florida Keys, USA, with other *C. thomasi* subspecies occurring elsewhere in the Caribbean. Formerly abundant in suitable habitats, the Miami blue butterfly declined in the 20th century due to a variety of anthropogenic and natural factors; at the time this study was conducted, the subspecies occurred in small populations on Bahia Honda in the middle Keys and several islands in Key West National Wildlife Refuge [40]. All butterflies used in this research were part of a captive colony originating from wild stock collected at Bahia Honda State Park, where the larvae apparently feed exclusively on gray nickerbean (*Caesalpinia bonduc*).

Females oviposit on leaf or flower buds or young leaves, usually singly as they walk along the plant. The first and second instar larvae feed cryptically in flower or leaf buds and are not usually tended by ants. However, after molting into the third instar, Miami blue butterfly larvae are commonly tended by ants while feeding on the host plant (Figure 1); *Camponotus* species are the most frequent ant associates [37, 38, 41]. Saarinen and Daniels [37] reported that all late-instar larvae found in the wild were tended by ants, but this may be due to the high detectability of larvae when surrounded by large and active ants. Larvae reared with *Camponotus floridanus* in laboratory conditions are rarely untended and regularly evert paired tentacular organs and secrete nectar from their dorsal nectar organ in response to ant presence [38].

2.2. Larval Performance Experiment. To generate variation in ant tending frequency, we reared Miami blue butterfly larvae in plastic trays (27 × 19 × 9.5 cm) in which we placed 0, 10, 20 or 30, workers of *Camponotus floridanus*, one of the ant species that most commonly tends larvae in the wild [37, 38]. We placed two larvae in 10 replicate trays for each ant abundance treatment for a total of 80 larvae. At the beginning of the experiment the larvae were all the same age (8 d) and similar in wet mass (1.43 mg ± 0.04 SE, n = 80). Each larva was placed in an open 150 mL plastic vial with cuttings of new growth from the host plant, gray nickerbean (*Caesalpinia bonduc*), to allow access by the ants but reduce larval wandering. We replaced the host plant and vial daily and transferred larvae from the old to the new cuttings with a sterilized paintbrush. We measured the mass of each larva every morning with a digital analytic balance accurate to 0.01 mg (Denver Instruments SI-215D). We recorded ant tending of each larva twice each day (~900 h and ~1600 h). Behaviors recorded as tending included antennating the larvae, consuming secretions from the dorsal nectary organ or running on or around the larvae. Upon pupation, we measured pupal mass and placed the pupae in individual closed vials without ants in the laboratory until adult emergence. When the adult butterflies emerged, we calculated the duration of the pupal stage, measured the forewing length, and noted the sex of each adult.

We tested the effects of ant abundance treatment (10, 20, or 30 ants), larval age, and time of observation (morning or afternoon) on frequency of ant tending with a generalized linear mixed-effects model with a binomial error and logit link that included larva within tray as a random variable. The age-specific probability of tending by ants for each of the three ant abundance treatments was estimated with the lmer function in the R package lme4 [42].

We tested the effects of ant tending frequency (measured as the overall proportion of observations in which tending was observed) and sex of larvae on the following growth parameters: maximum larval mass, age at maximum larval mass, percent loss of mass from the final larval instar to the pupa, pupal mass, age at pupation, duration of the pupal stage, and total combined duration of the immature stages. For the analysis of duration of the pupal stage, we included pupal mass as a predictor variable, and the analyses that tested sex as a predictor variable only included the 73 individuals (38 males and 35 females) that survived to adulthood. These relationships were tested with linear mixed-effects models in the R package nlme [42] and *P* values were evaluated against a table-wide false discovery rate of 0.05 for each predictor variable to control for multiple comparisons [43]. Although this approach resulted in multiple related tests, controlling for the false discovery rate allowed us to make inferences about stage-specific differences in growth due to sex and ant tending without overly inflating Type I error.

Our previous work found a strong positive relationship between forewing chord length and egg production for females but found no effect of size on male reproductive performance [39]. To test the explanatory power of pupal mass for adult size, thereby linking larval growth to a correlate of female fitness, we analyzed a linear model testing the effects of pupal mass and sex on forewing chord length. For this analysis, we included all of the individuals from this study and those from an earlier experiment [38].

2.3. Oviposition Preference Experiment. The oviposition experiments were conducted in small screen flight cages in which we placed two plastic trays containing cuttings of the host plant either with or without *C. floridanus* ants. Cages were placed under incandescent lights on a timer set to a 14 hr light/10 hr dark cycle with some intermittent dark periods during the day to stimulate activity and prevent overheating. Each day, we replaced the host plant cuttings, counted the previous day's eggs, alternated the position of the ant treatments, and provided fresh nectar sources for the female butterflies. Preliminary trials and observations showed that *C. floridanus* did not consume or damage Miami blue butterfly eggs.

We first tested the effects of ant interactions during the larval stage on ant-related oviposition preferences of adult females. Four mated female Miami blue butterflies from one of the larval treatments (i.e., tended by *C. floridanus* or not) were placed in each of 6 screen cages and eggs were counted on host plant cuttings with and without *C. floridanus* ants every morning for 7 days. We then conducted a similar oviposition preference trial to test if butterflies responded to ant pheromones rather than the physical presence of ants on the host plant. For this experiment, we tested oviposition preference between one control cutting of the host plant and one cutting that had been exposed to *C. floridanus* for 30 min prior to placement in the flight cage but from which ants were absent during the trial. We conducted this experiment in 5 flight cages, each containing 4 naïve female Miami blue butterflies, for 5 days.

The number of eggs varied substantially among days and cages within experiments, so we used a standardized measure of preference (the number of eggs on the control plant minus number of eggs on the plant with ants) as the response variable for analysis. We calculated this daily difference in egg number for each cage every day that eggs were present on at least one plant. To evaluate the effects of natal experience (i.e., naïve or ant tended) on oviposition preference, we tested a mixed-effects model with larval experience with ants and day as fixed effects and cage as a random grouping variable. For the test of the effects of ant pheromones on oviposition preference, we analyzed a mixed-effects model with day as a fixed effect and cage as a random grouping variable. We complemented these more detailed linear mixed-effects analyses with chi-squared tests on the cumulative egg distributions for the entire experiment.

3. Results

3.1. Larval Performance Experiment. The probability of tending by ants increased with larval age, and among treatment contrasts showed that tending was less frequent in the 10-ant treatment than in the 20- and 30-ant treatments throughout larval development (Figure 2, Table 1). Over the course of the study, larvae in these three treatments were tended during 49% (±5% SE), 79% (±2% SE), and 74% (±3% SE) of

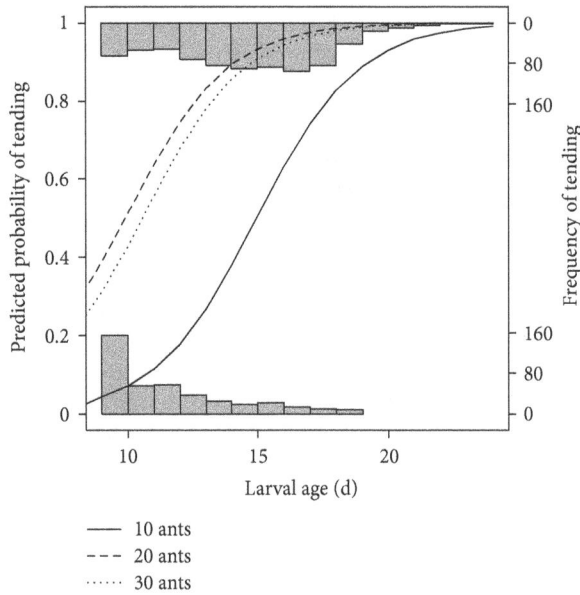

FIGURE 2: Fitted logistic regression curves showing the predicted probability of ant tending (left y-axis) as a function of larval age for the three ant abundance treatments in this study. The histograms show the frequency of observations in which tending occurred (upper right y-axis) or did not occur (lower right y-axis) each day for the three ant abundance treatments combined. Results of statistical analyses on the effects of ant abundance treatment and larval age are in Table 1.

TABLE 1: Results of generalized linear mixed-effects model testing the effects of ant abundance treatment (10, 20 or 30 ants), larval age (d) and time of observation (morning or afternoon) on the probability that ants would be tending Miami blue butterfly larvae. β is the coefficient describing the shape of the logistic curve (Figure 2). The table below presents the tests of contrasts among levels of ant abundance (the P-value of the 10 ant treatment is from a test against the null of $\beta = 0$ and the P-values associated with the 20 and 30 ant treatments indicate the difference between those treatments and the 10 ant treatment). The interaction term between ant abundance treatment and age of larvae was not significant so was removed to more accurately assess the main effects.

Variable	β (±SE)	z-value	P
Ant abundance treatment			
10 ants	−7.69 (0.59)	−13.10	<0.0001
20 ants	−5.09 (0.45)	5.84	<0.0001
30 ants	−5.43 (0.45)	5.02	<0.0001
Age of larvae	0.51 (0.034)	15.16	<0.0001
Time of observation	0.065 (0.16)	0.41	0.68

observations, respectively; larvae in the no ant treatment were obviously not tended.

The results of models testing ant tending frequency and sex on larval growth parameters are in Table 2. Note that many of the larval growth parameters were highly correlated, so the tests for sex and ant tending as predictor variables at a given stage are also related to those for other growth responses. For example, maximum larval mass strongly

TABLE 2: Results of linear mixed-effects model testing the effects of ant tending frequency and sex on growth parameters of Miami blue butterfly larvae. Non-significant interaction terms were omitted from some models to better estimate the effects of tending frequency and sex of larvae on growth responses. Uncorrected P-values are presented; test results lower than the table-wide false discovery rate for the corresponding predictor variable are indicated in bold italic font.

	Growth parameter tested		
Source of variation	df	F-value	P
Age at maximum larval mass			
Tending frequency	1, 32	0.83	0.37
Sex	1, 32	0.13	0.72
Tending frequency ∗ Sex	1, 32	4.55	0.041
Maximum larval mass			
Tending frequency	1, 32	15.30	***0.0004***
Sex	1, 32	12.61	***0.0012***
Tending frequency ∗ Sex	1, 32	4.54	0.041
Percent mass lost during prepupal stage			
Tending frequency	1, 32	1.71	0.20
Sex of larvae	1, 32	0.084	0.77
Tending frequency ∗ Sex	1, 32	6.55	***0.015***
Pupal mass			
Tending frequency	1, 32	5.55	0.025
Sex	1, 32	7.09	***0.012***
Tending frequency ∗ Sex	1, 32	7.52	***0.0099***
Age at pupation			
Tending frequency	1, 32	1.10	0.30
Sex	1, 32	0.022	0.88
Tending frequency ∗ Sex	1, 32	4.52	***0.048***
Duration of pupal stage			
Sex	1, 33	4.64	0.039
Pupal mass	1, 33	0.32	0.57
Total duration of immature stages			
Tending frequency	1, 33	1.45	0.24
Sex	1, 33	0.53	0.47

predicted pupal mass regardless of other variables but testing both responses showed that the interaction of tending frequency and sex affected the percent mass lost between these stages.

Male and female larvae both reached a maximum larval mass of approximately 70 mg when untended, but as tending frequency increased the maximum mass of female larvae increased whereas the maximum mass of males decreased. Tending frequency had a similar effect on the age at maximum larval mass: untended females required approximately 1 d longer than untended males to attain their maximum mass (16.31 d v. 15.4 d, resp.) but tending frequency decreased female development time (difference between never tended and always tended = −0.55 d) whereas it increased male development time (difference between never tended and always tended = +1.61 d). Across all ant abundance treatments, Miami blue butterfly larvae lost 20.8% (±0.66 SE) of their mass during the prepupal stage (time from maximum

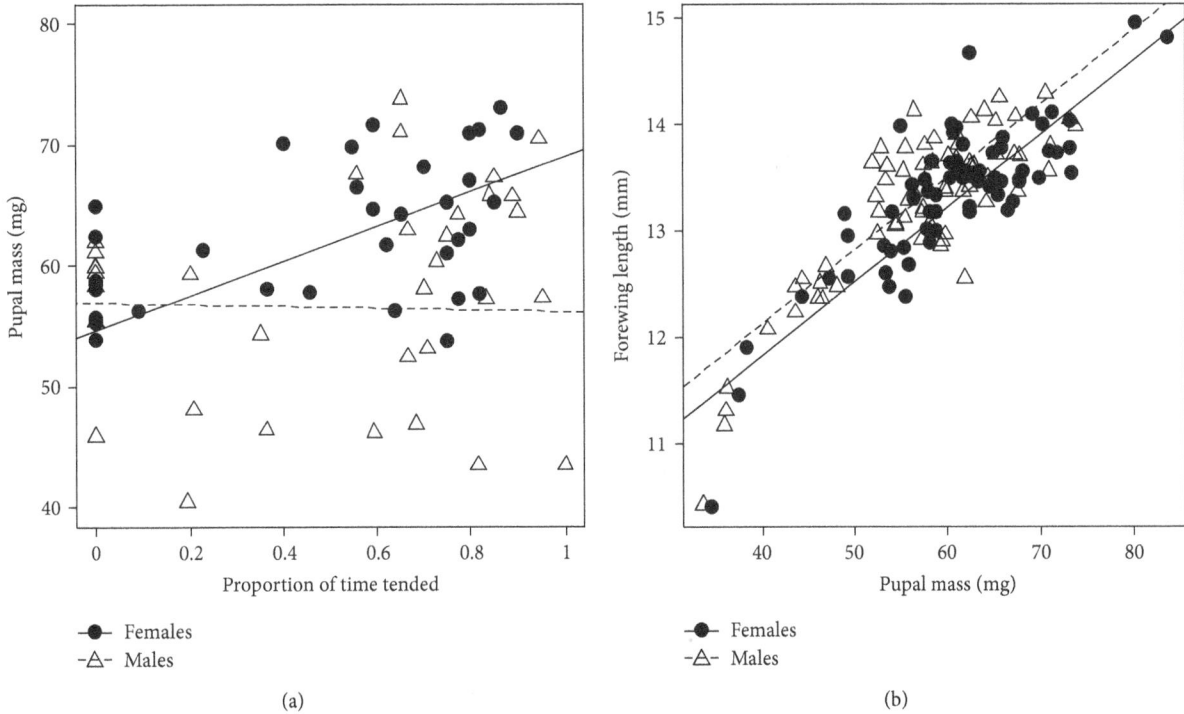

FIGURE 3: (a) Frequency of ant tending had sex-dependent effect on pupal mass, with female larvae pupating at a higher mass as tending increased whereas male larvae were unaffected by tending frequency. (b) Pupal mass and sex strongly predicted adult forewing chord length for Miami blue butterflies.

larval mass to pupation). This loss was influenced by sex and frequency of tending: the percent mass lost with no ant tending was significantly higher for females than for males (22.5 ± 1.7 SE and 17.4 ± 2.2 SE, resp.; $t = -2.27$, df = 32, $P = 0.03$) with the opposite relationship at very high levels of tending.

Female pupae were significantly larger than male pupae overall (61.9 mg ± 1.4 SE and 56.6 ± 2.0 SE, resp.; $t = 2.6$, df = 71, $P = 0.011$), but pupal mass of males and females was differentially affected by frequency of ant tending during the larval stage in a pattern similar to that for maximum larval mass. The mass of female and male pupae was not significantly different when they were untended as larvae (54.6 ± 2.5 SE and 57.0 ± 3.3 SE, resp.; $t = 0.73$, df = 32, $P = 0.47$); however, tending frequency increased pupal mass for females whereas the pupal mass of males was unaffected (Figure 3(a)). The frequency of ant tending also had sex-dependent effects on the age at which larvae pupated: females pupated earlier in response to higher tending frequency whereas males pupated later as tending frequency increased. Despite the numerous sex-specific larval growth parameters, the total time spent in the immature stages was invariant with respect to either sex or the frequency of ant tending (29.66 days ± 0.17 SE).

Both pupal mass and sex significantly predicted forewing length of adult butterflies ($F_{1,135} = 344.3$, $P < 0.0001$ and $F_{1,135} = 18.9$, $P < 0.001$, resp.). There was no difference in the slope of the relationship between the sexes (i.e., no significant interaction term), but males had a longer forewing chord length for a given pupal mass (Figure 3(b)).

3.2. Oviposition Preference Experiment. Naïve female Miami blue butterflies raised without ants as larvae laid 8.03 (±2.44 SE) fewer eggs per day per cage on plants with ants present (df = 9, $t = 2.47$, $P = 0.035$). However, for females tended by ants as larvae, the mean difference between the number of eggs on control plants and plants with ants was only 1.99 (±2.65 SE), which was not significantly different from no preference (df = 41, $t = 0.75$, $P = 0.46$). Although neither treatment produced an absolute preference for ovipositing where ants were present, females reared with ants as larvae laid significantly more eggs on plants with *C. floridanus* ants present than naïve females (Table 3). Ant-related oviposition patterns did not change as the butterflies senesced. The cumulative frequency of eggs across all days and cages was consistent with the results of the linear mixed effects model analyzing the daily results: naïve female Miami blue butterflies deposited only 30.5% of their eggs on host plant cuttings with *C. floridanus* present, whereas female Miami blue butterflies that were tended by *C. floridanus* as larvae laid 45.2% of their eggs on host plant cuttings with *C. floridanus* present (Figure 4; $\chi^2 = 15.26$, df = 1, $P < 0.0001$).

There was no effect of ant pheromones on oviposition preference when ants were not physically present. The mean difference between the number of eggs on control plants and those on plants exposed to ant trail pheromones was 1.13 (±15.53 SE) eggs per cage per day, which was not significantly different from no preference (df = 16, $F = 0.042$, $P = 0.84$). This result was corroborated by the cumulative test: of 772 eggs laid throughout the course of this experiment, 371 (48.1%) were on plants exposed to *C. floridanus* pheromones

TABLE 3: Results of linear mixed-effects analysis testing the influence of larval experience with ants and day on mean daily difference in egg number between host plants without and with *C. floridanus* ants present.

Source of variation	df	F	P
Ant interaction as larvae	1, 9	5.93	0.038
Day	1, 41	0.22	0.65

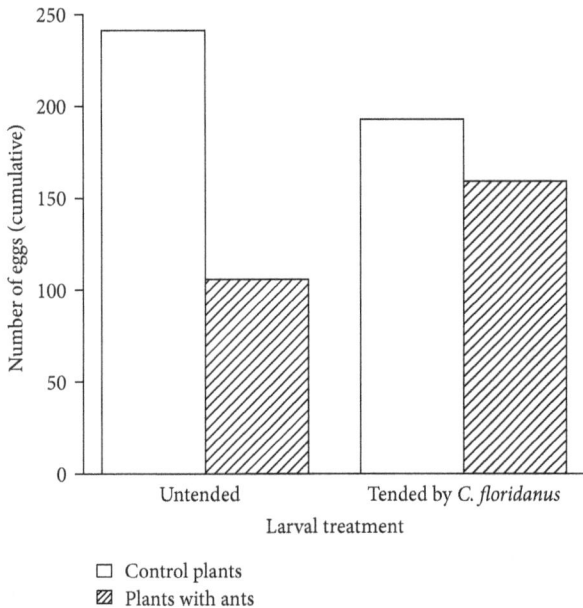

FIGURE 4: The total number of eggs laid by untended (naïve) females and females tended by *C. floridanus* as larvae on host plants with ants either present or absent. Although neither category of females preferred to oviposit on plants with *C. floridanus* present, the oviposition preference of those raised with ants did not differ from random egg distribution whereas the naïve females laid significantly fewer eggs on plants with ants.

and 401 (51.9%) were on control plants not exposed to ants ($\chi^2 = 1.17$, df = 1, $P = 0.28$).

4. Discussion

Identifying the costs and benefits of mutualistic relationships and investigating the conditions under which they manifest is invaluable for understanding the consequences of behavioral variation for life history strategies and fitness [44]. In the Lycaenidae, ant association may have a range of direct and indirect effects on development and survival [8, 45], and even relatively small differences in life history parameters due to interactions with ants can affect the evolution of lycaenid-ant mutualisms [46].

Our previous research demonstrated that wing length was positively related to higher lifetime fecundity for female Miami blue butterflies [39]. The present study showed that pupal mass increased with frequency of ant tending for female Miami blue butterfly larvae (but not for males) and that pupal mass and sex were reliable predictors of wing

length. By combining these results with previous findings, we estimate that always tended female larvae would pupate at 14.4 mg higher mass than untended larvae, a difference that would result in an additional 1 mm of wing length, which in turn would predict an increase in lifetime fecundity of 53 (±17.8) eggs. This provides clear evidence that ant-mediated variation in female larval growth has direct consequences for an important component of fitness. Patterns in the other size parameters (i.e., maximum larval mass, mass lost during the prepupal stage) largely supported the conclusion that female larvae benefitted from ant tending, whereas ants had neutral or slightly negative effects on male development.

If interacting with ants increases growth of female Miami blue butterfly larvae, then they should generally engage in behaviors that increase the probability of attracting and retaining ants. However, the proportion of observations in which tending occurred did not differ between male and female larvae. This apparent contradiction may have a relatively simple explanation in the broader perspective of the mutualistic relationship between the Miami blue butterfly and ants. If the primary benefit for lycaenids associating with ants is protection from predators, male and female Miami blue butterfly larvae may be equally likely to elicit ant tending. Attaining higher pupal mass is more beneficial for females than for males, so perhaps female larvae have evolved overcompensatory feeding behavior when tended by ants that results in faster growth and larger size [16, 47]. This explanation is consistent with the hypothesis that selection should reduce the costs of mutualism, particularly when the cost reduction mechanism (i.e., altered larval feeding and movement) does not directly conflict with the benefit gained by the other partner (i.e., dorsal nectar secretions that ants consume) [44].

Maximizing the larval growth objectives of large size and fast development generally involves substantial tradeoffs [15] or requires relaxation of other limitations such as food resources or predation risk [48]. We did not examine predation and antipredator protection in this study, but anecdotal evidence suggests that *C. floridanus* ants are effective at deterring natural enemies of Miami blue butterfly larvae [38]. Together with the sex-specific effects of ant tending on pupal mass, it appears that female Miami blue butterfly larvae largely mitigate the common tradeoff between size and age at pupation by interacting with ant mutualists.

Pierce and Elgar [28] proposed a suite of traits correlated with ant-related oviposition among lycaenid butterflies, but the Miami blue butterfly only meets some of the criteria. First, although the interaction with ants is beneficial for larval performance in this system, it is facultative rather than the obligate relationships found in some ant tended lycaenids and the first two instars feed within host plant buds and are not tended by ants. That said, later-instar larvae in the wild are usually guarded by presumably mutualistic ants, most commonly *C. floridanus* and *C. planatus* [38]. Second, there are both predictable and unpredictable components of the ant environment that could affect females' perception of site suitability for oviposition. Ants do not constantly patrol the host plant in large numbers and *C. floridanus* is primarily nocturnal, so presence at any given time may not

41

be a reliable indication of whether ants will tend Miami blue butterfly larvae at that potential oviposition site. Third, field surveys suggest that ants are likely to find larvae on the host plant regardless if they are present or absent at the time of oviposition [37, 38]. In short, the interaction between larvae of the Miami blue butterfly and ants features strong association with ants in the later instars but does not meet any of the other characteristics listed by Pierce and Elgar [28] as traits common among lycaenids thought to use ants as oviposition cues. Taken together, these aspects of Miami blue butterfly ecology suggest that there may be little selective pressure for ovipositing females to select sites based on ant presence.

The results of our study were only partially consistent with this prediction. We found no evidence that Miami blue butterflies preferentially oviposited on host plants where mutualistic ants were present. On the contrary, naïve females oviposited less frequently on plants occupied by the ant species that most commonly tends Miami blue butterfly larvae. It is not uncommon for ants that tend lycaenid larvae to behave antagonistically toward ovipositing females of the same species [28, 29, 49, 50], so there may be benefits to adult survival from avoiding interactions with potentially aggressive ants. However, it is not clear from our study why female butterflies tended by ants as larvae would respond differently to ant proximity or potential aggressive behaviors during oviposition.

The most provocative result of the oviposition experiment was that Miami blue butterflies tended by *C. floridanus* ants as larvae were significantly more likely to oviposit on plants with ants than were naïve females. This suggests that larval history of ant interaction can reduce the apparent avoidance of ants at potential oviposition sites, although it resulted in a relative preference rather than an absolute one. Natal habitat preference induction has been convincingly demonstrated in relatively few systems [33, 51, 52]. However, both behavioral and neurological studies have shown that larval exposure to odors or suboptimal foods can induce preference for or reduce aversion to otherwise deterrent chemical stimuli among both herbivorous and parasitoid insect taxa [34, 35, 53, 54]. We did not evaluate potential mechanisms for conditioning for ants by Miami blue butterfly larvae, but high-quality mutualists can have substantial positive effects on survival and performance, so it is not surprising that there may be selection for such behavior in ant-lycaenid mutualisms.

Previous research on lycaenid butterflies has suggested ant-associated oviposition for dozens of species with the presumed benefit of increased larval performance [8, 28]. Oviposition preference for host plants where ants are present has been subject to rigorous experimentation in some systems (e.g., *Jalmenus evagoras* [28, 55], *Anthene emolus* [29, 50], and *Ogyris amaryllis* [31]). However, other studies suggesting ant-associated oviposition have probably actually measured the correlation between ant presence and suitable host plants rather than strictly demonstrating oviposition preference for ants [56–58].

The results of our study lead to three suggestions for developing and testing hypotheses related to ant-related oviposition and larval growth among lycaenids. First, it is important to recognize that relative preference may be adaptively significant even if butterflies do not show an absolute preference for ovipositing in sites associated with ant mutualists. Second, researchers would benefit from considering the broader literature on oviposition preference [1]. Most tests of ant-related oviposition among lycaenids note the importance of the relationship with ants but do not consider the reliability of cues at the time of oviposition for future interaction or the likelihood that larvae will encounter ants even if they are initially absent. For example, oviposition on a high-quality host plant may be more important for many lycaenids than ant presence at the time of oviposition, particularly when larvae are likely to encounter ants later in life as is true for the Miami blue butterfly. In such cases, we would not expect females to use ants as oviposition cues even if ant tending has substantial effects on fitness. Third, identifying the mechanisms that allow ovipositing females to distinguish among host plants with and without ants should be a priority for future work [59, 60]. Other studies on lycaenids have suggested that visual [28, 29, 50], tactile [31], and chemical [57] cues associated with ant presence may stimulate oviposition. Further examination of the physiological and behavioral components of such mechanisms would greatly enrich the existing theory and empirical evidence of ant-related oviposition among lycaenids.

Acknowledgments

This study was conducted under Florida Fish and Wildlife Conservation Commission permit WX02525f. Funding was provided by the Florida Fish and Wildlife Conservation Commission, E.O. Dunn Foundation, U.S. Fish and Wildlife Service, and the National Fish and Wildlife Foundation; M.D. Trager was funded by a National Science Foundation Graduate Research Fellowship.

References

[1] W. J. Resetarits, "Oviposition site choice and life history evolution1," *American Zoologist*, vol. 36, no. 2, pp. 205–215, 1996.

[2] J. Bernardo, "Maternai effects in animal ecology," *American Zoologist*, vol. 36, no. 2, pp. 83–105, 1996.

[3] J. F. Rieger, C. A. Binckley, and W. J. Resetarits, "Larval performance and oviposition site preference along a predation gradient," *Ecology*, vol. 85, no. 8, pp. 2094–2099, 2004.

[4] K. A. Angelon and J. W. Petranka, "Chemicals of predatory mosquitofish (*Gambusia affinis*) influence selection of oviposition site by *Culex mosquitoes*," *Journal of Chemical Ecology*, vol. 28, no. 4, pp. 797–806, 2002.

[5] P. Ballabeni, M. Wlodarczyk, and M. Rahier, "Does enemy-free space for eggs contribute to a leaf beetle's oviposition preference for a nutritionally inferior host plant?" *Functional Ecology*, vol. 15, no. 3, pp. 318–324, 2001.

[6] T. H. Oliver, I. Jones, J. M. Cook, and S. R. Leather, "Avoidance responses of an aphidophagous ladybird, *Adalia bipunctata*, to aphid-tending ants," *Ecological Entomology*, vol. 33, no. 4, pp. 523–528, 2008.

[7] M. A. Morales, "Ant-dependent oviposition in the membracid *Publilia concava*," *Ecological Entomology*, vol. 27, no. 2, pp. 247–250, 2002.

[8] N. E. Pierce, M. F. Braby, A. Heath et al., "The ecology and evolution of ant association in the Lycaenidae (Lepidoptera)," *Annual Review of Entomology*, vol. 47, pp. 733–771, 2002.

[9] N. E. Pierce, R. L. Kitching, R. C. Buckley, M. F. J. Taylor, and K. F. Benbow, "The costs and benefits of cooperation between the Australian lycaenid butterfly, *Jalmenus evagoras*, and its attendant ants," *Behavioral Ecology and Sociobiology*, vol. 21, no. 4, pp. 237–248, 1987.

[10] M. Baylis and N. E. Pierce, "Lack of compensation by final instar larvae of the myrmecophilous lycaenid butterfly, *Jalmenus evagoras*, for the loss of nutrients to ants," *Physiological Entomology*, vol. 17, no. 2, pp. 107–114, 1992.

[11] A. M. Fraser, A. H. Axén, and N. E. Pierce, "Assessing the quality of different ant species as partners of a myrmecophilous butterfly," *Oecologia*, vol. 129, no. 3, pp. 452–460, 2001.

[12] J. H. Cushman, V. K. Rashbrook, and A. J. Beattie, "Assessing benefits to both participants in a lycaenid-ant association," *Ecology*, vol. 75, no. 4, pp. 1031–1041, 1994.

[13] D. Wagner, "Species-specific effects of tending ants on the development of lycaenid butterfly larvae," *Oecologia*, vol. 96, no. 2, pp. 276–281, 1993.

[14] K. Fiedler and B. Hölldobler, "Ants and *Polyommatus icarus* immatures (Lycaenidae)—sex-related developmental benefits and costs of ant attendance," *Oecologia*, vol. 91, no. 4, pp. 468–473, 1992.

[15] L. A. Kaminski and D. Rodrigues, "Species-specific levels of ant attendance mediate performance costs in a facultative myrmecophilous butterfly," *Physiological Entomology*, vol. 36, no. 3, pp. 208–214, 2011.

[16] K. Fiedler and C. Saam, "Does ant-attendance influence development in 5 European Lycaenidae butterfly species? (Lepidoptera)," *Nota Lepidopterologica*, vol. 17, no. 1-2, pp. 5–24, 1994.

[17] L. Rowe and D. Ludwig, "Size and timing of metamorphosis in complex life cycles: time constraints and variation," *Ecology*, vol. 72, no. 2, pp. 413–427, 1991.

[18] K. Gotthard, "Adaptive growth decisions in butterflies," *BioScience*, vol. 58, no. 3, pp. 222–230, 2008.

[19] D. Schluter, T. D. Price, and L. Rowe, "Conflicting selection pressures and life history trade-offs," *Proceedings of the Royal Society B: Biological Sciences*, vol. 246, no. 1315, pp. 11–17, 1991.

[20] P. A. Abrams, O. Leimar, S. Nylin, and C. Wiklund, "The effect of flexible growth rates on optimal sizes and development times in a seasonal environment," *American Naturalist*, vol. 147, no. 3, pp. 381–395, 1996.

[21] E. A. Bernays, "Feeding by lepidopteran larvae is dangerous," *Ecological Entomology*, vol. 22, no. 1, pp. 121–123, 1997.

[22] W. U. Blanckenhorn, "The quarterly review of biology: the evolution of body size: what keeps organisms small?" *Quarterly Review of Biology*, vol. 75, no. 4, pp. 385–407, 2000.

[23] W. U. Blanckenhorn, "Adaptive phenotypic plasticity in growth, development, and body size in the yellow dung fly," *Evolution*, vol. 52, no. 5, pp. 1394–1407, 1998.

[24] B. J. Danner and A. Joern, "Stage-specific behavioral responses of *Ageneotettix deorum* (Orthoptera: Acrididae) in the presence of lycosid spider predators," *Journal of Insect Behavior*, vol. 16, no. 4, pp. 453–464, 2003.

[25] K. Gotthard, "Increased risk of predation as a cost of high growth rate: an experimental test in a butterfly," *Journal of Animal Ecology*, vol. 69, no. 5, pp. 896–902, 2000.

[26] S. Nylin and K. Gotthard, "Plasticity in life-history traits," *Annual Review of Entomology*, vol. 43, pp. 63–83, 1998.

[27] P. R. Atsatt, "Lycaenid butterflies and ants—selection for enemy-free space," *American Naturalist*, vol. 118, no. 5, pp. 638–654, 1981.

[28] N. E. Pierce and M. A. Elgar, "The influence of ants on host plant selection by *Jalmenus evagoras*, a myrmecophilous lycaenid butterfly," *Behavioral Ecology and Sociobiology*, vol. 16, no. 3, pp. 209–222, 1985.

[29] P. Seufert and K. Fiedler, "The influence of ants on patterns of colonization and establishment within a set of coexisting lycaenid butterflies in a south-east Asian tropical rain forest," *Oecologia*, vol. 106, no. 1, pp. 127–136, 1996.

[30] A. M. Fraser, T. Tregenza, N. Wedell, M. A. Elgar, and N. E. Pierce, "Oviposition tests of ant preference in a myrmecophilous butterfly," *Journal of Evolutionary Biology*, vol. 15, no. 5, pp. 861–870, 2002.

[31] P. R. Atsatt, "Ant-dependent food plant selection by the mistletoe butterfly *Ogyris amaryllis* (Lycaenidae)," *Oecologia*, vol. 48, no. 1, pp. 60–63, 1981.

[32] A. L. Ward and D. J. Rogers, "Oviposition response of scarabaeids: does 'mother knows best' about rainfall variability and soil moisture?" *Physiological Entomology*, vol. 32, no. 4, pp. 357–366, 2007.

[33] J. M. Davis and J. A. Stamps, "The effect of natal experience on habitat preferences," *Trends in Ecology and Evolution*, vol. 19, no. 8, pp. 411–416, 2004.

[34] S. S. Liu, Y. H. Li, Y. Q. Liu, and M. P. Zalucki, "Experience-induced preference for oviposition repellents derived from a non-host plant by a specialist herbivore," *Ecology Letters*, vol. 8, no. 7, pp. 722–729, 2005.

[35] Y. Caubet, P. Jaisson, and A. Lenoir, "Preimaginal induction of adult behavior in insects," *Quarterly Journal of Experimental Psychology Section B-Comparative and Physiological Psychology*, vol. 44, no. 3-4, pp. 165–178, 1992.

[36] H. Sadeghi and F. Gilbert, "Oviposition preferences of aphidophagous hoverflies," *Ecological Entomology*, vol. 25, no. 1, pp. 91–100, 2000.

[37] E. V. Saarinen and J. C. Daniels, "Miami blue butterfly larvae (Lepidoptera: Lycaenidae) and ants (Hymeoptera: Formicidae): new information on the symbionts of an endangered taxon," *Florida Entomologist*, vol. 89, no. 1, pp. 69–74, 2006.

[38] M. D. Trager and J. C. Daniels, "Ant tending of Miami blue butterfly larvae (Lepidoptera: Lycaenidae): partner diversity and effects on larval performance," *Florida Entomologist*, vol. 92, no. 3, pp. 474–482, 2009.

[39] M. D. Trager and J. C. Daniels, "Size effects on mating and egg production in the Miami blue butterfly," *Journal of Insect Behavior*, vol. 24, no. 1, pp. 34–43, 2011.

[40] Florida Fish and Wildlife Conservation Commission, "Management plan: Miami blue *Cyclargus* (=*Hemiargus*) *thomasi bethunebakeri*," 2003.

[41] S. P. Carroll and J. Loye, "Invasion, colonization, and disturbance; historical ecology of the endangered Miami blue butterfly," *Journal of Insect Conservation*, vol. 10, no. 1, pp. 13–27, 2006.

[42] R Development Core Team, *R: A Language and Environment for Statistical Computing*, vol. 2, R Foundation for Statistical Computing, 2011.

[43] T. A. Waite and L. G. Campbell, "Controlling the false discovery rate and increasing statistical power in ecological studies," *Ecoscience*, vol. 13, no. 4, pp. 439–442, 2006.

[44] J. L. Bronstein, "The costs of mutualism," *American Zoologist*, vol. 41, no. 4, pp. 825–839, 2001.

[45] B. Stadler, K. Fiedler, T. J. Kawecki, and W. W. Weisser, "Costs and benefits for phytophagous myrmecophiles: when ants are not always available," *Oikos*, vol. 92, no. 3, pp. 467–478, 2001.

[46] K. H. Keeler, "A model of selection for facultative non-symbiotic mutualism," *American Naturalist*, vol. 118, no. 4, pp. 488–498, 1981.

[47] D. Wagner and C. Martínez del Rio, "Experimental tests of the mechanism for ant-enhanced growth in an ant-tended lycaenid butterfly," *Oecologia*, vol. 112, no. 3, pp. 424–429, 1997.

[48] G. Uhl, S. Schmitt, M. A. Schäfer, and W. Blanckenhorn, "Food and sex-specific growth strategies in a spider," *Evolutionary Ecology Research*, vol. 6, no. 4, pp. 523–540, 2004.

[49] N. Collier, "Identifying potential evolutionary relationships within a facultative lycaenid-ant system: ant association, oviposition, and butterfly-ant conflict," *Insect Science*, vol. 14, no. 5, pp. 401–409, 2007.

[50] K. Fiedler and U. Maschwitz, "The symbiosis between the weaver ant, *Oecophylla smaragdina*, and *Anthene emolus*, an obligate myrmecophilous lycaenid butterfly," *Journal of Natural History*, vol. 23, no. 4, pp. 833–846, 1989.

[51] A. B. Barron, "The life and death of Hopkins' host-selection principle," *Journal of Insect Behavior*, vol. 14, no. 6, pp. 725–737, 2001.

[52] N. Janz, L. Söderlind, and S. Nylin, "No effect of larval experience on adult host preferences in *Polygonia c-album* (Lepidoptera: Nymphalidae): on the persistence of Hopkins' host selection principle," *Ecological Entomology*, vol. 34, no. 1, pp. 50–57, 2009.

[53] S. Ray, "Survival of olfactory memory through metamorphosis in the fly *Musca domestica*," *Neuroscience Letters*, vol. 259, no. 1, pp. 37–40, 1999.

[54] Y. Akhtar and M. B. Isman, "Larval exposure to oviposition deterrents alters subsequent oviposition behavior in generalist, *Trichoplusia ni* and specialist, *Plutella xylostella* moths," *Journal of Chemical Ecology*, vol. 29, no. 8, pp. 1853–1870, 2003.

[55] M. Baylis and N. E. Pierce, "The effect of host-plant quality on the survival of larvae and oviposition by adults of an ant-tended lycaenid butterfly, *Jalmenus evagoras*," *Ecological Entomology*, vol. 16, no. 1, pp. 1–9, 1991.

[56] D. Jordano, J. Rodríguez, C. D. Thomas, and J. Fernández Haeger, "The distribution and density of a lycaenid butterfly in relation to *Lasius* ants," *Oecologia*, vol. 91, no. 3, pp. 439–446, 1992.

[57] S. F. Henning, "Biological groups within the Lycaenidae (Lepidoptera)," *Journal of the Entomological Society of Southern Africa*, vol. 46, no. 1, pp. 65–85, 1983.

[58] A. S. Seymour, D. Gutiérrez, and D. Jordano, "Dispersal of the lycaenid *Plebejus argus* in response to patches of its mutualist ant *Lasius niger*," *Oikos*, vol. 103, no. 1, pp. 162–174, 2003.

[59] D. Wagner and L. Kurina, "The influence of ants and water availability on oviposition behaviour and survivorship of a facultatively ant-tended herbivore," *Ecological Entomology*, vol. 22, no. 3, pp. 352–360, 1997.

[60] M. Musche, C. Anton, A. Worgan, and J. Settele, "No experimental evidence for host ant related oviposition in a parasitic butterfly," *Journal of Insect Behavior*, vol. 19, no. 5, pp. 631–643, 2006.

5

Habitat Choice and Speciation

Sophie E. Webster,[1] Juan Galindo,[1,2] John W. Grahame,[3] and Roger K. Butlin[1]

[1] *Department of Animal and Plant Sciences, The University of Sheffield, Western Bank, Sheffield S10 2TN, UK*
[2] *Departamento de Bioquímica, Genética e Inmunología, Facultad de Biología, Universidad de Vigo, 36310 Vigo, Spain*
[3] *Institute of Integrative and Comparative Biology, Faculty of Biological Sciences, University of Leeds, Leeds LS2 9JT, UK*

Correspondence should be addressed to Sophie E. Webster, sophie.webster@sheffield.ac.uk

Academic Editor: Rui Faria

The role of habitat choice in reproductive isolation and ecological speciation has often been overlooked, despite acknowledgement of its ability to facilitate local adaptation. It can form part of the speciation process through various evolutionary mechanisms, yet where habitat choice has been included in models of ecological speciation little thought has been given to these underlying mechanisms. Here, we propose and describe three independent criteria underlying ten different evolutionary scenarios in which habitat choice may promote or maintain local adaptation. The scenarios are the result of all possible combinations of the independent criteria, providing a conceptual framework in which to discuss examples which illustrate each scenario. These examples show that the different roles of habitat choice in ecological speciation have rarely been effectively distinguished. Making such distinctions is an important challenge for the future, allowing better experimental design, stronger inferences and more meaningful comparisons among systems. We show some of the practical difficulties involved by reviewing the current evidence for the role of habitat choice in local adaptation and reproductive isolation in the intertidal gastropod *Littorina saxatilis*, a model system for the study of ecological speciation, assessing whether any of the proposed scenarios can be reliably distinguished, given current research.

1. Introduction

The role of divergent natural selection in speciation has been widely studied in recent years [1]. There is now broad acceptance that selection of this type can lead to the evolution of reproductive isolation, even in the face of gene flow [2]. Nevertheless, significant controversy remains. Is "ecological speciation" really distinct from other modes of speciation [3]? Why does reproductive isolation remain incomplete in some cases but not in others [4]? Do chromosomal rearrangements [5] or divergence hitchhiking [6] help to overcome the antagonism between selection and recombination? What is the role of the so-called "magic traits" [7]?

"Habitat isolation" is one part of the ecological barrier to gene exchange between species that includes effects due to local adaptation, competition, and choice [8]. In this paper, we will focus our attention on habitat choice, discussing the nature of its role in ecological speciation and the potential contribution towards reproductive isolation of various forms of habitat choice. We define habitat choice as any behaviour that causes an individual to spend more time in one habitat type than another compared with the expectation based on random dispersal (see "habitat selection", p. 184, [9]). On the basis of this definition, a simple reduction in dispersal distance would not constitute habitat choice. Examples of mechanisms that might underlie choice include active movement into a preferred habitat; reduced dispersal in the preferred habitat relative to a nonpreferred habitat; preferential settling of propagules after a dispersal phase or a change in the timing of dispersal that influences the probability of arriving in a different habitat. Habitats may be spatially separated on various scales, from abutting distributions to a fine-scale mosaic, even different parts of the same host plant [10]. They also need not be separated in space at all: temporal or seasonal separation is possible. In the case of seasonal separation, allochronic emergence or reproduction [11] effectively constitutes forms of habitat choice but will not be considered here.

Habitat heterogeneity can lead to ecological speciation in the presence of gene flow. It requires divergent selection, which results in the establishment of a multiple-niche polymorphism [1] (Figure 1). This might arise *in situ* or on secondary contact between previously allopatric populations. A mechanism for nonrandom mating must then become associated with this polymorphism [12]. As a result of this two-step process, habitat choice can influence the probability of speciation or the degree of reproductive isolation achieved in one of two ways; firstly, habitat choice may increase the range of parameters under which a stable polymorphism can be maintained by selection in a heterogeneous environment [13]. This effect can be independent of any effect on mating pattern if, for example, individuals feed in two contrasting environments but mate away from the food resources. Secondly, habitat choice may cause assortative mating if it results in partially independent mating pools. Habitat choice of this type may be favoured by selection against hybrids and so constitutes a form of reinforcement [14]. Choice of mating habitat may cause assortment without influencing local adaptation but, clearly, habitat choice may have both effects (i.e., on polymorphism and assortative mating) simultaneously in some systems, strongly influencing the likelihood and speed of speciation [15].

Additionally, the trait responsible for habitat choice and the locally adapted trait—responsible for adaptations to local conditions—must generally become associated, typically requiring the build-up of linkage disequilibrium but potentially facilitated by pleiotropy (see preference-performance correlations in [16]), as is the case for other components of reproductive isolation [2]. However, it is possible for habitat choice to be under direct selection and contribute to nonrandom mating, thus constituting a "magic trait" (*sensu* Gavrilets [17]). "Magic traits" are usually discussed in the context of locally adaptive traits that also contribute to assortative mating, such as signals and preferences, though the concept can clearly be applied to any trait that promotes reproductive isolation. A recent attempt at clarification of the "magic trait" definition suggested "a trait subject to divergent selection and a trait contributing to nonrandom mating that are pleiotropic expressions of the same gene(s)" [7] but this view unfortunately confounds two distinct ideas: the impact of a single trait on more than one component of reproductive isolation and the effect of a gene on more than one trait. Here, we follow the proposal by Smadja and Butlin [2] to distinguish "single-effect" and "multiple-effect" traits and avoid the use of the, now confusing, term "magic."

Multiple-effect traits facilitate the evolution of reproductive isolation by reducing or removing the need for linkage disequilibrium and so avoiding the negative effects of recombination [2, 18]. Their contribution to speciation still depends on the magnitude of their effects [19]. A trait that contributes to reproductive isolation through habitat choice may also contribute to reproductive isolation in other ways, including but not exclusively through effects that lead to direct selection, and so may be a multiple-effect trait. Other things being equal, we expect such traits to increase the probability or speed of speciation.

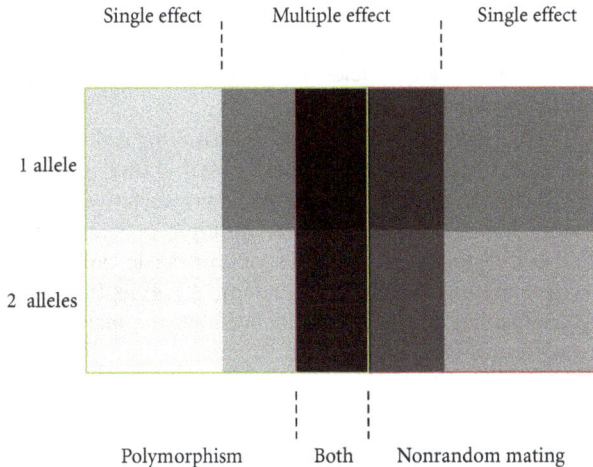

FIGURE 1: Habitat choice and the probability of speciation. Darker colours represent increased probability of speciation. See main text for further explanation.

Habitat choice may also evolve by either a "one-allele" or a "two-allele" mechanism [18], as in the "single-variation" and "double-variation" models of de Meeûs et al. [20]. A "one-allele" scenario might involve "habitat matching" (e.g., [21]), for example, causing an animal to move until it is cryptic against its background and then to remain stationary. Indirect selection would favour the spread of such an allele in an environment with two backgrounds where predation maintained a polymorphism for alternate cryptic colours, without the need for linkage disequilibrium because the matching allele is favoured in association with both colour morphs. Alternatively, a "two-allele" mechanism might involve one allele favouring upward movement and another favouring downward movement. Here, the evolution of habitat choice, and so reproductive isolation, relies on linkage disequilibrium between the upward movement allele and alleles conferring increased fitness in the high habitat and between the downward allele and the alternate alleles at the fitness loci. Note that the "one-allele" versus "two-allele" distinction still holds for polygenic traits and can be made without knowing the genetic basis of the trait—the primary issue is whether the trait has to change in the same direction (one-allele) or in different directions (two-allele) in the diverging populations. "One-allele" mechanisms increase the probability of speciation and can potentially be more effective in limiting gene flow between subpopulations than two-allele traits. As with multiple-effect traits, this is because they remove the need for linkage disequilibrium.

Habitat choice may be plastic, including the possibility of learning, and plasticity can be an important factor in speciation [22]. The extent of plastic response may vary, up to the point where no genetic difference needs to exist between individuals showing behavioural preferences for contrasting environments. However, plasticity itself has a genetic basis; evolution can act upon the degree of plasticity, both the ability to learn and biases in the way individuals learn can evolve. Evolution of plasticity or learning may be considered examples of a "one-allele" mechanism, where the ability to modify the learning phenotype in a way that results in advantageous habitat choice is genetically determined and the same alleles, for effective learning, are beneficial everywhere.

These three different criteria for classifying traits responsible for habitat choice (whether habitat choice influences the maintenance of polymorphism or assortative mating; whether habitat choice traits are multiple or single effect; whether they follow a "one-allele" or "two-allele" inheritance mechanism) are largely independent. An exception arises because habitat choice may influence both the maintenance of polymorphism and assortative mating, and this is an example of a multiple-effect trait. If all possible combinations between the three criteria are considered, they may therefore lead, at least theoretically, to ten different habitat choice scenarios (Figure 1). Here we will discuss the implications of each one of these criteria for habitat choice and speciation, including examples of empirical studies and/or models. We will also review and discuss the current evidence for the role of habitat choice in local adaptation, including its possible effect on ecological speciation in an intertidal gastropod, the rough periwinkle *Littorina saxatilis*. We use this example to illustrate some of the practical difficulties involved in distinguishing among habitat choice scenarios empirically. This system also emphasises the generality of the distinctions by comparison with the more widely studied insect models in which host plants form the contrasting habitats.

2. Habitat Choice Scenarios

Figure 1 provides a conceptual classification of habitat choice scenarios, illustrating the likelihood of progression towards speciation under each scenario (a scenario consisting of a combination of criteria). It suggests that, at one extreme, habitat choice may contribute to the maintenance of polymorphism without directly contributing towards reproductive isolation (as indicated by the lightest area) whereas, at the other extreme, habitat choice evolution may be a primary driver of speciation (as indicated by the darkest, central areas). For simplicity we have erected a categorical system (our "criteria"). However, due to the dynamic nature of the evolutionary process, one or more habitat choice scenarios from our framework may contribute to speciation at different stages in the process. For example, stabilisation of multiple-niche polymorphism may often be an early stage of speciation whereas the evolution of divergent preferences for mating habitat could be a form of reinforcement at a late stage of the process. Below, we will discuss the relevance of each criterion of this classification with reference to illustrative examples. However, it will quickly become clear that there is often insufficient empirical information to be certain how individual case studies fit into our classification, partly because the role of habitat choice in speciation has not been addressed in a strong conceptual framework. Additionally, we were unable to identify good empirical examples that satisfactorily demonstrate some of our proposed scenarios—this highlights both the limited range of the studies that have been undertaken in this area and the difficulty in

identifying the mechanisms involved. We illustrate some of these difficulties in a final section dealing with one particular example, the periwinkle *Littorina saxatilis*.

2.1. Polymorphism or Nonrandom Mating. The first of our categories deals with the role of polymorphism and nonrandom mating in habitat choice scenarios: the establishment of polymorphism generated by divergent selection is a necessary step in ecological speciation.

Ecological speciation in the presence of gene flow requires divergent selection, which results in the establishment of polymorphism [1]. The maintenance of genetic variation within populations in heterogeneous environments has been widely discussed in the past, from theoretical models to experimental studies (reviewed in [23]). One of the first models, by Levene [24], showed that the maintenance of polymorphism in an environment with two habitats was possible. In this model there is random migration of individuals into the habitats, selection favours one of the genotypes in each habitat, each habitat contributes a constant number of individuals to the next generation, and there is random mating. The range of the parameters necessary to maintain polymorphism is rather restricted; nevertheless, this model was the basis for Maynard Smith's [25] classic analysis of sympatric speciation. It is clear that under this model habitat choice is favoured by indirect selection (local adaptation to one of the habitats will secondarily promote preference for that habitat) and that habitat choice considerably expands the range of parameters under which the polymorphism is stable. Many subsequent models have shown that frequency- or density-dependent selection makes polymorphism more likely (e.g., [26, 27]) and so extends the range of situations in which an initial polymorphism will create conditions for the evolution of habitat choice, or where preexisting habitat choice will favour the establishment of multiple-niche polymorphism, reinforcing local adaptation (e.g., [20, 28]). The interplay between habitat choice and local adaptation has recently been reviewed by Ravigné et al. [13, 29]. From the point of view of speciation, the maintenance of local adaptation creates opportunities for the further evolution of reproductive isolation [25]. Therefore, if habitat choice helps to maintain polymorphism and if polymorphism reinforces local adaptation, it also increases the likelihood of speciation. This is true even if habitat choice has no influence on mating pattern. However, if habitat choice also generates assortative mating, its contribution to speciation will be greater (Figure 1). Habitat choice of this type is favoured by indirect selection when a polymorphism is established and this constitutes a form of reinforcement [14]. Finally, it is possible that habitat choice may apply only to the mating habitat and so influence mating patterns without being directly connected to a source of divergent selection. To understand the role of habitat choice in speciation, it is important to distinguish these categories and to determine the stage in the speciation process at which choice evolves.

Natural cases where both polymorphism and habitat choice are present in the population (but the choice does not influence mating pattern) are uncommon in the literature.

From the point of view of speciation, these examples should represent initial stages of speciation or divergence, but the outcome of this process is highly dependent on the reigning conditions (e.g., habitat distribution, habitat size, and selection coefficients) and there may actually be no progression towards greater isolation [4]. The most clear-cut examples are likely to come from species where mating occurs in a different habitat from the majority of the life-cycle, such as aquatic insects with a brief aerial mating phase. One interesting possible case concerns the aphid genus *Cryptomyzus* in which sibling species, which still occasionally hybridise in the field, utilise either dead nettle (*Lamium galeobdolon*) or hemp-nettle (*Galeopsis tetrahit*) as summer host. On these hosts they reproduce asexually, but their sexual generations occur on redcurrant (*Ribes rubrum*) regardless of the summer host [30]. Here, there is assortative mating but it is not due to the strong preference for distinct summer hosts. Rather, it seems to be mediated by differences in the diurnal cycle of pheromone release by females. This situation, where multiple races or species share the same primary host while utilising distinct summer hosts, seems to be common in aphids. The scenario where habitat choice can maintain a polymorphism but have no influence on nonrandom mating appears to be rare but may simply be under-represented in the literature. It is not likely to be favourable for progression to complete reproductive isolation.

The opposite situation, where habitat choice influences mating alone, is also not widely documented but is also most likely where mating occurs in a distinct habitat from the majority of the life cycle. A possible example may occur in the mosquito *Anopheles gambiae* whose larvae develop in small, often temporary water bodies but whose adults mate in aerial swarms. The M and S molecular forms of *A. gambiae* cooccur in many parts of Africa and show strong, but incomplete reproductive isolation. A major contributor to pre-zygotic isolation appears to be the choice of distinct habitats in which to form mating swarms [31]. This scenario is also unfavourable for speciation because there is no close connection between assortative mating and a source of divergent selection.

In the majority of cases habitat choice is likely to influence both maintenance of polymorphism and assortative mating, as mating usually takes place in the same habitat as the life-cycle phase in which selection occurs. A trait responsible for habitat choice of this type may be considered a multiple-effect trait because it has effects on reproductive isolation both through assortative mating and through enhancing local adaptation. This combination may generate a higher probability of progressing to complete reproductive isolation than cases where habitat choice influences only one component of isolation. The situation is typical of many phytophagous insects, which remain on their host to mate, and has been very widely studied in this context [32], including important early models (e.g., [28]) and classic research on the apple maggot fly, *Rhagoletis pomonella*.

Sympatric host races of the apple maggot fly in North America have evolved in the last 150 years, with a host shift occurring from hawthorn to apple around 1860 [33]. Life cycles in these two host races are very similar; mating occurs on the plant and females oviposit in ripening fruit. Larvae complete their development in the ripe fallen fruits, pupate in the soil, and undergo a facultative diapause until spring. Adults congregate again on the host plant for mating. Host races are differentially adapted, primarily through their diapause characteristics which match the timing of the life-cycles to differences in phenology of their hosts [34, 35]. There are also host-associated differences in larval survival [36].

Host fidelity in *Rhagoletis* is partly a result of limited dispersal but there is clear evidence for active habitat choice involving fruit size and colour and, especially, volatile chemical signatures [37, 38]. Because there are clear fitness differences and mating occurs on the host fruit, host choice clearly contributes to both stabilisation of the coexistence of the races and to assortative mating between them.

The historical role of host choice is less easily determined. Feder [39] suggested that individuals with a genetic preference for apples, the derived host, may have gained an immediate selective advantage, perhaps involving the use of an empty niche or escape from parasitism ("apple race flies have less parasitoids than hawthorn race flies because parasitoids use plant cues when searching for their hosts," [40]). In this case, habitat choice may have evolved first and facilitated subsequent host adaptation, rather than evolving in response to the fitness costs of oviposition on the wrong host. Habitat choice would then be a multiple-effect trait since it would be under direct selection, as well as contributing to assortative mating (see below). A further complication to this hypothesis is the association of some host-specific traits with chromosomal inversions. These inversions appear to predate the introduction of apples to North America and may have evolved during a period of allopatry [41]. Their presence in the population may also have facilitated the host shift, interacting with changes in preference.

2.2. Habitat Choice: "Multiple-Effect Trait" versus "Single-Effect Trait". The *Rhagoletis* example nicely illustrates the distinction between multiple-effect and single-effect traits, as defined in Section 1. We can envisage two possible historical sequences. In the first, multiple-effect, scenario, an allele arose in the ancestral hawthorn population which increased the likelihood of females ovipositing on apple. This habitat choice allele was favoured by direct selection because larvae developing on apple had higher survival but at the same time contributed to isolation through its impact on assortative mating. This led to the establishment of a population on apple trees, which further adapted to the new habitat, including divergence in diapause timing. Johnson et al. [42] modelled a scenario of this type. An alternative, single-effect, scenario would begin with a proportion of females ovipositing on apple by chance. This favoured alleles for high survival on apple leading to the establishment of a multiple-niche polymorphism. Indirect selection could then have favoured habitat choice through its impact on assortative mating alone, requiring linkage disequilibrium between survival and choice alleles to be established in the face of gene flow and recombination (a form of reinforcement). Distinguishing such alternatives retrospectively is likely to be very difficult.

The importance of phenology in *Rhagoletis* suggests a further option. Apple and hawthorn fruit are temporally separated habitats. Therefore, any change in diapause timing early in the evolution of the apple race would have constituted a multiple-effect trait under direct selection, because of the benefits of matching timing to the host and also influencing habitat choice. Since the timing difference could have generated assortative mating, this trait would have made three distinct contributions to reproductive isolation.

Interestingly, disrupted host finding in *Rhagoletis* hybrids also contributes to postzygotic isolation [43], a neglected potential contribution of habitat choice to speciation. This constitutes a distinct pathway by which a habitat choice trait can have multiple effects on reproductive isolation, which is independent of the polymorphism versus nonrandom mating criterion. For this reason, we consider the single-effect/multiple-effect distinction separately, because the impact of a habitat choice trait on both polymorphism and nonrandom mating is not the only way in which it can act as a multiple-effect trait.

Another good example of a direct link between habitat choice and adaptation is the pea aphid, *Acyrthosiphon pisum*. In this phytophagous insect, feeding, mating, and oviposition occur only on the host plant. In the northeastern United States, host races of the subspecies *A. pisum pisum* live on alfalfa (*Medicago*) and clover (*Trifolium*), crops that are sometimes grown in adjacent fields, and some gene flow persists between the races [6]. The aphids acceptance of the host plant is one of the main reasons for assortative mating: pea aphids can distinguish between their preferred host and the alternate host by probing with their stylets. When they detect the alternate host they do not feed but will move in search of another plant in order to increase their probabilities of reproductive success [44]. Thus, habitat choice and habitat-associated fitness are aspects of the same underlying trait of host acceptance, which can be considered a multiple-effect trait. As with the apple maggot fly, host acceptance also influences assortative mating, and this situation greatly facilitates the evolution of reproductive isolation [2]. It is likely that the key genetic changes of host acceptance involve the aphids chemical senses, and the recent characterisation of chemosensory gene families in the pea aphid [45] opens the way to identification of the responsible genes.

Hawthorne and Via [46] showed that, for the traits they defined as host acceptance and host-associated performance, QTL mapped close together in the pea aphid genome. They suggest that there may be either close linkage between genes for the different traits or alleles with pleiotropic effects on the traits. Following Smadja and Butlin [2], we suggest that it is more instructive to view host acceptance as a multiple-effect trait with direct effects on fitness. Of course, there may be other traits that also adapt the aphids to different host environments, such as mechanisms for tackling host defensive compounds.

2.3. "One-Allele" versus "Two-Allele" Mechanisms. "One-allele" and "two-allele" mechanisms have been widely discussed in speciation research since the distinction was introduced by Felsenstein [18] and the distinction has been considered one of the most useful ways to categorise speciation [47]. In principle, habitat choice may evolve by either mechanism [20] but making the distinction for empirical examples is not straightforward. Therefore, we begin with conceptual examples to illustrate the ways in which "one-allele" and "two-allele" mechanisms might apply to habitat choice before suggesting possible case studies.

In "one-allele" mechanisms, a single allele present in a population under divergent or disruptive selection generates habitat preference independently of the direction of selection. One possible way for this to work would be through natal habitat preference induction (reviewed in [48]), in which experience with a natal habitat shapes the preferences of individuals for that habitat. Experience with particular stimuli increases subsequent preference for a habitat that contains those same stimuli, which might help dispersing individuals to locate a suitable habitat quickly and efficiently. Because assessing habitat quality involves time, risk, and energy invested in sampling potential habitats, selection should favour mechanisms that help individuals to select and use habitats that best suit their phenotypes. Any allele that spreads in response to such selection would enhance divergent habitat choice in a population showing local adaptation to multiple niches. Habitat matching, for example in cryptically coloured species, would have a similar effect without the need to condition on the natal habitat (matching habitat choice, [21]), promoting local adaptation and even leading to speciation. On the other hand, in response to environmental change, adaptation could involve change or relaxation of habitat choice instead of adaptation to modified conditions, resulting in more variable habitat use [21], which could lead to breakdown of reproductive isolation. Under these conditions the "two-allele" mechanism would offer greater resistance to hybridisation than the "one-allele" mechanism, although "two-allele" habitat choice would be less likely to initiate the process of speciation.

"Two-allele" habitat choice is more likely to involve innate preferences for specific habitat features, such as substrate colour or odour or the presence of particular resources. Divergent or disruptive selection is required to establish such distinct preferences.

Differentiating between "one-allele" (e.g., habitat matching) and "two-allele" (distinct preferences in different subpopulations) scenarios is not an easy task, as previously mentioned. As an example, we consider colour morphs of another phytophagous insect, the walking stick *Timema cristinae* (see [49] for a review). Two host species with highly divergent leaf colour and shape, *Ceanothus spinosus* and *Adenostoma fasciculatum,* are utilised. Insects found on different hosts have different cryptic colour patterns because of selection due to predation. These wingless insects feed, mate, and reproduce on the same host individual and movement between plants is restricted (12 m per generation, [50]). They show host preference [51] and partial assortative mating [52]. Immigrant inviability (selection against ecotypes from the contrasting habitat: i.e., host plant) is also an important process maintaining the ecotype divergence [53]. It is clear that habitat choice contributes to the maintenance of *Timema* ecotypes but the mechanistic basis of this preference has not yet been characterised. The preference could, for example, involve detection of different chemical compounds on the host plant surfaces that are unrelated to crypsis: a "two-allele" mechanism. Alternatively, the insects may match their own colouration to the background, a mechanism that could fall into the "one-allele" category. These possibilities could be distinguished experimentally, for example by allowing insects to choose between backgrounds of different colour/pattern in the absence of plant material.

Preferences for divergent chemical signals are likely to underlie host-plant preference in many phytophagous insects [54]. These are likely to involve "two-allele" mechanisms where different alleles for positive or negative responses [55] to particular stimuli have to be fixed in the diverging subpopulations and have to be associated, through linkage disequilibrium or pleiotropy, with traits that underlie local adaptation. Because this is less favourable to speciation than matching mechanisms, it is important to make the distinction in more case studies.

Particularly interesting possible cases of "one-allele" habitat choice involve learning, including imprinting. Such cases are sometimes described as "nongenetic" but clearly the ability to imprint or the strength of imprinting can have a genetic basis and the spread of a single allele can then cause divergent habitat preferences by promoting imprinting on different signals. Obligate brood parasitic birds represent potential examples of this process. In the continuum of divergence and/or speciation [4, 56, 57], different races or species pairs represent different stages in the process of ecological speciation, and this is also the case for brood parasitic birds. Indigobirds (genus *Vidua*) are obligate, host-specific brood parasites of firefinches (genus *Lagonosticta*) and other estrildid finches (family Estrildidae). Assortative mating is due to song learning and mimicry of the host song by the males [58, 59], and this allows a host switch to occur in a single generation. Females also learn the song of their foster parents and choose their mates and the nests they parasitise using song. Different degrees of divergence are found depending on the species under study, for example, some of them are morphologically indistinguishable [60, 61] and they lack genetic differentiation at neutral markers [62]. When host species of different indigobirds have overlapping distributions, hybridisation can occur due to egg-laying mistakes (e.g., a female lays an egg in a nest parasitised by another indigobird species). Individuals of different species would then learn the same song, through imprinting, and are likely to hybridise. However, in most of the cases, indigobird species have evolved several other polymorphisms, such as different male plumage colour and nestling mouth markings that match those of their respective hosts [63], thereby enabling young indigobirds to better compete for parental care in host nests. These polymorphisms represent different axes of divergence (greater "ecological niche dimensionality," [64]) promoting increased divergence, despite the possibility of accidental hybridisation in regions where host ranges overlap. This suggests that, following colonisation of a new host, host fidelity due to imprinting can be sufficient for divergent natural and/or sexual selection on morphology, ecology, and/or behaviour to generate progress on the speciation continuum. Imprinting causes habitat choice, through its influence on female choice of host nests, as well as mate choice and so is a multiple-effect as well as a "one-allele" trait [2, 7].

In these sections, we have selected empirical examples that illustrate each of our classifying criteria. Figure 1 shows the ways in which these criteria may combine to create conceptual scenarios with varying probabilities of progression to speciation. Working from the framework, how easily can these conceptual scenarios be applied to real-world systems? Our brief review of the literature suggests that habitat choice studies pertaining to ecological speciation are biased towards phytophagous insects. This is not surprising, because "host race" formation seems to be a common route to speciation, which has been widely studied. However, in an attempt to expand the scope of habitat choice studies in an ecological speciation context, we discuss below the evidence for habitat choice in the intertidal gastropod genus *Littorina*. We examine the current evidence for habitat choice, discussing which scenarios are likely to be involved and the difficulties in trying to distinguish them.

3. Habitat Choice in *Littorina*

Intertidal gastropods present ideal systems for studies of habitat choice: the littoral zone can create extreme environmental gradients and highly heterogeneous habitats within relatively short distances, and generally the animals are easy to locate and manipulate for both *in situ* and lab-based trials. Heterogeneous habitats of this type can lead to differential survival and generate divergent selection in polymorphic populations. Microhabitat use in this landscape has been identified as strongly influencing survival in intertidal gastropods [65–67], so habitat choice presents itself as a likely trait to respond to this selection.

Large-scale transplant experiments have indicated habitat preference behaviour in *Littorina* species, such as *L. keenae* [68], *L. angulifera* [69], and *L. unifasciata* [70]. All show that the snails tend to return to the approximate tidal height from which they were displaced, exhibiting directional movement towards the shore level of origin. However, these transplant experiments may also be influenced by effects of differential survival that are hard to separate from behavioural effects. We will examine this problem below.

Littorina saxatilis, the rough periwinkle, is a marine gastropod that is emerging as a model system for studying ecological speciation. It is widely distributed across rocky shorelines in the North Atlantic, extremely polymorphic (shell colour, shell shape, and behaviour), and prone to ecotype formation due to local adaptation because of its low average dispersal [71]. Pairs of phenotypically divergent ecotypes occupying different niches in the intertidal zone are found over scales of tens of metres or shorter across different shores along its distribution and are maintained through divergent natural selection [72]. These ecotypes of *L. saxatilis* have been studied in detail on shores from three geographical regions (Sweden, UK, and Spain), and a process of parallel ecological speciation between them has been suggested ([72], but see [73, 74]). However, despite displaying phenotypic divergence, the ecotypes are not completely reproductively isolated, with gene flow still occurring (Sweden: [75], UK: [76], and Spain: [77]).

The ecotype pairs on each of these shores are separated on a microgeographic scale, exhibiting adaptations to the prevailing habitat. On Swedish shores, the habitat is composed of a mosaic of cliff habitat punctuated with boulder fields, whereas the UK and Spanish ecotype pairs are found on the same shores, but at different levels of the littoral zone. The ecotypes in the UK are known as H and M (high-shore and midshore), those in Sweden are termed E and S (exposed and sheltered), and the Spanish pair are termed RB and SU (ridged-banded and smooth-unbanded) (see [73]). The M, S, and RB ecotypes are morphologically congruent, exhibiting thick shells, relatively small shell apertures, and large body size. These features are considered to be adaptations to an important selection pressure: predation by crabs. The H, E, and SU ecotypes from these three shores also share similar morphological characters: smaller size, thinner shells, and a larger shell aperture. In avoiding the hazards of crab predation by their position on the shore (low in Spain, high in Britain and Sweden), these ecotypes are free to develop larger shell apertures, increasing foot area and thus grip on the substrate to minimise dislodgement. Nonrandom mating is also observed in each population of ecotype pairs [78–81], primarily due to assortative mating by size (see [72] for a review). L. saxatilis lacks a pelagic larval stage, instead exhibiting direct development where females retain their brood internally and release fully formed young [71]. Many other littorinid species (such as L. littorea) produce a pelagic larval stage, allowing dispersal over a wide range and maintaining gene flow between populations [71]. The low dispersal range of L. saxatilis (1–4 m, [82, 83]) limits gene flow, and this facilitates much greater local adaptation in this species than in many of its congeners [72].

Since selection drives differential adaptation to closely adjacent habitats, habitat choice mechanisms could easily be imagined to play a role in population divergence. Random dispersal combined with selection against less fit phenotypes may superficially look like habitat choice as the phenotypes are segregated into divergent habitats, as noted above. This is particularly true where dispersal distance is short allowing selection to produce sharp phenotypic transitions at habitat boundaries. Selection for reduced postnatal dispersal [84] may accentuate this effect. However, as there is no active behavioural mechanism, this does not constitute habitat choice as we define it.

A possible exception, where habitat choice can be inferred from phenotypic distributions, is where habitat heterogeneity is on a scale much smaller than the dispersal distance. It is then not possible for selection to maintain genetic differentiation between patches [85] independent from habitat choice, although phenotypic plasticity could still result in strong phenotype-habitat associations. Morphological and AFLP (amplified fragment length polymorphism, see [86] for a review) clines have been identified, which are too steep to have been generated by selection alone [87, 88]. In these cases, additional mechanisms such as habitat choice may contribute to the genesis and maintenance of

the gradient by strengthening barriers to gene flow. In the middle of the shore gradient in Spain, mussel and barnacle dominated patches are intermingled on a scale of a few centimetres and the RB and SU L. saxatilis ecotypes are associated with these patches [89]. This is strongly suggestive of active habitat choice [90]. The heterogeneous nature of this connecting habitat may be particularly important in the maintenance of the hybrid zone and the segregation of the ecotypes, as has been demonstrated with Bombina toads [91]. Nevertheless, in order to determine the role of habitat choice in maintaining divergent populations, in L. saxatilis and in other comparable systems, it is necessary to utilise manipulative experiments (e.g., using mark-recapture approaches).

Clear evidence for home-site advantages in littorinid species has been documented [83, 89, 92] along with evidence of selection on shell characters. In this context, we consider a home-site advantage to be where individuals are likely to have increased fitness in the habitat or microhabitat to which their ecotype may be presumed to be adapted. This advantage may vary at different stages of the life history. Is there also good evidence for habitat choice in the L. saxatilis ecotypes? Has L. saxatilis evolved habitat choice in response to divergent selection, or did nonrandom mating and adaptive polymorphism evolve in the presence of preexisting habitat choice? Habitat choice can be an adaptive behaviour, increasing fitness in the "home" habitat even when only a single habitat type is occupied [93] and so could have been present before ecotype differentiation began. Is habitat choice a multiple-effect trait in Littorina and is it based on "one-allele" or "two-allele" genetic variation? We discuss the evidence for the presence of habitat choice in L. saxatilis and consider whether it is possible to make any of these distinctions.

Work on Swedish populations indicates that morphological adaptation to the contrasting environment has a strong genetic basis but has an element of plasticity which can improve local adaptation [94, 95]. However, the E (exposed) ecotype displayed significantly lower levels of plasticity than the S (sheltered) ecotype, indicating differential plasticity within local populations. This leads to the question of the effect of plasticity on the role of habitat choice: it is feasible that lower phenotypic plasticity might favour the evolution of genetically based habitat choice, to increase occupation of the environment in which individuals are more fit. Increased plasticity may decrease selection for habitat choice, since individuals would be better able to adapt their phenotype to local conditions. This might be tested if the degree of morphological plasticity varies among regions, leading to a prediction of varying habitat choice.

Experimental evidence for nonrandom dispersal in L. saxatilis ecotype populations has been obtained in both Spain and Sweden [78, 82, 83]. In the Swedish populations, displaced snails exhibited greater average dispersal distances than nondisplaced ones and dispersal differed between E and S ecotypes, in addition to a tendency to recapture snails in their own habitat more often than expected from random dispersal [83]. However, this tendency to recapture snails in their own habitat may be a function of differential survival in native and nonnative habitats. Additionally, although survival rates and migration distances were measured, direction of movement was not. Erlandsson et al. [82] expanded on this study to determine whether the dispersal was directional in the Spanish population. They detected random dispersal when snails were placed at their native shore level (with overall dispersal distances averaging less than 2 m), whereas when animals were transplanted to their nonnative shore level they moved further and more directionally, with the Spanish RB morph exhibiting the greatest directional response. Although this hinted at habitat choice in this ecotype, the recapture rate was low (<20%) and conclusions were drawn only from recaptured individuals (which are likely to be a strongly selected sample), therefore it is difficult to make any meaningful conclusions about habitat choice from these simpler experimental studies.

On the Galician shore in Spain, Cruz et al. [78] tried to separate survival and habitat preference in the two ecotypes of *L. saxatilis* using two reciprocal transplant experiments. In the first experiment, sample groups of each ecotype were transplanted both to their native and nonnative shore level at each of two sites. Snails were then recaptured and their movement recorded from two days after transplant. The study compared the recapture positions of the transplanted snails and the recapture positions of the corresponding control snails, to correct for movements that may been induced via prevailing climatic conditions. In addition to the confounding effect that such forced migration might have had on the snails survival, transplanted snails may have dispersed beyond the study area leading to reduced recapture. One way in which Cruz et al. [78] avoided these complications was to argue that only directional movement could result in more than 50% of the released snails in the treatment group being recaptured in the direction of their preferred habitat (up shore for RB, down shore for SU). With this stringent criterion, habitat choice was only observed in one site.

The second experiment involved collection of snails from five intertidal levels on each of two sites and reciprocal transplants across sites for release. This destroyed the correlation between shore level and snail phenotype (measured as the first axis of shell shape variation). Over a period of two weeks, they observed the reformation of the shell-shape cline and measured the relative contributions, for the cline reformation, of the snails migrating movements and survival. Using a clever comparison between the change in average positions of all snails recaptured and the change in position of those that were known to have survived, they separated the contributions of differential survival and habitat choice to the changing cline. Differential movement contributed between one-third and one-half of the change in cline at one site and hardly at all at the other site, leading to the conclusion that habitat choice was present but less important than differential viability in the maintenance of the phenotypic cline.

This important study illustrates the difficulties associated with demonstrating habitat choice in the field. Despite considerable effort and thoughtful design, the experiments were still hampered by low recapture rates (around 50%)

and could be criticised for releasing snails at high densities in potentially unnatural positions. The analyses do not provide quantitative estimates of survival, dispersal, or their ecotype-habitat interactions, the sort of variables that would be needed to model likely evolutionary scenarios. A recent theoretical model of ecotype formation in *Littorina saxatilis* [96] did not include habitat choice as a parameter. Due to the currently unknown contribution of habitat choice to reproductive isolation in this species, it would be interesting to see how the potential to evolve choice might influence model outcomes.

Cruz et al. [78] also discussed the possible behavioural basis of habitat choice. The observation that shell morphology provided the best predictor of habitat-specific viability but that sampling location best explained the pattern of movement led them to suggest that shell shape and habitat choice are genetically independent. Therefore, in the terms we use here, shell shape is not a multiple-effect trait in the sense that a change in shape alters the fitness profile but does not automatically alter habitat preference (as it might if snails had a pre-existing tendency to move to a habitat that was favourable for their shell shape). Note, however, that shell size does seem to be multiple-effect in that it influences both differential survival and assortative mating. A "two-allele" mechanism (or "double-variation": [20, 78]) seems more likely than a "one-allele" mechanism for the same reason. Movement to the optimum habitat could be a "one-allele" mechanism but would result in a strong association between shell shape and differential movement, which may occur prior to local adaptation. An upward bias in RB and a downward bias in SU would most likely be a two-allele mechanism, dependent on linkage disequilibrium and so less tightly linked to shape. Under this assumption, habitat choice could evolve after local adaptation. Lack of preference in hybrids [90] tends to support this conclusion, suggesting that habitat choice evolved after local adaptation whereas a tendency toward matching position to optimal habitat preceded local adaptation. However, direct behavioural tests and genetic analysis are needed to confirm these speculations and will be difficult if habitat choice is as weak in other regions as it seems to be in Spain.

Other littorinid studies highlight the role of chemoreception in influencing the behaviour of individuals (such as trail following: [97, 98]). It has been determined that *L. saxatilis* E ecotype males (S males were not studied) are able to discriminate between mucous trails of the female of each ecotype [99] and show a clear preference for trails of females within the size range of the E ecotype females. In addition to its role in assortative mating, trail following could play a role in habitat choice. This has been studied to an extent in *L. littorea* [100]. When chemical cues were removed from the "home" boulder and substrate, *L. littorea* displayed a significantly impaired ability to navigate back to the boulder from which they had been displaced. In *L. saxatilis* it would be interesting to separate the role of assortative mating from habitat choice. Trail following could impact dispersal experiments by making individual movements nonindependent [78] and, if it forms the basis of philopatry, could represent a "one-allele" habitat choice mechanism. More studies are needed to unveil the role of chemoreception in habitat choice in this species.

Did size-assortative mating evolve after ecological partitioning and evolution of habitat choice? Or did habitat choice facilitate ecotype formation after the development of assortative mating? The model by Sadedin et al. [96] suggested that assortative mating may be considered ancestral. However, although a number of ecological factors were modelled, habitat choice was not included as a parameter. Dispersal was included, but this was not directional. Dispersal was an important consideration though: frequent long-range random dispersal eliminated spatial genetic structure and did not lead to ecotype formation. Although the role and mechanism of habitat choice in *Littorina* have not yet been explicitly modelled, we may draw inferences from models developed for other organisms. Early models suggested that when fitness, mating, and habitat choice are all based on the same character, speciation with gene flow may result—the degree of reproductive isolation is determined by the strength of assortative mating and the strength of disruptive selection. For moderate selection, habitat-based nonrandom mating also facilitates reproductive isolation. However, in simulations, the size-related mate choice mechanism in *L. saxatilis* could not explain more than a small part of the sexual isolation between morphs [101]. This implies that size-related mate choice, although considered a multiple-effect trait, may only be important in a speciation context if it evolves in parallel with other ecological traits, including habitat choice.

These studies highlight the difficulties in connecting theoretical evolutionary scenarios with existing empirical data. However, if future habitat choice studies are carried out with an explicit conceptual framework in mind and across a wide range of study systems, they will contribute more effectively to our understanding of speciation.

4. Conclusion

The influence of habitat choice on ecological speciation clearly varies in both magnitude and mechanism, and in many cases we cannot be sure about the contribution it makes to reproductive isolation or at what stage it evolved. The empirical examples discussed for some of our projected scenarios provide an indication of which evolutionary scenarios have been observed in natural systems. We would expect those scenarios where habitat choice does not strongly favour progress towards speciation to be detected in studies of within-species polymorphism, whereas those promoting speciation may be more prevalent among studies of ecological speciation. A more exhaustive review is needed to test this prediction but may be premature since many case studies do not yet provide enough information to distinguish among scenarios for the evolution of habitat choice.

Although there have been some valuable habitat choice studies on *Littorina saxatilis*, there are still a lot of unanswered questions regarding its role in the maintenance of both the phenotypic and genetic clines. As a candidate system for ecological speciation, the understanding of the role of habitat choice prior to complete reproductive isolation in

L. saxatilis is an important facet in our overall understanding of the processes and mechanisms leading to species formation.

Describing the role of habitat choice within the conceptual framework that we propose represents an important step in understanding speciation. It shows how habitat choice can affect reproductive isolation in very different manners, influencing the likelihood of speciation and potentially leading to different stages along the continuum of speciation. Empirical studies of habitat choice in divergent populations or closely related species, representing different stages of speciation and different evolutionary scenarios, should form a focus for future research. When analysed within such a conceptual framework, we believe these studies will give more insight into the part that habitat choice plays in ecological speciation than if they are considered in isolation.

Acknowledgments

The authors would like to thank Rui Faria, Steven Parratt, and the three anonymous referees for their useful comments on the paper. All authors are supported by the Natural Environment Research Council (NERC). J. Galindo is currently funded by "Isidro Parga Pondal" research fellowship from Xunta de Galicia (Spain).

References

[1] D. Schluter, "Evidence for ecological speciation and its alternative," *Science*, vol. 323, no. 5915, pp. 737–741, 2009.

[2] C. M. Smadja and R. K. Butlin, "A framework for comparing processes of speciation in the presence of gene flow," *Molecular Ecology*, vol. 20, no. 24, pp. 5123–5140, 2011.

[3] J. M. Sobel, G. F. Chen, L. R. Watt, and D. W. Schemske, "The biology of speciation," *Evolution*, vol. 64, no. 2, pp. 295–315, 2010.

[4] P. Nosil, L. J. Harmon, and O. Seehausen, "Ecological explanations for (incomplete) speciation," *Trends in Ecology and Evolution*, vol. 24, no. 3, pp. 145–156, 2009.

[5] M. Kirkpatrick and N. Barton, "Chromosome inversions, local adaptation and speciation," *Genetics*, vol. 173, no. 1, pp. 419–434, 2006.

[6] S. Via, "Natural selection in action during speciation," *Proceedings of the National Academy of Sciences of the United States of America*, vol. 106, pp. 9939–9946, 2009.

[7] M. R. Servedio, G. S.V. Doorn, M. Kopp, A. M. Frame, and P. Nosil, "Magic traits in speciation: magic but not rare?" *Trends in Ecology and Evolution*, vol. 26, no. 8, pp. 389–397, 2011.

[8] J. A. Coyne and H. A. Orr, *Speciation*, Sinauer, Sunderland, Mass, USA, 2004.

[9] D. J. Futuyma, "Ecological specialization and generalization," in *Evolutionary Ecology*, C. W. Fox, D. A. Roff, and D. J. Fairbairn, Eds., pp. 177–189, Oxford University Press, New York, NY, USA, 2001.

[10] R. F. Denno, M. S. McClure, and J. R. Ott, "Interspecific interactions in phytophagous insects: competition reexamined and resurrected," *Annual Review of Entomology*, vol. 40, pp. 297–331, 1995.

[11] E. Tauber, H. Roe, R. Costa, J. M. Hennessy, and C. P. Kyriacou, "Temporal mating isolation driven by a behavioral gene in *Drosophila*," *Current Biology*, vol. 13, no. 2, pp. 140–145, 2003.

[12] M. Kirkpatrick and V. Ravigné, "Speciation by natural and sexual selection: models and experiments," *American Naturalist*, vol. 159, no. 3, pp. S22–S35, 2002.

[13] V. Ravigné, U. Dieckmann, and I. Olivieri, "Live where you thrive: joint evolution of habitat choice and local adaptation facilitates specialization and promotes diversity," *American Naturalist*, vol. 174, no. 4, pp. E141–E169, 2009.

[14] M. R. Servedio and M. A. F. Noor, "The role of reinforcement in speciation: theory and data," *Annual Review of Ecology, Evolution, and Systematics*, vol. 34, pp. 339–364, 2003.

[15] A. P. Hendry, P. Nosil, and L. H. Rieseberg, "The speed of ecological speciation," *Functional Ecology*, vol. 21, no. 3, pp. 455–464, 2007.

[16] J. Jaenike and R. D. Holt, "Genetic variation for habitat preference: evidence and explanations," *American Naturalist*, vol. 137, pp. S67–S90, 1991.

[17] S. Gavrilets, *Fitness landscapes and the Origin of Species*, Princeton University Press, Princeton, NJ, USA, 2004.

[18] J. Felsenstein, "Skepticism towards Santa Rosalia, or why are there so few kinds of animals?" *Evolution*, vol. 35, pp. 124–138, 1981.

[19] B. C. Haller, L. F. De Léon, G. Rolshausen, K. M. Gotanda, and A. P. Hendry, "Magic traits: distinguishing the important from the trivial," *Trends in Ecology and Evolution*, vol. 27, no. 1, pp. 4–5, 2012.

[20] T. de Meeûs, Y. Michalakis, F. Renaud, and I. Olivieri, "Polymorphism in heterogeneous environments, evolution of habitat selection and sympatric speciation: soft and hard selection models," *Evolutionary Ecology*, vol. 7, no. 2, pp. 175–198, 1993.

[21] P. Edelaar, A. M. Siepielski, and J. Clobert, "Matching habitat choice causes directed gene flow: a neglected dimension in evolution and ecology," *Evolution*, vol. 62, no. 10, pp. 2462–2472, 2008.

[22] D. W. Pfennig, M. A. Wund, E. C. Snell-Rood, T. Cruickshank, C. D. Schlichting, and A. P. Moczek, "Phenotypic plasticity's impacts on diversification and speciation," *Trends in Ecology and Evolution*, vol. 25, no. 8, pp. 459–467, 2010.

[23] P. W. Hedrick, "Genetic-polymorphism in heterogeneous environments: a decade later," *Annual Review of Ecology and Systematics*, vol. 17, pp. 535–566, 1986.

[24] H. Levene, "Genetic equilibrium when more than one ecological niche is available," *American Naturalist*, vol. 87, pp. 331–333, 1953.

[25] J. Maynard Smith, "Sympatric speciation," *American Naturalist*, vol. 100, pp. 637–650, 1966.

[26] D. Udovic, "Frequency-dependent selection, disruptive selection, and the evolution of reproductive isolation," *American Naturalist*, vol. 116, pp. 621–641, 1980.

[27] D. Sloan Wilson and M. Turelli, "Stable underdominance and the evolutionary invasion of empty niches," *American Naturalist*, vol. 127, no. 6, pp. 835–850, 1986.

[28] S. R. Diehl and G. L. Bush, "The role of habitat preference in adaptation and speciation," in *Speciation and Its Consequences*, D. Otte and J. Endler, Eds., pp. 345–365, Sinauer Associates, Sunderland, Mass, USA, 1989.

[29] V. Ravigné, I. Olivieri, and U. Dieckmann, "Implications of habitat choice for protected polymorphisms," *Evolutionary Ecology Research*, vol. 6, no. 1, pp. 125–145, 2004.

[30] J. A. Guldemond and A. F. G. Dixon, "Specificity and daily cycle of release of sex pheromones in aphids: a case of reinforcement," *Biological Journal of the Linnean Society*, vol. 52, no. 3, pp. 287–303, 1994.

[31] A. Diabaté, A. Dao, A. S. Yaro et al., "Spatial swarm segregation and reproductive isolation between the molecular forms of *Anopheles gambiae*," *Proceedings of the Royal Society B*, vol. 276, no. 1676, pp. 4215–4222, 2009.

[32] K. W. Matsubayashi, I. Ohshima, and P. Nosil, "Ecological speciation in phytophagous insects," *Entomologia Experimentalis et Applicata*, vol. 134, no. 1, pp. 1–27, 2010.

[33] G. L. Bush, J. L. Feder, S. H. Berlocher, B. A. McPheron, D. C. Smith, and C. A. Chilcote, "Sympatric origins of *R. pomonella*," *Nature*, vol. 339, no. 6223, p. 346, 1989.

[34] J. L. Feder, J. B. Roethele, B. Wlazlo, and S. H. Berlocher, "Selective maintenance of allozyme differences among sympatric host races of the apple maggot fly," *Proceedings of the National Academy of Sciences of the United States of America*, vol. 94, no. 21, pp. 11417–11421, 1997.

[35] D. C. Smith, "Heritable divergence of *Rhagoletis* pomonella host races by seasonal asynchrony," *Nature*, vol. 336, no. 6194, pp. 66–67, 1988.

[36] D. Schwarz and B. A. McPheron, "When ecological isolation breaks down: sexual isolation is an incomplete barrier to hybridization between *Rhagoletis* species," *Evolutionary Ecology Research*, vol. 9, no. 5, pp. 829–841, 2007.

[37] C. Linn, J. L. Feder, S. Nojima, H. R. Dambroski, S. H. Berlocher, and W. Roelofs, "Fruit odor discrimination and sympatric host race formation in *Rhagoletis*," *Proceedings of the National Academy of Sciences of the United States of America*, vol. 100, no. 20, pp. 11490–11493, 2003.

[38] C. E. Linn, H. Dambroski, S. Nojima, J. L. Feder, S. H. Berlocher, and W. L. Roelofs, "Variability in response specificity of apple, hawthorn, and flowering dogwood-infesting *Rhagoletis* flies to host fruit volatile blends: implications for sympatric host shifts," *Entomologia Experimentalis et Applicata*, vol. 116, no. 1, pp. 55–64, 2005.

[39] J. L. Feder, "The apple maggot fly, *Rhagoletis* pomonella: flies in the face of conventional wisdom about speciation?" in *Endless Forms*, D. J. Howard and S. H. Berlocher, Eds., pp. 130–144, Oxford University Press, New York, NY, USA, 1998.

[40] J. L. Feder, "The effects of parasitoids on sympatric host races of *Rhagoletis pomonella* (Diptera: Tephritidae)," *Ecology*, vol. 76, no. 3, pp. 801–813, 1995.

[41] J. L. Feder, J. B. Roethele, K. Filchak, J. Niedbalski, and J. Romero-Severson, "Evidence for inversion polymorphism related to sympatric host race formation in the apple maggot fly, *Rhagoletis pomonella*," *Genetics*, vol. 163, no. 3, pp. 939–953, 2003.

[42] P. A. Johnson, F. C. Hoppensteadt, J. J. Smith, and G. L. Bush, "Conditions for sympatric speciation: a diploid model incorporating habitat fidelity and non-habitat assortative mating," *Evolutionary Ecology*, vol. 10, no. 2, pp. 187–205, 1996.

[43] C. E. Linn, H. R. Dambroski, J. L. Feder, S. H. Berlocher, S. Nojima, and W. L. Roelofs, "Postzygotic isolating factor in sympatric speciation in *Rhagoletis* flies: reduced response of hybrids to parental host-fruit odors," *Proceedings of the National Academy of Sciences of the United States of America*, vol. 101, no. 51, pp. 17753–17758, 2004.

[44] M. C. Caillaud and S. Via, "Specialized feeding behavior influences both ecological specialization and assortative mating in sympatric host races of pea aphids," *American Naturalist*, vol. 156, no. 6, pp. 606–621, 2000.

[45] C. Smadja, P. Shi, R. K. Butlin, and H. M. Robertson, "Large gene family expansions and adaptive evolution for odorant and gustatory receptors in the pea aphid, *Acyrthosiphon*

pisum," *Molecular Biology and Evolution*, vol. 26, no. 9, pp. 2073–2086, 2009.

[46] D. J. Hawthorne and S. Via, "Genetic linkage of ecological specialization and reproductive isolation in pea aphids," *Nature*, vol. 412, no. 6850, pp. 904–907, 2001.

[47] M. R. Servedio, "The role of linkage disequilibrium in the evolution of premating isolation," *Heredity*, vol. 102, no. 1, pp. 51–56, 2009.

[48] J. M. Davis and J. A. Stamps, "The effect of natal experience on habitat preferences," *Trends in Ecology and Evolution*, vol. 19, no. 8, pp. 411–416, 2004.

[49] P. Nosil, "Divergent host plant adaptation and reproductive isolation between ecotypes of *Timema cristinae* walking sticks," *American Naturalist*, vol. 169, no. 2, pp. 151–162, 2007.

[50] C. P. Sandoval, *Geographic, ecological and behavioral factors affecting spatial variation in color or morph frequency in the walking-stick Timema cristinae*, Ph.D. thesis, University of California, Santa Barbara, Calif, USA, 1993.

[51] P. Nosil, C. P. Sandoval, and B. J. Crespi, "The evolution of host preference in allopatric vs. parapatric populations of *Timema cristinae* walking-sticks," *Journal of Evolutionary Biology*, vol. 19, no. 3, pp. 929–942, 2006.

[52] P. Nosil, B. J. Crespi, and C. P. Sandoval, "Host-plant adaptation drives the parallel evolution of reproductive isolation," *Nature*, vol. 417, no. 6887, pp. 440–443, 2002.

[53] P. Nosil, T. H. Vines, and D. J. Funk, "Perspective: reproductive isolation caused by natural selection against immigrants from divergent habitats," *Evolution*, vol. 59, no. 4, pp. 705–719, 2005.

[54] C. Smadja and R. K. Butlin, "On the scent of speciation: the chemosensory system and its role in premating isolation," *Heredity*, vol. 102, no. 1, pp. 77–97, 2008.

[55] J. Feder, S. Egan, and A. A. Forbes, "Ecological adaptation and speciation: the evolutionary significance of habitat avoidance as a post-zygotic reproductive barrier to gene flow," *International Journal of Ecology*. In press.

[56] A. P. Hendry, "Ecological speciation! Or the lack thereof?" *Canadian Journal of Fisheries and Aquatic Sciences*, vol. 66, no. 8, pp. 1383–1398, 2009.

[57] J. Mallet, "Hybridization, ecological races and the nature of species: empirical evidence for the ease of speciation," *Philosophical Transactions of the Royal Society B*, vol. 363, no. 1506, pp. 2971–2986, 2008.

[58] R. B. Payne, L. L. Payne, and J. L. Woods, "Song learning in brood-parasitic indigobirds *Vidua chalybeata*: song mimicry of the host species," *Animal Behaviour*, vol. 55, no. 6, pp. 1537–1553, 1998.

[59] R. B. Payne, L. L. Payne, J. L. Woods, and M. D. Sorenson, "Imprinting and the origin of parasite-host species associations in brood-parasitic indigobirds, *Vidua chalybeata*," *Animal Behaviour*, vol. 59, no. 1, pp. 69–81, 2000.

[60] C. N. Balakrishnan, K. M. Sefc, and M. D. Sorenson, "Incomplete reproductive isolation following host shift in brood parasitic indigobirds," *Proceedings of the Royal Society B*, vol. 276, no. 1655, pp. 219–228, 2009.

[61] R. B. Payne, C. R. Barlow, C. N. Balakrishnan, and M. D. Sorenson, "Song mimicry of black-bellied Firefinch *Lagonosticta* rara and other finches by the brood-parasitic Cameroon indigobird *Vidua camerunensis* in West Africa," *Ibis*, vol. 147, no. 1, pp. 130–143, 2005.

[62] K. M. Sefc, R. B. Payne, and M. D. Sorenson, "Genetic continuity of brood-parasitic indigobird species," *Molecular Ecology*, vol. 14, no. 5, pp. 1407–1419, 2005.

[63] R. B. Payne, "Nestling mouth markings and colors of Old World finches Estrildidae: mimicry and coevolution of nestling finches and their brood parasite," *Miscellaneous Publications Museum of Zoology University of Michigan*, vol. 194, pp. 1–45, 2005.

[64] P. Nosil and C. P. Sandoval, "Ecological niche dimensionality and the evolutionary diversification of stick insects," *PLoS ONE*, vol. 3, no. 4, Article ID e1907, 2008.

[65] E. G. Boulding, "Mechanisms of differential survival and growth of two species of *Littorina* on wave-exposed and on protected shores," *Journal of Experimental Marine Biology and Ecology*, vol. 169, no. 2, pp. 139–166, 1993.

[66] K. M. M. Jones and E. G. Boulding, "State-dependent habitat selection by an intertidal snail: the costs of selecting a physically stressful microhabitat," *Journal of Experimental Marine Biology and Ecology*, vol. 242, no. 2, pp. 149–177, 1999.

[67] R. P. Kovach and D. A. Tallmon, "Strong influence of microhabitat on survival for an intertidal snail, *Nucella lima*," *Hydrobiologia*, vol. 652, no. 1, pp. 49–56, 2010.

[68] L. P. Miller, M. J. O'Donnell, and K. J. Mach, "Dislodged but not dead: survivorship of a high intertidal snail following wave dislodgement," *Journal of the Marine Biological Association of the United Kingdom*, vol. 87, no. 3, pp. 735–739, 2007.

[69] D. A. Antwi and C. Ameyaw-Akumfi, "Migrational orientation in two species of littoral gastropods (*Littorina angulifera* and *Nerita senegalensis*)," *Marine Biology*, vol. 94, no. 2, pp. 259–263, 1987.

[70] M. G. Chapman, "Assessment of variability in responses of intertidal periwinkles to experimental transplantations," *Journal of Experimental Marine Biology and Ecology*, vol. 236, no. 2, pp. 171–190, 1999.

[71] D. G. Reid, *Systematics and Evolution of Littorina*, Ray Society, London, UK, 1996.

[72] E. Rolán-Alvarez, "Sympatric speciation as a by-product of ecological adaptation in the Galician *Littorina saxatilis* hybrid zone," *Journal of Molluscan Studies*, vol. 73, no. 1, pp. 1–10, 2007.

[73] R. K. Butlin, J. Galindo, and J. W. Grahame, "Sympatric, parapatric or allopatric: the most important way to classify speciation?" *Philosophical Transactions of the Royal Society B*, vol. 363, no. 1506, pp. 2997–3007, 2008.

[74] K. Johannesson, M. Panova, P. Kemppainen, C. André, E. Rolan-Alvarez, and R. K. Butlin, "Repeated evolution of reproductive isolation in a marine snail: unveiling mechanisms of speciation," *Philosophical Transactions of the Royal Society B*, vol. 365, no. 1547, pp. 1735–1747, 2010.

[75] M. Panova, J. Hollander, and K. Johannesson, "Site-specific genetic divergence in parallel hybrid zones suggests non-allopatric evolution of reproductive barriers," *Molecular Ecology*, vol. 15, no. 13, pp. 4021–4031, 2006.

[76] C. S. Wilding, R. K. Butlin, and J. Grahame, "Differential gene exchange between parapatric morphs of *Littorina saxatilis* detected using AFLP markers," *Journal of Evolutionary Biology*, vol. 14, no. 4, pp. 611–619, 2001.

[77] J. Galindo, P. MorÁn, and E. RolÁn-Alvarez, "Comparing geographical genetic differentiation between candidate and noncandidate loci for adaptation strengthens support for parallel ecological divergence in the marine snail *Littorina saxatilis*," *Molecular Ecology*, vol. 18, no. 5, pp. 919–930, 2009.

[78] R. Cruz, C. Vilas, J. Mosquera, and C. García, "Relative contribution of dispersal and natural selection to the maintenance of a hybrid zone in *Littorina*," *Evolution*, vol. 58, no. 12, pp. 2734–2746, 2004.

[79] J. Hollander, M. Lindegarth, and K. Johannesson, "Local adaptation but not geographical separation promotes assortative mating in a snail," *Animal Behaviour*, vol. 70, no. 5, pp. 1209–1219, 2005.

[80] S. L. Hull, "Assortative mating between two distinct microallopatric populations of *Littorina saxatilis* (Olivi) on the northeast coast of England," *Hydrobiologia*, vol. 378, no. 1–3, pp. 79–88, 1998.

[81] A. R. Pickles and J. Grahame, "Mate choice in divergent morphs of the gastropod mollusc *Littorina saxatilis* (Olivi): speciation in action?" *Animal Behaviour*, vol. 58, no. 1, pp. 181–184, 1999.

[82] J. Erlandsson, E. Rolán-Alvarez, and K. Johannesson, "Migratory differences between ecotypes of the snail *Littorina saxatilis* on Galician rocky shores," *Evolutionary Ecology*, vol. 12, no. 8, pp. 913–924, 1998.

[83] K. Janson, "Selection and migration in two distinct phenotypes of *Littorina saxatilis* in Sweden," *Oecologia*, vol. 59, no. 1, pp. 58–61, 1983.

[84] B. J. Balkau and M. W. Feldman, "Selection for migration modification," *Genetics*, vol. 74, no. 1, pp. 171–174, 1973.

[85] M. Slatkin, "Gene flow in natural populations," *Annual Review of Ecology and Systematics*, vol. 16, pp. 393–430, 1985.

[86] S. Bensch and M. Åkesson, "Ten years of AFLP in ecology and evolution: why so few animals?" *Molecular Ecology*, vol. 14, no. 10, pp. 2899–2914, 2005.

[87] N. H. Barton and G. M. Hewitt, "Analysis of hybrid zones," *Annual Review of Ecology and Systematics*, vol. 16, pp. 113–148, 1985.

[88] J. W. Grahame, C. S. Wilding, and R. K. Butlin, "Adaptation to a steep environmental gradient and an associated barrier to gene exchange in *Littorina saxatilis*," *Evolution*, vol. 60, no. 2, pp. 268–278, 2006.

[89] E. Rolán-Alvarez, K. Johannesson, and J. Erlandsson, "The maintenance of a cline in the marine snail *Littorina saxatilis*: the role of home site advantage and hybrid fitness," *Evolution*, vol. 51, no. 6, pp. 1838–1847, 1997.

[90] M. Carballo, A. Caballero, and E. Rolán-Alvarez, "Habitat-dependent ecotype micro-distribution at the mid-shore in natural populations of *Littorina saxatilis*," *Hydrobiologia*, vol. 548, no. 1, pp. 307–311, 2005.

[91] C. J. MacCallum, B. Nürnberger, N. H. Barton, and J. M. Szymur, "Habitat preference in the *Bombina hybrid* zone in Croatia," *Evolution*, vol. 52, no. 1, pp. 227–239, 1998.

[92] D. G. Reid, "The comparative morphology, phylogeny and evolution of the gastropod family littorinidae," *Philosophical Transactions B*, vol. 324, no. 1220, pp. 1–110, 1989.

[93] H.D. Rundle and J. W. Boughman, "Behavioural ecology and speciation," in *Evolutionary Behavioral Ecology*, D. F. Westneat and C. W. Fox, Eds., Oxford University Press, New York, NY, USA, 2010.

[94] J. Hollander and R. K. Butlin, "The adaptive value of phenotypic plasticity in two ecotypes of a marine gastropod," *BMC Evolutionary Biology*, vol. 10, no. 1, article 333, 2010.

[95] J. Hollander, M. L. Collyer, D. C. Adams, and K. Johannesson, "Phenotypic plasticity in two marine snails: constraints superseding life history," *Journal of Evolutionary Biology*, vol. 19, no. 6, pp. 1861–1872, 2006.

[96] S. Sadedin, J. Hollander, M. Panova, K. Johannesson, and S. Gavrilets, "Case studies and mathematical models of ecological speciation. 3: ecotype formation in a Swedish snail," *Molecular Ecology*, vol. 18, no. 19, pp. 4006–4023, 2009.

[97] G. Chelazzi, P. Dellasantina, and M. Vannini, "Long-lasting substrate marking in the collective homing of the gastropod *Nerita textilis*," *Biological Bulletin*, vol. 168, pp. 214–221, 1985.

[98] M. S. Davies and P. Beckwith, "Role of mucus trails and trail-following in the behaviour and nutrition of the periwinkle *Littorina littorea*," *Marine Ecology Progress Series*, vol. 179, pp. 247–257, 1999.

[99] K. Johannesson, J. N. Havenhand, P. R. Jonsson, M. Lindegarth, A. Sundin, and J. Hollander, "Male discrimination of female mucous trails permits assortative mating in a marine snail species," *Evolution*, vol. 62, no. 12, pp. 3178–3184, 2008.

[100] C. Chapperon and L. Seuront, "Cue synergy in Littorina littorea navigation following wave dislodgement," *Journal of the Marine Biological Association of the United Kingdom*, vol. 89, no. 6, pp. 1133–1136, 2009.

[101] E. Rolán-Alvarez, J. Erlandsson, K. Johannesson, and R. Cruz, "Mechanisms of incomplete prezygotic reproductive isolation in an intertidal snail: testing behavioural models in wild populations," *Journal of Evolutionary Biology*, vol. 12, no. 5, pp. 879–890, 1999.

6

The Effect of Timing of Grassland Management on Plant Reproduction

Tommy Lennartsson, Jörgen Wissman, and Hanna-Märtha Bergström

Swedish Biodiversity Centre, Swedish University of Agricultural Sciences, P.O. Box 7007, 750 07 Uppsala, Sweden

Correspondence should be addressed to Tommy Lennartsson, tommy.lennartsson@slu.se

Academic Editor: Andrew Denham

Seminatural grasslands are maintained by regular anthropogenic disturbance, usually grazing or mowing. Management action late in the growing season was historically more common than today. Two experimental grazing regimes, continuous stocking from May to September and late-onset grazing from mid-July, were compared in two Swedish grasslands. Effects on flowering and fruit production were studied and related to plant functional traits. Change in vegetation composition over six years was analysed in one grassland. Delayed onset of grazing enhanced fruit production up to four times. Phenology of reproduction was the most important plant trait explaining differences in reproduction among species. Diversity of vascular plant species was higher after six years of late-onset grazing. No differences in vegetation height or proportion of grazed shoots were found by the end of the season. The results suggest that early reproduction may function as an escape from damage and that late onset of grazing may be used as a substitute for labour-intense traditional mowing.

1. Introduction

Nonwooded habitats harbour a large proportion of the biodiversity in many temperate regions. High light influx favours small species of plants and species-rich vegetation [1, 2]. This vegetation together with favourable microclimate forms a rich fauna of invertebrates, and both the vegetation and the lower fauna serve as a resource base for higher animals. One group of nonwooded habitats, particularly important to biodiversity, is semi-natural grasslands, that is, unfertilized, uncultivated grasslands, in which anthropogenic disturbances, usually grazing or mowing, are essential or important for keeping the habitats open [3]. In the historical landscape, such grasslands constituted a necessary nutrient base for agriculture and were managed for production of hay and pasture. In the modern landscape, they are usually managed for conservation purposes.

Abandonment of semi-natural grasslands has been identified as a major threat to the European flora and fauna [4, 5]. Most countries in Western Europe have lost more than 95% of their original grassland areas, for example, Sweden [6] and the UK [7]. The unfavourable conservation status of semi-natural grasslands is indicated by the high numbers of red-listed species confined to such ecosystems [8]. Grassland species are threatened both by ceased management and by insufficient management quality. The latter is partly due to the fact that the present management methods differ considerably from the historical management that has formed the rich grassland biodiversity, for example, in terms of type, timing, and intensity of management [9]. While cessation of management has been generally acknowledged as an important cause of the decline of grassland biodiversity [4], suboptimal management of the remaining grasslands has attracted less attention [10]. An increased focus on management quality may be motivated since as grassland areas become reduced and fragmented, it becomes increasingly important to optimise management of the remaining patches.

In order to obtain such optimal management, we can manipulate the type, timing, and intensity of management, and to some extent the abundance of certain spatial habitat structures which affect the disturbance and the vegetation, for example, shrubs and trees. In this study, we focus on one of these management tools, the timing of management, and we investigate its importance for plant reproduction in Swedish semi-natural grasslands.

Historically, large grassland areas in Europe were left undisturbed until late in the growing season (July-September) when mowing or late grazing took place [9]. The use of late management in the remaining grassland patches has however been reduced drastically, being replaced by grazing during most of the season [11, 12]. For production reasons, grazing is today often started as early as possible because the nutrient content of the pasture is highest early in the summer [13, 14]. The shift from late to early management has been shown to have strong, usually negative, effects on grassland biodiversity [15, 16], especially when grazing is intense [17, 18]. The timing of management strongly affects the reproduction of plants, which in turn affects both plant population viability and the resource for populations of pollen eaters, nectar eaters, seed predators, and several invertebrate herbivores.

In ecological terms, grazing and mowing exert disturbance and stress to the vegetation by damaging plants and removing biomass. This disturbance is necessary for grassland ecosystems since it maintains openness and counteracts succession towards tall vegetation dominated by few competitive species. Plants in grassland habitats are adapted to the disturbance, for example, by mechanisms related to palatability and mechanical defence [19, 20], growth form [21], and phenology of reproduction [22–24]. Phenology can be assumed to be a particularly important trait in mown and other late managed habitats since it determines whether a plant can produce seeds before the vegetation is disturbed [25].

In this study we experimentally compared two cattle grazing regimes, continuous stocking from May to September and late onset of grazing (grazing from mid-July to September) in terms of effects on plant reproduction. Late onset of grazing was chosen as an alternative to the prevailing continuous stocking regime both because it is a traditional grazing regime and because it may potentially function as a more practical substitute for labour-intense traditional mowing. The study addresses the following questions: (1) what are the differences between continuously and late grazed pastures in terms of production of flowers and mature fruits? (2) How is plant reproductive success affected by species-specific traits such as phenology of reproduction? (3) Can cattle graze the old vegetation late in the summer, or does late-onset grazing have potentially negative effects in terms of ungrazed vegetation by the end of the season? (4) Is late grazing a possible substitute for labour-intense mowing for conservation management?

2. Methods

2.1. Study System and Experimental Setup.

Two, slightly different grazing experiments are analysed in this study. Both were performed in semi-natural grasslands in south-central Sweden, normally grazed annually by 12–18-month-old steers from mid-May to late September. One grassland (Harpsund) is situated in the county of Södermanland (59°05′ N 16°29′ E), the other (Pustnäs) in the county of Uppland (59°48′ N 17°40′ E), in south-central Sweden. Both sites are flat and without notable slope and have nutrient-poor soils as the result of long history of grazing without fertilisation other than the reallocation of nutrient through dung and urine from the grazers. The vegetation at both sites is mainly of dry-mesic herb-rich *Agrostis capillaris* meadow type [26]. Particularly common species are shown in Table 2 for Harpsund, and the same dominant species are found in Pustnäs. The late and continuously stocked areas in Harpsund were 4 and 6 hectares, resp., and in Pustnäs 1 and 3 hectares, respectively.

Within each grassland, one area was separated from the continuously stocked pasture by fencing until 20 July, when the fence was opened and the cattle were allowed to move freely between the two treatment areas. The grazing pressure was c. 1.8 steers (12 months of age in May) per hectare, corresponding to c. 630 kg live-weight per hectare. When the pasture area increased at late onset of grazing, the stocking density was maintained by adding extra steers. In Pustnäs three exclosure cages per treatment were used in order to examine if the plant reproduction differed between treatment areas independently of treatment.

The use of two large treatment areas per site, instead of a number of small, interspersed areas, may raise statistical problems [27, 28]. Due to this, and as a result of the differences in data sampling design between the two grasslands, all analyses were performed for each grassland separately. The results are interpreted acknowledging the possible area effects. Large treatment areas are, on the other hand, necessary to study natural grazing effects of the cattle. The grazing protected cages in Pustnäs were treated as two random samples, controlling for treatment area.

2.2. Data Sampling.

The sampling design differs slightly between the two grasslands. Data on four main response variables were collected: (1) production, development, and grazing of reproductive units (see below) during the grazing season (both sites); (2) height and grazing of vegetation during the season (both sites); (3) frequency of grazed shoots of vascular plants in mid-August (Harpsund); (4) number of vascular plant species (Pustnäs). The experiments started in 1997 in Pustnäs and in 2001 in Harpsund.

A reproductive unit was defined for each species as the smallest unit of reproductive organs (buds, flowers, fruits) that could be readily recognized and counted in the field. For most herbs, the reproductive unit was defined as a single bud, flower, and fruit (gynoecium). In *Asteraceae* a reproductive unit was defined as a flower head, and in *Plantago*, spike-forming grasses and sedges were defined as a spike. In small-flowered herbs and in sedges with panicles, cymes, or composed umbels or racemes, reproductive units were defined as the smallest possible assemblage of reproductive organs, usually the second- or third-level branch in the inflorescence. In grasses with panicles, spikelets were used as reproductive units.

Production and development of reproductive units, from bud to mature fruit, were monitored in twenty randomly distributed 1×1 m plots per grazing regime in Harpsund, and in seven 50×50 cm plots per grazing regime in Pustnäs.

The plots were examined 5–7 times during the season at c. 20-day intervals (see Figure 2 for exact dates). At each occasion the reproductive units of each occurring species were counted and assigned to one of the following classes: grazed (grazed reproductive shoot or branch), bud, flowering, postflowering, immature fruit, and mature/dehisced fruit. For composite reproductive units (see above), the unit was assigned to a certain class if approximately 75% of the flowers in the reproductive unit belonged to that class. Thus, reproduction was measured at the level of reproductive unit, with no estimates of seed production.

The height of the vegetation in the plots was measured eight times from May 24 to September 17 using a rising plate meter [29]. The vegetation height provides an estimate of the grazing intensity. In Harpsund, grazing intensity was estimated also by counting the numbers of grazed and ungrazed shoots 20 days after late onset of grazing in 2001 and 2002, in 25 0.1×0.1 m plots per grazing regime, located in a fixed pattern in one 0.5×25 m area per grazing regime.

In Pustnäs, three 1×2 m coarse-grid metal net cages were randomly located in each treatment in order to indicate area-specific differences between the two treatment areas, in absence of the grazing effect. Data on vegetation height and the production and development of reproductive units were collected as described above.

Plant reproduction was estimated as relative fruit production, defined as the proportion of the reproductive units that developed mature fruits. This estimate was calculated per species in Harpsund, for all species that occurred in a minimum of five plots per grazing treatment. The species were analyzed in order to investigate if some species-specific traits could explain how different species responded to the grazing regimes. This analysis was done by correlating relative fruit production with three species-specific plant traits, with two estimates of species-specific preference by the grazers, and with spontaneous abortion of reproductive units during the growing season. The following plant traits were analyzed: plant height according to Lid [30], Ellenberg light index according to Ellenberg et al. [31], and date of fruit maturation (the approximate date when 50% of the fruits had reached mature stage in late grazing). As estimates of species-specific grazing preference, we used the proportion of reproductive units that were grazed during the first three weeks of grazing, and during the entire season. The level of herbivory during the first weeks of grazing indicates whether the plant species is a "first choice" of the grazers, or whether it is less preferred. Species experiencing low herbivory during the entire season can be regarded as low-preference species [32]. Spontaneous abortion of reproductive units, finally, indicates whether fruit set of the species is restricted by other factors than herbivory, for example, competition in tall vegetation. Twenty-eight species occurred in a minimum of five plots per grazing treatment (see Table 2) and were used in comparison of relative fruit production between grazing treatments.

In Pustnäs the frequencies of all species present in 20 randomly distributed 0.5×0.5 m plots were monitored to examine species change over time. This was initiated in 1997, and the same plots were reinventoried in 2003 (except two plots in both treatments which were not found). The taxonomic nomenclature follows Mossberg and Stenberg [33].

2.3. Statistical Treatment. Since most of the data deviated considerably from the normal distribution and in most cases were count data, Spearman's rank correlation, Mann-Whitney U-test, and Poisson's regressions were used [34]. Differences in the number of grazed shoots between grazing regimes and years, diversity indexes, and species number per plot were however approximately normally distributed and were therefore analyzed using repeated measures in ANOVA (SAS mixed models procedure) and t-tests, respectively. Analysis of differences in the relative fruit production between grazing regimes at different dates was performed using repeated measures in generalized linear models, Poisson's distribution, and log link function (SAS Genmod procedure, GEE-analysis). Bonferroni's corrections of multiple comparisons among species in the two grazing regimes were not necessary since only the number of significance was discussed, not the single cases.

The change of vegetation during six years of grazing experiment was estimated using the Shannon diversity index H [35] in Pustnäs in 1997, the initial year of the experiment, and again in 2003.

3. Results

The number of species did not differ significantly between the two treatment areas in any of the two grassland sites. In Harpsund, on average 26.0 species of vascular plants occurred per m2 plot in the continuously stocked and 28.6 species in the late grazed area (Mann-Whitney $U = 132$, $N = 40$, $P = 0.07$). In Pustnäs, each 50×50 cm plot had on average 13.4 species with continuous stocking and 12.3 species in late-onset grazing (Mann-Whitney $U = 13$, $N = 14$, $P = 0.12$).

3.1. Reproduction. Grazing regime strongly affected plant reproduction, the average production of reproductive units, both in total and in terms of mature reproductive units (Figure 1). Also the average relative fruit production was higher in late grazing compared to continuous stocking, both at plot level and at species level (Figure 1).

The mean density of reproductive units was higher in the late grazing treatment compared to continuous stocking at all observation dates except for the first date in spring ($P < 0.05$, generalized linear model, Poisson's errors, and log link function, with individual plots as repeated measures between observation dates, $df = 1$ for all analyses).

The exclosure cages in Pustnäs showed that no significant differences between treatment areas were found, neither for production of reproductive units (Mann-Whitney $U = 2.0$, $N = 6$, $P = 0.28$) nor for maturation ($U = 3.5$, $N = 6$, $P = 0.66$). Ungrazed plots inside the cages produced on average 451 reproductive units per square meter in the continuously stocked area and 442 in the late grazed area. Of these reproductive units, 401 ± 33 (89 ± 1.6 per cent) matured

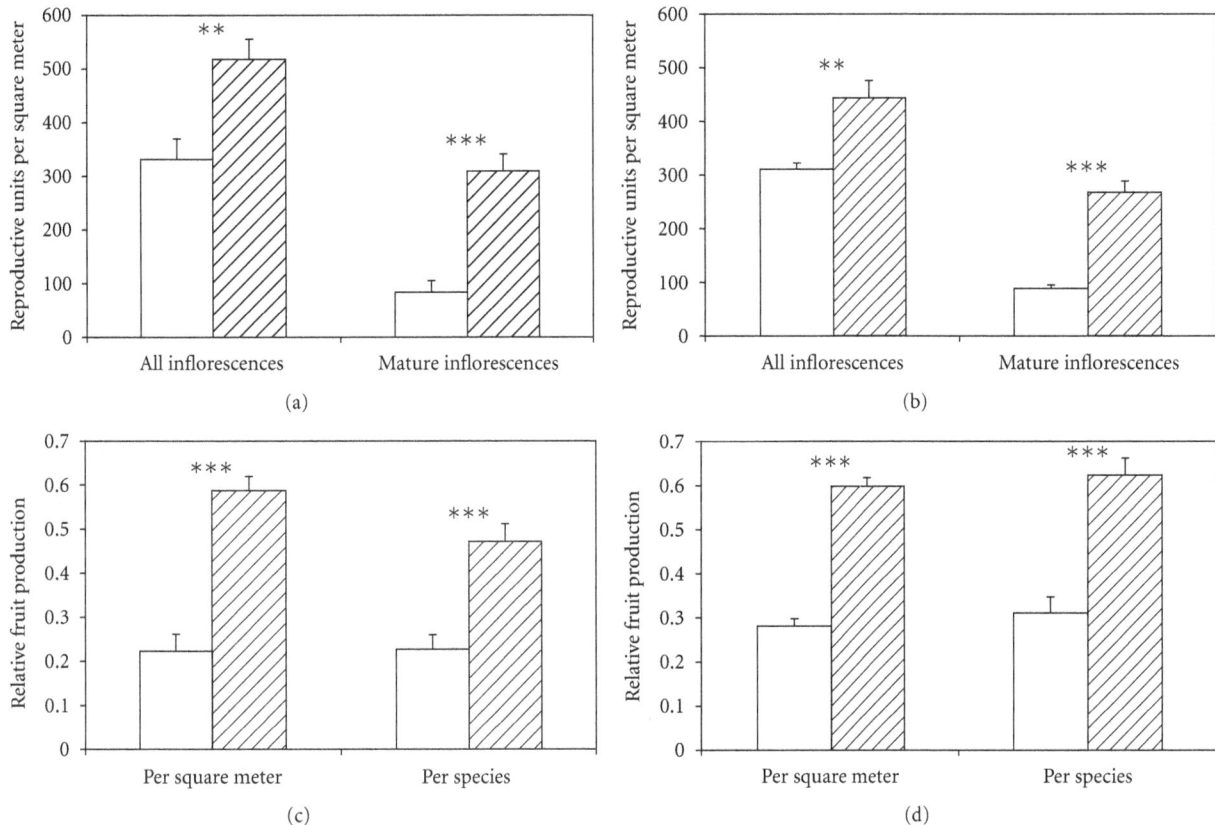

FIGURE 1: Mean production of reproductive units (see text for explanation) in two grazed semi-natural grasslands, Harpsund (a) and Pustnäs (b), and mean (per m^2 and per species) relative fruit production (proportion of inflorescences developing to mature fruit stage) in Harpsund (c) and Pustnäs (d). Two grazing regimes were compared, continuous stocking from mid-May to early October (white bars) and late onset of grazing (grazing from mid-July to early October, dashed bars). Error bars show one SE. Asterisks indicate significant differences between grazing regimes within site (*$P < 0.05$, **$P < 0.01$, ***$P < 0.001$, Mann-Whitney U-test, N (Harpsund) = 20 per grazing regime, N (Pustnäs) = 7 per grazing regime).

to fruit stage in the continuously stocked area and 401 ± 61 (91 ± 2.0 per cent) in the late grazed area.

3.2. Species-Specific Responses. In Harpsund, relative fruit production was significantly higher in late grazing compared to continuous stocking for 15 of 28 analyzed plant species. No species had higher relative fruit production in continuous stocking (Figure 2, Table 2). Six species produced no mature fruits at all in the 20 plots in continuous stocking, compared to one species in late grazing.

In Pustnäs, there were too few sampling plots to allow for testing of species-specific differences between grazing regimes. Sixteen species occurred in three plots or more per treatment. For seven of the 16 species, relative fruit production in late grazing exceeded that of continuous stocking by a factor > 2. No species had higher fruit set in continuous stocking.

The mean relative fruit production of single species was negatively correlated with date of fruit maturation for the species. This effect was more prominent in late grazing (Figure 3).

Grazing preference, in terms of degree of herbivory (see Methods), had no significant effect on species-specific relative fruit production, because the relative fruit production was not correlated with the species-specific herbivory of all reproductive units, neither over the entire season (Spearman's rank correlation $P > 0.10$) nor during the first weeks of grazing ($P > 0.15$). No correlation was found between relative fruit production and any of the other tested species-specific parameters: the Ellenberg light index, plant height, or abortion of reproductive units.

Grasses reached significantly higher relative fruit production than herbs and sedges both in late grazing (Mann-Whitney $U = 47$, $N = 34$, $P = 0.002$) and continuous stocking ($U = 33$, $N = 26$, $P = 0.013$, Figure 3).

3.3. Diversity and Species Richness. The H (the Shannon diversity index, [34]) in plots of the vegetation change experiment in Pustnäs did not differ between treatments in 1997, the initial year of the experiment (mean for continuous stocking and late grazing, 1.37 and 1.39, t-test, $t = -60$, $df = 38$, $P = 0.555$). In 2003, after six years, the diversity in late grazed plots had increased compared to the continuously stocked plots (mean for continuous stocking and late grazing, 1.41 and 1.58, resp., t-test, $t = -6.68$, $df = 36$, $P = < 0.001$). H-indices were not significantly different between

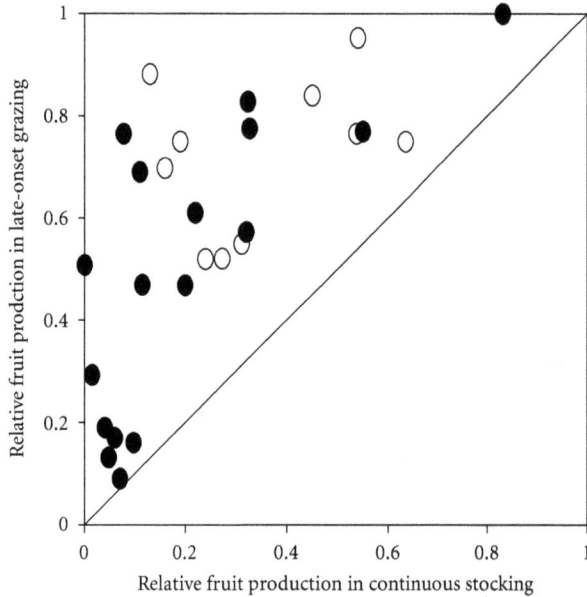

FIGURE 2: Mean per-species relative fruit production (proportion of inflorescences developing to mature fruit stage) of 27 plant species of grasses (open circles) and herbs/sedges (filled circles) in two grazing regimes, late onset of grazing (grazing from mid-July to early October), and continuous stocking (grazing from mid-May to early October), in a semi-natural grassland at Harpsund, see text for explanations.

TABLE 1: The mean proportion of grazed shoots in mid-August in two grazing regimes (continuous stocking May-September and late grazing mid-July to September), at two sites (Harpsund and Pustnäs), over two years, 2001 and 2002. $N = 20$ plots per grazing regime in Harpsund and $N = 7$ in Pustnäs.

Year	Site	Grazing regime	Mean proportion of grazed shoots (S.E.)
2001	Harpsund	Continuous	0.83 (0.033)
		Late	0.71 (0.050)
	Pustnäs	Continuous	0.98 (0.009)
		Late	0.82 (0.036)
2002	Harpsund	Continuous	0.40 (0.022)
		Late	0.46 (0.055)
	Pustnäs	Continuous	0.46 (0.018)
		Late	0.49 (0.019)

8.1, $P = 0.006$) but not of grazing regime ($P = 0.862$) or the interaction between grazing regime and year ($P = 0.069$), whereas the proportion of grazed shoots in Pustnäs was affected by the interaction ($\chi^2 = 6.7$, $P = 0.011$) but neither by grazing regime ($P = 0.084$) nor year ($P = 0.140$).

4. Discussion

This study shows that the timing of grassland management, here grazing, strongly affects the fruit production of vascular plants which may affect the viability of populations of both plants and other organisms depending on plant reproduction, for example, insects depending on nectar and pollen. The response to the grazing regime, in terms of relative fruit production, varied between species, mainly depending on each species' phenology of reproduction. The plant species diversity increased over time with the grazing regime tested here, late-onset grazing. Late onset of grazing carried no obvious costs in terms of ungrazed vegetation by the end of the season.

In the two studied semi-natural pastures subject to continuous stocking from May to September, the vegetation height was kept almost constantly low throughout the season. As a consequence, the production of fruits was reduced to a fifth compared to undisturbed conditions. By delaying the onset of grazing from mid-May to 20 July, the production of buds and flowers was increased 2–2.7 times compared to continuous stocking and the production of mature fruits c. 4 times. The results were very similar in the two studied grasslands. The increased plant reproduction in late grazing was not an effect of more shoots and reproductive organs escaping herbivory completely, but of more shoots escaping herbivory long enough to produce mature fruits. No difference in number of grazed shoots was found between treatments by the end of the season, but the vegetation was on average 0.3–1.2 cm taller in late than in continuous stocking. The difference of 1.2 cm was statistically significant, but most likely has no ecological significance, for example, in terms of litter accumulation [36].

years in continuous stocking (t-test, $t = -1.53$, $df = 36$, $P = 0.134$).

The number of species per plot showed a similar pattern. The mean number of species per plot was initially 15.1 in continuous stocking and 14.9 in late grazing (t-test, $t = 0.16$, $df = 37$, $P = 0.875$). After six years, species number was 14.9 in continuous stocking compared to an increase to 19.4 species per plot in late grazing (t-test, $t = -4.53$, $df = 34$, $P = <0.001$).

3.4. Vegetation Height and Grazing of Shoots. Grazing regime affected the vegetation height during the early part of the season, but the difference was negligible in late summer and autumn (Figure 4). In continuous stocking, the average vegetation height was low throughout the season, with a slight decrease from c. 4.5 to c. 2.5 cm (Figure 4). Late-onset grazing allowed the vegetation to grow to a maximum average of 8.8 and 8.5 cm of height in Harpsund and Pustnäs, respectively, before the onset of grazing in mid-July. After one month, the late grazed vegetation was reduced to on average 3.5 cm in both grasslands, that is, 1.2 cm taller vegetation than in continuous stocking in Harpsund (Mann-Whitney $U = 119.0$, $n = 20$, $P = 0.025$) and 0.3 cm taller in Pustnäs ($P = 0.32$, Figure 4). Inside the exclosure cages in Pustnäs, the vegetation height never differed significantly between the two treatment areas ($P > 0.35$).

The proportion of grazed shoots was measured in mid-August, that is, one month after the onset of late grazing (Table 1). In Harpsund, the proportion of grazed shoots was affected by year (ANOVA, repeated measures: $df = 1$, $\chi^2 = $

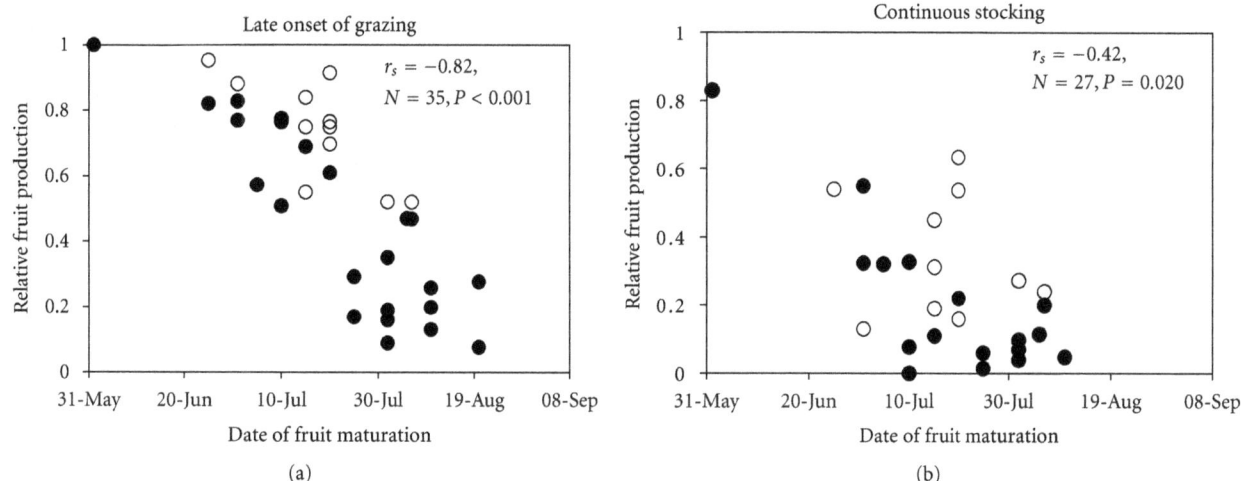

FIGURE 3: Mean per-species relative fruit production (proportion of inflorescences developing to mature fruit stage) of grasses (open circles) and herbs/sedges (filled circles), as a function of each species' date of fruit maturation (see text for explanation). Two grazing regimes were compared, late onset of grazing (grazing from mid-July to early October), and continuous stocking (grazing from mid-May to early October) in a semi-natural grassland at Harpsund, see text for explanations. Spearman's rank correlation coefficients, r_s, are shown, together with significance tests. For clarity, no error bars are shown.

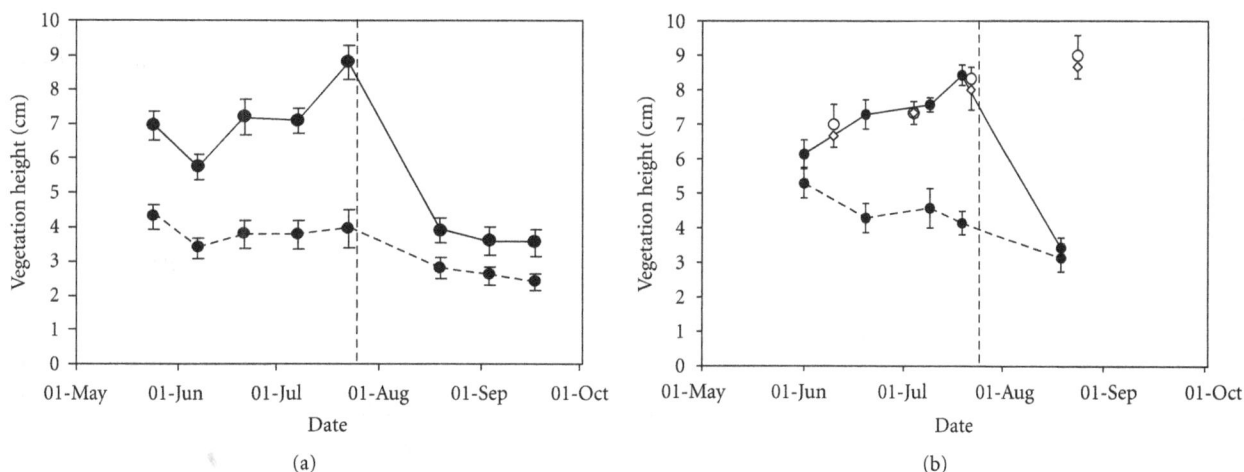

FIGURE 4: Average vegetation height during the growing season in continuously stocked (dashed line) and late (solid line) grazed areas, in two semi-natural grasslands, Harpsund (a) and Pustnäs (b). Vegetation height inside exclosure cages in Pustnäs is shown by open circles (cages in continuous stocking) and open diamonds (cages in late grazing). Error bars show 1 standard error. Onset of late grazing is indicated with a vertical dotted line. $N = 20$ plots per treatment in Harpsund, 7 in Pustnäs; 3 exclosure cages per treatment in Pustnäs. A dry period in early June caused a temporal drop in turgor and therefore also in the measure of vegetation height.

Tall vegetation before late onset of grazing can be expected to affect the reproduction of small plant species negatively, through competition for light [37]. However, no such effects were found, as plant height was not correlated with relative fruit production, neither was the Ellenberg light index [31], which would have been expected if fruit set was affected by competition for light. Furthermore, the abortion of reproductive units was not correlated with relative fruit production, neither at species nor plot levels. This indicates that effects of spontaneous abortion due to, for example, competition did not contribute significantly to the differences in relative fruit production between species and grazing regimes.

In both grazing regimes, comparison of species showed that relative fruit production was positively correlated with early reproduction. Early flowering has been suggested to be an adaptation to predictably late disturbances such as mowing [24, 38, 39]. This study shows that also late onset of grazing favoured early reproducing vascular plants in a similar way. By 20 July approximately half of the most abundant species had 50% or more of their fruits mature. Although the vegetation was rapidly grazed after onset of grazing in 20 July, some fruits also successfully matured after that date, and a total of 50–100% (median 77%) of the reproductive units of early reproducing species produced mature fruits. In the other, later reproducing half of the species,

TABLE 2: Mean relative fruit production of 28 species occurring in five or more plots per the two grazing regimes: continuous stocking and late grazing, in the grassland Harpsund, see text for details. Differences between grazing regimes are tested using the Mann-Whitney U-test. N-values show the number of plots in which the species were present.

Species	Continuous		Late		U	P
	N	Mean (SE)	N	Mean (SE)		
Agrostis capillaris	19	0.24 (0.07)	20	0.52 (0.06)	91.0	0.005
Alchemilla sp.	9	0.01 (0.01)	13	0.29 (0.09)	24.0	0.014
Anthoxanthum odoratum	9	0.54 (0.15)	20	0.95 (0.04)	48.0	0.048
Helictotrichon pubescens	14	0.54 (0.09)	13	0.77 (0.11)	50.5	0.048
Carex caryophyllea	5	0.11 (0.07)	8	0.49 (0.25)	12.5	0.250
Cerastium fontanum	15	0.08 (0.04)	16	0.77 (0.11)	29.5	<0.001
Dactylis glomerata	5	0.27 (0.12)	9	0.52 (0.15)	15.0	0.310
Deschampsia caespitosa	5	0.31 (0.14)	5	0.55 (0.20)	8.50	0.400
Festuca pratensis	12	0.45 (0.13)	13	0.84 (0.14)	45.5	0.073
Festuca rubra	19	0.19 (0.05)	19	0.75 (0.086)	50.5	<0.001
Filipendula vulgaris	7	0.20 (0.14)	6	0.47 (0.18)	13.5	0.260
Lotus corniculatus	14	0.04 (0.02)	18	0.19 (0.059)	74.0	0.039
Luzula campestris	15	0.55 (0.10)	19	0.77 (0.04)	99.5	0.140
Phleum pratense	9	0.63 (0.15)	10	0.75 (0.09)	40.5	0.710
Plantago lanceolata	16	0.22 (0.07)	20	0.61 (0.05)	37.0	<0.001
Poa pratensis	20	0.16 (0.06)	11	0.70 (0.10)	31.0	0.001
Potentilla erecta	6	0.05 (0.05)	17	0.13 (0.04)	30.0	0.120
Primula veris	5	0 (0)	10	0.51 (0.13)	7.50	0.020
Ranunculus auricomus	13	0.32 (0.11)	8	0.57 (0.17)	33.0	0.160
Ranunculus polyanthemos	17	0.33 (0.09)	14	0.78 (0.08)	39.0	0.001
Rumex acetosa	16	0.32 (0.10)	12	0.83 (0.09)	40.0	0.008
Stellaria graminea	16	0.08 (0.04)	7	0 (0)	38.5	0.100
Taraxacum sp.	9	0.83 (0.11)	6	1.00 (0)	6.0	0.013
Trifolium medium	8	0.11 (0.06)	17	0.47 (0.09)	28.5	0.018
Trifolium pratense	17	0.11 (0.04)	20	0.69 (0.07)	23.0	<0.001
Trifolium repens	17	0.07 (0.02)	18	0.09 (0.03)	148.0	0.860
Veronica chamaedrys	19	0.06 (0.23)	17	0.17 (0.06)	128.5	0.230
Vicia cracca	6	0.10 (0.06)	8	0.16 (0.12)	23.0	0.870

a median of 26% of the reproductive units produced mature fruits. In comparison with late grazing, mowing in mid-July would have considerably reduced fruit set of all species except for the very early ones. This is because mowing results in immediate removal of reproductive organs, whereas removal by grazing is more extended in time. Much of the fruit maturation in the studied grasslands took place during 20 July–5 August, which suggests that a few days difference in mowing time or onset of grazing would make a large difference in plant fruit set.

Early reproduction may also be an adaptation to more continuous disturbance such as grazing, since the risk of reproductive organs being grazed before fruit maturation is proportional to the time a plant is exposed to a grazed environment [40]. In continuous stocking, the early reproducing half of the species produced a median of 0.32 mature fruits per reproductive unit, compared to 0.08 in the late reproducing species.

Escape from herbivory may also be achieved by mechanisms related to grazing preference [41, 42] or growth-form

[43]. Grasses had higher fruit set than herbs and sedges in both grazing regimes, partly as an effect of early reproduction, but to some extent also as an effect of less herbivory on grasses. The degree of herbivory experienced by a species is an estimate of species-specific grazing preference. Here, preference was not correlated with species-specific fruit set, indicating that differences in preference explains little of the variation in fruit set between species in this system. Plant traits that affect preference may however be a more important mechanism under less intensive grazing, because opportunities for herbivore selection between plant species increase with decreasing grazing intensity [44]. No obvious growth-form-related defence mechanisms, such as thorns, occurred among the analyzed species. Low-growing plants can be expected to be able to escape cattle herbivory to some extent [22], but plant height was not correlated with fruit set in this study.

The species diversity increased during six years of late grazing treatment while the diversity in the continuously stocked plots remained at a constant level. Also, the species number increased by 30 per cent in late grazing, but remained constant after six years in continuous stocking. Similar observations are made by, for example, Pavlů et al. [45], who reported an increase of forbs in the vegetation after delayed onset of grazing on old leys. An increase of seed production in late grazed areas was pointed out as one of the possible explanations.

In full-scale grazing experiments, the different treatments will always be located in adjacent areas because of the otherwise biased effects of fences (e.g., [45]). Yet, this may cause problems in separating treatment effects from area effects. In this study, however, several factors indicate that the observed differences between grazing regimes were not area effects. Firstly, the two grasslands showed very similar differences between grazing regimes, which indicates a grazing regime effect rather than a site effect. Secondly, the control cages protected from grazing in Pustnäs did not differ between treatment areas in any measured parameter. Thirdly, similarity between treatment areas was indicated by similar number of species. Finally, the differences in plant reproduction between treatments were large and could clearly be attributed to observed grazing of shoots and reproductive organs.

Apart from the increased fruit set and plant species diversity, the undisturbed period in early summer can also be assumed to be positive for phytophagous, nectar- and pollen-feeding insects. Low amounts of pollen and nectar resources have been suggested to be an important threat to, for example, wild bees [46–48] and lepidopterans [49], of which several species are red listed [8].

Acknowledging that only two grasslands were studied, the results clearly indicate that if continuous stocking from spring to autumn is applied in grasslands with a history of late management, plant species diversity can be assumed to decrease as well as the resource for organisms depending on plant reproduction. This is alarming since continuous stocking from May to September is the most common management regime in Swedish and northern European semi-natural grasslands, largely irrespective of their management history [9]. For example, the widespread shift from mowing to grazing may lead to considerable changes of vegetation composition and plant population viability in semi-natural grasslands. The study suggests an increased use of analyses of management history in grassland conservation and management, which would most likely result in an increased use of late management. The study also shows that late onset of grazing may well function as a substitute for labour-intense mowing of semi-natural grassland.

Acknowledgments

The authors thank J. Bengtsson, Å. Berg, and C. Glynn for discussions, K. Lehtlä, A. Denham, and two anonymous reviewers for comments on the manuscript, S. Overud and E. Sjödin for valuable discussions and help with the field work, and E. Spörndly for being responsible for cattle handling. They further thank Per Rudengren and the Harpsund farm and the Kungsängen experimental farm at SLU for putting their land and grazers at our disposal. The study was funded by the Foundation for Strategic Environmental Research (HagmarksMISTRA) and the Swedish Council for Forestry and Agricultural Research (Award 34.0297/98 to T. Lennartsson).

References

[1] H. Sjörs, *Ekologisk Botanik. Biologi 10*, Almqvist & Wiksell, Uppsala, Sweden, 1971.

[2] M. Pärtel and M. Zobel, "Small-scale plant species richness in calcareous grasslands determined by the species pool, community age and shoot density," *Ecography*, vol. 22, no. 2, pp. 153–159, 1999.

[3] U. Emanuelsson, *The Rural Landscapes of Europe—How man has Shaped Europe's Nature*, Formas, 2009.

[4] R. M. Fuller, "The changing extent and conservation interest of lowland grasslands in England and Wales: a review of grassland surveys 1930–1984," *Biological Conservation*, vol. 40, no. 4, pp. 281–300, 1987.

[5] D. Stanners and P. Bourdeau, *Europe's Environment. The Dobris Assessment*, European Environment Agency, Copenhagen, Denmark, 1995.

[6] Statistics Sweden, "Betesmarker—historiska data," Statistiska Meddelanden 36:9001, Statistics Sweden, Borås, Sweden, 1990.

[7] Nature Conservancy Council, *Nature Conservation in Great Britain*, Nature Conservancy Council, Peterborough, UK, 1984.

[8] U. Gärdenfors, *The 2010 Red List of Swedish Species*, ArtDatabanken, Uppsala, Sweden, 2010.

[9] E. Gustavsson, A. Dahlström, M Emanuelsson, J. Wissman, and T. Lennartsson, "Combining historical and ecological knowledge to optimise biodiversity conservation in semi-natural grasslands," in *The Importance of Biological Interactions in the Study of Biodiversity*, J. L. Pujol, Ed., pp. 173–196, 2011.

[10] G. L. A. Fry, "Conservation in agricultural ecosystems," in *The Scientific Management of Temperate Communities for Conservation*, I. F. Spellerberg, F. B. Goldsmith, and M. G. Morris, Eds., pp. 415–443, Blackwell, London, UK, 1991.

[11] G. Beaufoy, D. Baldock, and J. Clark, *The Nature of Farming. Low Intensity Farming Systems in Nine European Countries*, Institute for European Environmental Policy, London, UK, 1995.

[12] A. García, "Conserving the species-rich meadows of Europe," *Agriculture, Ecosystems and Environment*, vol. 40, no. 1–4, pp. 219–232, 1992.

[13] K. H. Patriksson, *Skötselhandbok för Gårdens Natur-och Kulturvärden*, Swedish Board of Agriculture, Jönköping, Sweden, 1998.

[14] I. Pehrson, *Bete och Betesdjur*, Swedish Board of Agriculture, Jönköping, Sweden, 2001.

[15] T. Lennartsson and J. G. B. Oostermeijer, "Demographic variation and population viability in *Gentianella campestris*: effects of grassland management and environmental stochasticity," *Journal of Ecology*, vol. 89, no. 3, pp. 451–463, 2001.

[16] A. C. Linusson, G. A. I. Berlin, and E. G. A. Olsson, "Reduced community diversity in semi-natural meadows in southern Sweden, 1965–1990," *Plant Ecology*, vol. 136, no. 1, pp. 77–94, 1998.

[17] E. M. Veenendaal, W. H. O. Ernst, and G. S. Modise, "Reproductive effort and phenology of seed production of savanna grasses with different growth form and life history," *Vegetatio*, vol. 123, no. 1, pp. 91–100, 1996.

[18] B. Dumont, A. Farruggia, J. P. Garel, P. Bachelard, E. Boitier, and M. Frain, "How does grazing intensity influence the

diversity of plants and insects in a species-rich upland grassland on basalt soils?" *Grass and Forage Science*, vol. 64, no. 1, pp. 92–105, 2009.

[19] R. M. Callaway, Z. Kikvidze, and D. Kikodze, "Facilitation by unpalatable weeds may conserve plant diversity in overgrazed meadows in the Caucasus Mountains," *Oikos*, vol. 89, no. 2, pp. 275–282, 2000.

[20] L. H. Fraser and J. P. Grime, "Interacting effects of herbivory and fertility on a synthesized plant community," *Journal of Ecology*, vol. 87, no. 3, pp. 514–525, 1999.

[21] I. Noymeir, M. Gutman, and Y. Kaplan, "Responses of Mediterranean grassland plants to grazing and protection," *Journal of Ecology*, vol. 77, no. 1, pp. 290–310, 1989.

[22] R. Fernandez Ales, J. M. Laffarga, and F. Ortega, "Strategies in Mediterranean grassland annuals in relation to stress and disturbance," *Journal of Vegetation Science*, vol. 4, no. 3, pp. 313–322, 1993.

[23] T. Lennartsson, P. Nilsson, and J. Tuomi, "Induction of overcompensation in the field gentian, *Gentianella campestris*," *Ecology*, vol. 79, no. 3, pp. 1061–1072, 1998.

[24] H. J. Zopfi, "Ecotypic variation in *Rhinanthus alectorolophus* (Scopoli) Pollich (Scrophulariaceae) in relation to grassland management. I. Morphological delimitations and habitats of seasonal ecotypes," *Flora*, vol. 188, no. 1, pp. 15–39, 1993.

[25] M. Akhalkatsi and J. Wagner, "Reproductive phenology and seed development of *Gentianella caucasea* in different habitats in the Central Caucasus," *Flora*, vol. 191, no. 2, pp. 161–168, 1996.

[26] L. Påhlsson, *Vegetationstyper i Norden*, Nordic Council of Ministers, Copenhagen, Denmark, 1994.

[27] S. H. Hurlbert, "Pseudoreplication and the design of ecological field experiments," *Ecological Monographs*, vol. 54, pp. 187–211, 1984.

[28] L. Oksanen, "Logic of experiments in ecology: is pseudoreplication a pseudoissue?" *Oikos*, vol. 94, no. 1, pp. 27–38, 2001.

[29] M. E. Castle, "A simple disc instrument for estimating herbage yield," *Journal of the British Grassland Society*, vol. 31, pp. 37–40, 1976.

[30] J. Lid, *Norsk og Svensk Flora*, Det Norske Samlaget, Oslo, Norway, 1979.

[31] H. Ellenberg, H. E. Weber, R. Düll, V. Wirth, and W. Werner, *Zeigerwerte von Pflanzen in Mitteleuropa*, Scripta Geobotanica, Göttingen, Germany, 2001.

[32] V. G. Allen, C. Batello, E. J. Berretta et al., "An international terminology for grazing lands and grazing animals," *Grass and Forage Science*, vol. 66, no. 1, pp. 2–28, 2011.

[33] B. Mossberg and L. Stenberg, *Den Nya Nordiska Floran*, Wahlström & Widstrand, Stockholm, Sweden, 2003.

[34] G. P. Quinn and M. J. Keough, *Experimental Design and Data Analysis for Biologists*, Cambridge University Press, Cambridge, UK, 2002.

[35] M. Begon, J. L. Harper, and C. R. Townsend, *Ecology*, Blackwell Science, London, UK, 3rd edition, 1996.

[36] J. Wissman, *Grazing regimes and plant reproduction in semi-natural grasslands*, Ph.D. thesis, Swedish University of Agricultural Sciences, 2006.

[37] H. M. Jutila and J. B. Grace, "Effects of disturbance on germination and seedling establishment in a coastal prairie grassland: a test of the competitive release hypothesis," *Journal of Ecology*, vol. 90, no. 2, pp. 291–302, 2002.

[38] T. Karlsson, "Early-flowering taxa of *Euphrasia* (Scrophulariaceae) on Gotland, Sweden," *Nordic Journal of Botany*, vol. 4, pp. 303–326, 1984.

[39] I. S. Warwick and D. Briggs, "The genecology of lawn weeds. III. Cultivation experiments with *Achillea millefolium* L., *Bellis perennis* L., *Plantago lanceolata* L., *Plantago major* L. and *Prunella vulgaris* L. collected from lawns and contrasting grassland habitats," *New Phytologist*, vol. 83, pp. 509–536., 1979.

[40] T. Lennartsson, *Demography, reproductive biology and adaptive traits in Gentianella campestris and G. amarella—evaluating grassland management for conservation by using indicator plant species*, Ph.D. thesis, Swedish University of Agricultural Sciences, 1997.

[41] J. M. Bullock and C. A. Marriot, "Plant responses to grazing and opportunities for manipulation," in *Grazing Management*, A. J. Rook and P. D. Penning, Eds., pp. 17–26, British Grassland Society, London, UK, 2000.

[42] M. Fenner, M. E. Hanley, and R. Lawrence, "Comparison of seedling and adult palatability in annual and perennial plants," *Functional Ecology*, vol. 13, no. 4, pp. 546–551, 1999.

[43] P. Tiffin, "Mechanisms of tolerance to herbivore damage: what do we know?" *Evolutionary Ecology*, vol. 14, no. 4–6, pp. 523–536, 2000.

[44] L. Jerling and M. Andersson, "Effects of selective grazing by cattle on the reproduction of *Plantago maritima*," *Holarctic Ecology*, vol. 5, no. 4, pp. 405–411, 1982.

[45] V. Pavlů, M. Hejcman, L. Pavlů, J. Gaisler, P. Nežerková, and L. Meneses, "Changes in plant densities in a mesic species-rich grassland after imposing different grazing management treatments," *Grass and Forage Science*, vol. 61, no. 1, pp. 42–51, 2006.

[46] W. I. Linkowski, B. Cederberg, and L. A. Nilsson, *Vildbin och Fragmentering*, Swedish Board of Agriculture, Jönköping, Sweden, 2004.

[47] A. Pekkarinen, "Oligolectic bee species and their decline in Finland (Hymenoptera: Apoidea)," in *Proceedings of the Proceedings of the 24th Nordic Congress of Entomology*, pp. 151–156, 1999.

[48] M. Franzén and S. G. Nilsson, "How can we preserve and restore species richness of pollinating insects on agricultural land?" *Ecography*, vol. 31, no. 6, pp. 698–708, 2008.

[49] C. Sarin and K. O. Bergman, "Habitat utilisation of burnet moths (*Zygaena* spp.) in southern Sweden: a multi-scale and multi-stage perspective," *Insect Conservation and Diversity*, vol. 3, no. 3, pp. 180–193, 2010.

Determining Effective Riparian Buffer Width for Nonnative Plant Exclusion and Habitat Enhancement

Gavin Ferris,[1] Vincent D'Amico,[2] and Christopher K. Williams[1]

[1] Department of Entomology & Wildlife Ecology, University of Delaware, Newark, DE 19716, USA
[2] US Forest Service, Newark, DE 19716, USA

Correspondence should be addressed to Christopher K. Williams, ckwillia@udel.edu

Academic Editor: Patricia Mosto

Nonnative plants threaten native biodiversity in landscapes where habitats are fragmented. Unfortunately, in developed areas, much of the remaining forested habitat occurs in fragmented riparian corridors. Because forested corridors of sufficient width may allow forest interior specializing native species to retain competitive advantage over edge specialist and generalist nonnative plants, identifying appropriate corridor widths to minimize nonnative plants and maximize ecosystem integrity is of habitat management concern. We measured the occurrences of 4 species of nonnative plants across the widths of 31 forested riparian corridors of varying widths in the White Clay Creek watershed of Pennsylvania and Delaware. Using repeated measures ANOVA, Japanese honeysuckle (*Lonicera japonica*) and multiflora rose (*Rosa multiflora*) prevalence did not significantly decline across buffer widths. However, garlic mustard (*Alliaria petiolata*) and oriental bittersweet (*Celastrus orbiculatus*) declined strongly within the first 15–25 m. Managing for riparian corridor widths a minimum of 15–25 m has the potential to enhance habitat quality but no corridor width (≤ 55 m) will exclude all invasive plants.

1. Introduction

Fragmentation of formerly intact habitats is an unavoidable consequence of human utilization and has resulted in patchy natural habitats in most human-dominated landscapes, especially in the eastern United States [1]. For many species, fragmentation contributes to habitat and resource loss, reduced gene flow, edge effects, and increased time in an inhospitable matrix leading to higher mortality and decreased reproductive rates [2]. Forested riparian buffers may serve as de facto corridors in some of these fragmented landscapes, but in other cases they may be the only appreciable forested habitat remaining. If preserving native biodiversity in human-dominated landscapes is to be a priority, then these forested riparian areas provide a critical opportunity for conservation.

One way of improving the habitat value of forested riparian corridors, and thus protecting biodiversity in the fragmented habitats in which they exist, is to protect the integrity of the food web supported within them. A common reduction in this integrity occurs when nonnative plants dominate the foundation of a trophic pyramid. Herbaceous plants native to temperate forests are vulnerable to edge effects as a consequence of their low dispersal, slow growth rates, long prereproductive periods, and low reproductive outputs [3–5]. Alternatively, most nonnative plant species are generalists or edge-specialists, which often make them the dominant competitor in fragmented habitats [6]. Increased nutrient loads due to agricultural and residential runoff [7, 8] may further increase the competitiveness of nonnative plants, thus further endangering native plant biodiversity [9].

As the foundation of the trophic pyramid is altered by nonnative plants, the enemy release hypothesis postulates that native herbivores will be less likely to feed on these plants whose defenses they have not evolved to overcome [10, 11]. A logical extension of this concept suggests that a potential consequence of nonnative plant dominance is a decrease in native herbivore (especially arthropod) diversity and biomass that exist in these altered trophic pyramids [12–17]. Therefore, invasion by nonnative plants may trigger a trophic cascade [18] negatively affecting higher trophic levels.

It is typical that riparian buffer width recommendations are based on remediation of nutrient runoff (e.g., 30 m [8, 19]). Unfortunately, existing guidelines for riparian buffers widths intended to provide wildlife habitat or ecosystem health are generally based on expert opinion rather than empirical data, and nonnative plants are hardly mentioned except to say that intentional plantings should be of native species [8, 20]. The shade and lower rates of disturbance offered by forest interiors make them less prone to invasion by many nonnative plants than edge habitats [21, 22]. If managers can determine a distance from buffer edge at which nonnative plant prevalence is significantly reduced, then management guidelines can be established for making these mitigation efforts more valuable for conservation by preserving trophic integrity. It is the goal of this research to identify prevalence patterns of 4 common nonnative plants across different widths of riparian buffers to improve habitat management guidelines.

2. Study Area

Our research was focused within the White Clay Creek watershed of Chester County, Pennsylvania, and New Castle County, Delaware. The White Clay Creek watershed is part of the Piedmont Physiographic Region and drains into the Delaware Bay. The major land uses within the 27,923 ha watershed are agriculture (36%), wooded/open space (29%), and residential (25%) [23]. Some 60 tree and shrub species are found within the watershed, as well as more than 200 wildflowers. White Clay Creek itself has been declared a National Wild and Scenic River.

During the summers of 2007-2008, we used Geographic Information Systems to establish 1500 randomly placed points along first- and second-order streams in the White Clay Creek watershed. We further ground truthed these points to select study sites that met all of the following criteria in an effort to reduce bias: (1) relatively equal forested buffer width on both sides of the stream, (2) similar land use (agricultural versus residential) on both sides of the stream, (3) was not recently disturbed by human activity, and (4) had a forest structure and width similar to that of the rest of the corridor within 100 m up and downstream from the point. Thirty-one sites met these 4 criteria (Figure 1).

3. Methods

At each site, we established 3 line transects that were 5 m apart and were perpendicular to the buffer. Transects ran from the edge of the forested section of the corridor to the stream bank with the outside edge of the corridor being defined as even with the upland side of trunks of the outermost canopy trees (>10 m). Each measured corridor was considered the sampling unit, and we measured the response of distance from edge on nonnative prevalence within that sampling unit. The distance between the stream and the corridor edge measured 10 m in 7 sites, 20 m in 4 sites, 35 m in 8 sites, 55 m in 7 sites, and 80 m in 5 sites.

We recorded the occurrence of 4 common nonnative plants using a slightly modified version of the transect method described by Canfield [24]. The line interception method is based on a foundation that the sampling unit is

a line transect, which is visualized having length and vertical dimensions only with no width, and the researcher only measures the intercept of plants through the vertical plane. Along each transect, we laid a metric measuring tape and recorded the percent interception of oriental bittersweet (*Celastrus orbiculatus*), Japanese honeysuckle (*Lonicera japonica*), multiflora rose (*Rosa multiflora*), and garlic mustard (*Alliaria petiolata*) occurring underneath the tape within each 5 m of the length of the transect. We averaged the 3 transects to calculate site level percentages across the buffer. At both ends of each transect and at 5 m increments beginning at the outside edge, we estimated canopy coverage using a densitometer and recorded an index of basal area using a forester's 10X cruising prism. We used repeated measures ANOVA ($\alpha = 0.10$ (to account for a lower sample size of measured streams that met our sampling criteria), with Greenhouse-Geisser corrections to account for violations of sphericity) to determine if each nonnative species prevalence changed with increasing distance from forested riparian edge while considering between corridor effects and within-corridor covariates of canopy coverage, basal density, and total buffer width. We made the assumption (based on visual assessments) that vegetation composition and structure were spatially autocorrelated at increasing distances from corridor edges and therefore repeated measures ANOVA was a necessary statistic (as compared to the use of linear regression when data points are assumed independent of each other). Although repeated measures is often used for measuring autocorrelated temporal changes, its fundamental statistical structure allows for analysis of vegetation changes spatially changing at further distances from the edge. Repeated-measures analysis further analyzes the statistical probability ($\alpha = 0.10$) that within subject effects (nonnative prevalence away from corridor edge) are explained by linear, quadratic, or cubic relationships. If a significant relationship existed in any or all three trends, we only reported the one trend shape with the strongest statistical support. Because the forested riparian corridors measured were of different widths, we could not run a single repeated measures ANOVA on the whole data set (testing for trends over 80 m) because a repeated measures ANOVA is sensitive to missing data. Rather we conducted a series of step wise repeated measurements including all corridors that had data within set distance bands. For example, all 31 corridors could be used to determine if nonnative prevalence changed at 5 m increments within the first 10 m. However, only 24 sites could be used to test if nonnative prevalence changed at 5 m increments within the first 20 m. Additionally, only 20 sites could test for differences across 35 m, 12 sites could test for differences across 55 m, and 5 sites could test for differences across 80 m.

4. Results

Although the prevalence of Japanese honeysuckle appeared to show a decline away from riparian buffer edges, high variances produced no main effects on distance, buffer width, canopy coverage, or basal density (Table 1, Figure 2). The prevalence of garlic mustard at 10 m was affected by the positive main effects of distance and distance/canopy coverage

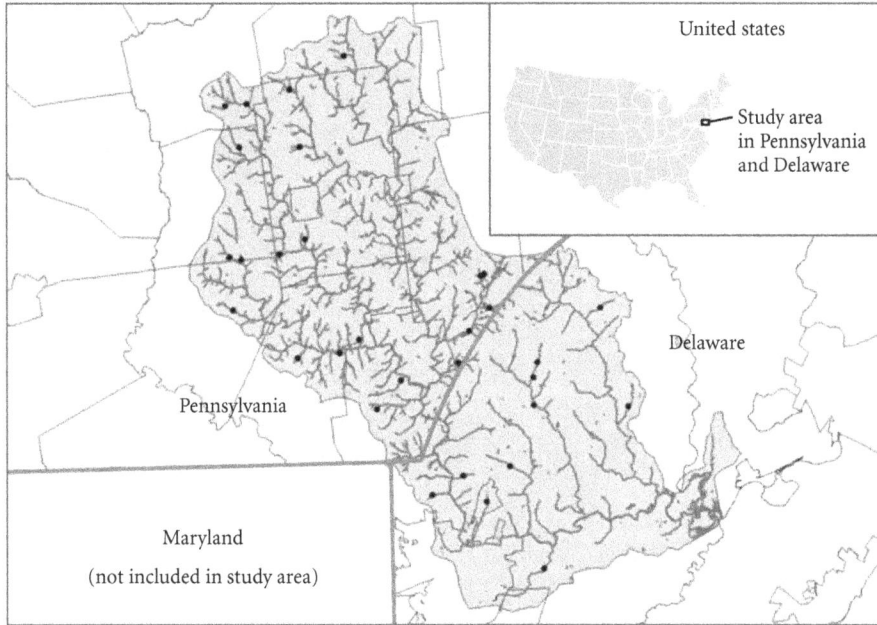

FIGURE 1: Thirty-one study sites within White Clay Creek watershed of Chester County, Pennsylvania, and New Castle County, Delaware, USA.

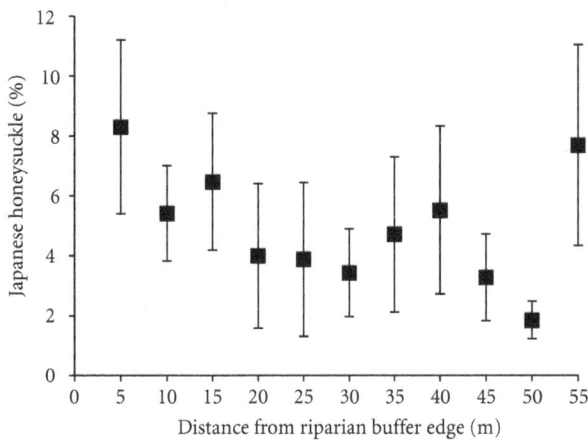

FIGURE 2: The prevalence of Japanese honeysuckle away from the edge of forested riparian buffers in Delaware and Pennsylvania, 2007-2008.

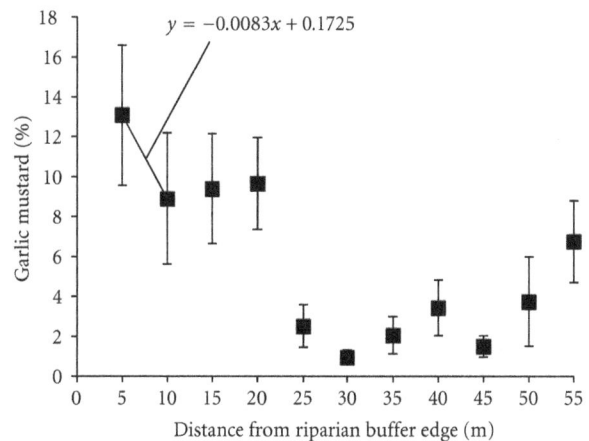

FIGURE 3: The prevalence of garlic mustard away from the edge of forested riparian buffers in Delaware and Pennsylvania, 2007-2008. Within-subject contrasts indicated a significant ($P < 0.10$) linear trend through 10 m.

interaction ($F_{1,17} = 4.96$, $P = 0.04$ and $F_{1,17} = 5.30$, $P = 0.03$, respectively, Table 1) following a linear trend ($F_{1,17} = 4.96$, $P = 0.04$, Figure 3). At 20 m, distance/total buffer width interaction showed significant reductions in garlic mustard ($F_{18.57,20.43} = 1.81$, $P = 0.10$, Table 1) but the within-subject contrast failed to detect any slope structure (Figure 3). Beyond 25 m, the mean prevalence of garlic mustard remained at low levels (1–7%).

The prevalence of multiflora rose showed no reduction at 10 m as a function of the main effects of distance, total buffer width, basal density, or canopy coverage. However, at 20 m distance, distance/total buffer width interaction, and distance/canopy coverage interaction showed significant

main effects on the reduction of multiflora rose ($F_{1.98,19.84} = 2.88$, $P = 0.08$; $F_{1.98,19.84} = 2.70$, $P = 0.09$; and $F_{21.83,19.84} = 1.88$, $P = 0.08$, respectively, Table 1) following a quadratic trend ($F_{1,10} = 10.86$, $P < 0.01$, Figure 4). No within-subject main effects were observed at 35 m; however, significant between-sample effects occurred ($F_{1,9} = 4.38$, $P = 0.07$, Table 1). At 55 m, distance and distance/canopy coverage interaction showed significant positive main effects on the reduction of multiflora rose ($F_{2.70,10.81} = 2.99$, $P = 0.08$ and $F_{2.70,10.81} = 3.00$, $P = 0.08$, respectively, Table 1) following an increasing linear trend ($F_{1,4} = 4.49$, $P = 0.10$, Figure 4).

TABLE 1: Summary of repeated measures ANOVA models on nonnative prevalence across multiple riparian buffer widths in Delaware and Pennsylvania, 2007-2008. We tested for within-sample effect of distance from buffer edge as well as the interaction between distance and total riparian buffer width, basal density at sampled distance, and canopy coverage at sampled distance. Stars indicate significance ($P \leq 0.10$). We further tested for trends across buffer width exhibiting significant ($P \leq 0.10$) linear, quadratic, or cubic relationships.

| Species | Repeated measures ANOVA model | Distance (m) | | | |
| | | 10 | 20 | 35 | 55 |
		$n = 31$	$n = 24$	$n = 20$	$n = 12$
Japanese honeysuckle	Distance	—	—	—	—
	Distance contrast	—	—	—	—
	Distance * total buffer width	—	—	—	—
	Distance * basal density	—	—	—	—
	Distance * canopy coverage	—	—	—	—
	Between sample effect	—	—	—	—
Garlic mustard	Distance	*	—	—	—
	Distance contrast	Linear	—	—	—
	Distance * total buffer width	—	*	—	—
	Distance * basal density	—	—	—	—
	Distance * canopy coverage	*	—	—	—
	Between sample effect	—	—	—	—
Multiflora rose	Distance	—	*	—	*
	Distance contrast	—	Quadratic	—	Linear
	Distance * total buffer width	—	*	—	—
	Distance * basal density	—	—	—	—
	Distance * canopy coverage	—	*	—	*
	Between sample effect	—	—	*	—
Bittersweet	Distance	*	*	—	—
	Distance contrast	Linear	Cubic	—	—
	Distance * total buffer width	—	*	—	—
	Distance * basal density	—	—	—	—
	Distance * canopy coverage	*	*	—	—
	Between sample effect	—	—	—	—

These trends indicate that multiflora rose was relatively consistent in all forested buffers despite their distance from edge.

Oriental bittersweet showed the most consistent response to increasing distance from the buffer edge. At 10 m, distance and distance/canopy coverage interaction both produced significant reductions in oriental bittersweet ($F_{1,17} = 3.77$, $P = 0.07$ and $F_{1,17} = 4.08$, $P = 0.06$, respectively, Table 1) following a linear trend ($F_{1,17} = 3.77$, $P = 0.07$, Figure 5). At 20 m, distance, distance/total buffer width interaction and distance/canopy coverage interaction showed main effects on the reduction of oriental bittersweet ($F_{1.86,18.64} = 2.80$, $P = 0.09$; $F_{20.50,18.64} = 1.90$, $P = 0.09$; and $F_{1.86,18.64} = 3.17$, $P = 0.07$, respectively, Table 1) following a cubic trend ($F_{1,10} = 4.50$, $P = 0.06$, Figure 5). No main effects were observed at 35 m or 55 m to affect oriental bittersweet prevalence; however, bittersweet levels continued to drop to low mean levels (~1-2%) at the furthest distances away from the edge.

5. Discussion

By definition, corridors exist in highly fragmented landscapes. Riparian corridors are especially prone to edge effects because the stream bank itself creates another source for repeated disturbance. Furthermore, as many riparian corridors exist as "buffer" strips intended to prevent sediment and nutrient runoff [8, 19], corridor soils are often nutrient rich. Increased resource availability and disturbance regimes contribute to enhanced invasive plant dominance [25]. The combination of frequent disturbance and increased resource availability makes it unsurprising that invasive plant occurrence is frequent in the riparian corridors of the White Clay Creek watershed. Although we set our alpha level at 0.10 to compensate for lower sample size, which increased the threshold for rejecting the null hypothesis as compared to other studies, we feel our results indicated observable trends that will aid future management decisions for corridor management. In particular, although multiflora rose and

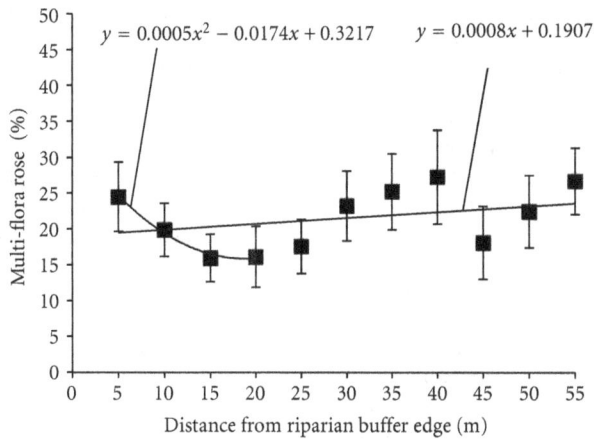

FIGURE 4: The prevalence of multiflora rose away from the edge of forested riparian buffers in Delaware and Pennsylvania, 2007-2008. Within-subject contrasts indicated a significant ($P < 0.10$) decreasing quadratic trend through 20 m but a linear increasing trend through 55 m.

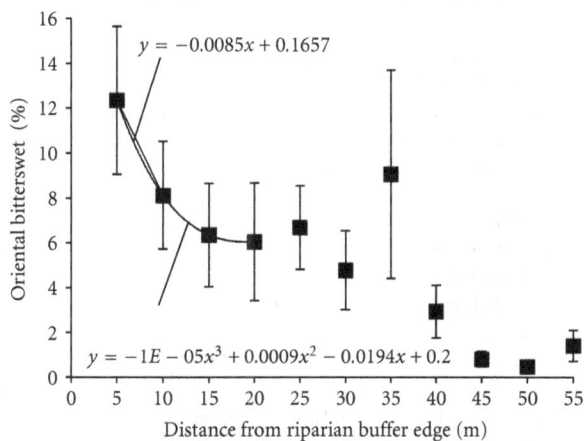

FIGURE 5: The prevalence of multiflora rose away from the edge of forested riparian buffers in Delaware and Pennsylvania, 2007-2008. Within-subject contrasts indicated a significant ($P < 0.10$) decreasing linear trend through 10 m and a decreasing cubic trend through 20 m.

Japanese honeysuckle showed little change in prevalence across different buffer widths, notable reductions occurred in oriental bittersweet occurred by 15 m and garlic mustard by 25 m. We will discuss possible reasons for these relationships in all four species.

Multiflora rose was the most commonly occurring invasive plant and showed no decline across corridor widths. Its frequency should not be surprising given the extent to which it was historically planted across the landscape. Multiflora rose was introduced from Japan and Korea in the 1860s and promoted across the eastern half of the United States as "living fences" and for enhancement of wildlife habitat [26]. Two of the owners of farms in the study area relayed memories of planting multiflora roses supplied by the government during WWII when wire fences were removed for metal drives. Its

long history in the study area may partially account for its establishment throughout the widths of most corridors. Also, there are multiple varieties of multiflora rose [27], leading to the possibility there could be some cultivars that are better suited to interior forest conditions and others that are more competitive at corridor edges. Although we did not note cultivars in this study, we encourage future researchers to study this possibility. Further, multiflora rose seeds are dispersed through bird scat [26], which may skew propagule pressure into corridors by disproportionately vectoring seeds into forested areas where birds roost and nest. Our study area also has a high population of white-tailed deer [23] and multiflora rose, garlic mustard, and Japanese honeysuckle has been shown to be less susceptible to deer herbivory than many native species and some other nonnative species [28, 29]. Preferred browsing on native plants by deer is one example of how enemy release can enhance the competitive abilities of nonnative plants in novel habitats.

In addition to being resistant to deer herbivory and having multiple varieties like multiflora rose [30], Japanese honeysuckle has also been shown to be shade-tolerant, and it benefits from an extended growing season as an evergreen [26]. A vining, climbing habit allows Japanese honeysuckle to make use of space even after another plant has become established. The combination of its growth pattern, shade tolerance, extended growing season, and deer resistance may explain the lack of response to corridor width.

While oriental bittersweet shares Japanese honeysuckle's vining habit and its seeds can germinate in partial to full shade [31], it is considered shade intolerant [32]. Bittersweet also does not share the resistance to deer herbivory displayed by other nonnative species, with at least one study showing a reduction in oriental bittersweet in response to deer browsing [28]. Therefore, the declines in oriental bittersweet as a function of buffer width and canopy coverage interactions were not surprising.

Unlike oriental bittersweet, garlic mustard is highly resistant to deer herbivory and has been shown to respond positively to deer browsing on its competitors [29]. While described as shade-tolerant [33], garlic mustard has also been shown to positively respond to increased light availability [34]. While its shade tolerance and deer herbivory may allow garlic mustard to exist in forest interiors, it has previously been shown to be more competitive in forest edges than in forest interiors [35]. Our results suggest that modest increases in corridor width along with canopy coverage interactions have the potential to limit garlic mustard recruitment.

Reactions to distance from buffer edge and canopy coverage in garlic mustard and oriental bittersweet support the assertion that because these species are less prevalent in forest interiors, expanding forested areas will decrease their competitive advantage over native species in riparian corridors. However, the disappearance of these significant trends after 20 m is troubling. It is possible that our sample sizes for larger corridor widths were simply too small to detect a decline in prevalence beyond 20 m—a consequence of the relative rarity of corridors beyond this width. It is also possible that the higher rates of disturbance within riparian buffers diminishes the interior forest character of these corridors such that they no longer cause a decline in invasive species within their interiors. We recommend further research to investigate why small populations of nonnative species can still exist at far distances from corridor or forest edges.

The extent to which expanding riparian corridors has the desired effect of restricting invasive plants and promoting native biodiversity will vary according to the invasive plants present in the area and the response of that particular ecosystem's ecologically important plants to forest interior conditions. The fact that we observed a lack of corridor-induced declines in Japanese honeysuckle and multiflora rose illustrates this point, and we encourage further research to clarify these results. However, because data still show that oriental bittersweet and garlic mustard showed a decline and leveling off by 15–25 m, this indicates that increasing buffer width will be more effective at excluding some invasive plants than others and that some native plants will likely respond better than others to similar conditions. This suggests that increasing corridor width may not only improve wildlife habitat by creating more of it, but by enhancing its quality by restraining the influx of some invasive vegetation and promoting plants native to the ecosystem in question and their associated herbivores. We further encourage future research to determine if a reduction in the nonnative plant assemblage, as compared to its complete elimination, will still improve trophic responses and ecosystem function.

References

[1] L. Fahrig, "Effects of habitat fragmentation on biodiversity," *Annual Review of Ecology, Evolution, and Systematics*, vol. 34, pp. 487–515, 2003.

[2] L. Fahrig, "Effect of habitat fragmentation on the extinction threshold: a synthesis," *Ecological Applications*, vol. 12, no. 2, pp. 346–353, 2002.

[3] P. Bierzychudek, "The demography of jack-in-the-pulpit, a forest perennial that changes sex," *Ecological Monographs*, vol. 52, no. 4, pp. 335–351, 1982.

[4] G. R. Matlack, "Plant species migration in a mixed-history forest landscape in eastern North America," *Ecology*, vol. 75, no. 5, pp. 1491–1502, 1994.

[5] O. Honnay, H. Jacquemyn, B. Bossuyt, and M. Hermy, "Forest fragmentation effects on patch occupancy and population viability of herbaceous plant species," *New Phytologist*, vol. 166, no. 3, pp. 723–736, 2005.

[6] K. MacQuarrie and C. Lacroix, "The upland hardwood component of Prince Edward Island's remnant Acadian forest: determination of depth of edge and patterns of exotic plant invasion," *Canadian Journal of Botany*, vol. 81, no. 11, pp. 1113–1128, 2003.

[7] J. E. Kundell and T. C. Rasmussen, "Recommendations of the 1995 Georgia Board of Regents'scientific panel on evaluating the erosion measurement standard defined by the Georgia Erosion and Sedimentation Act," in *Proceedings of the Georgia Water Resources Conference*, pp. 211–217, 1995.

[8] S. Wenger, *A Review of the Scientific Literature on Riparian Buffer Width, Extent, and Vegetation*, University of Georgia, Institute of Ecology, Office of Public Service and Outreach, Athens, Ga, USA, 1999.

[9] M. A. Rickey and R. C. Anderson, "Effects of nitrogen addition on the invasive grass Phragmites australis and a native com-

[10] J. H. Lawton and K. C. Brown, "The population and community ecology of invading biological sciences," *Philosophical Transactions of the Royal Society of London B*, vol. 314, no. 1167, pp. 607–617, 1986.

[11] J. M. Crawley, "The population biology of invaders," *Philosophical Transactions of the Royal Society of London B*, vol. 314, no. 1167, Article ID 10.1098/rstb.1986.0082, pp. 711–731, 1987.

[12] D. W. Tallamy, "Do alien plants reduce insect biomass?" *Conservation Biology*, vol. 18, no. 6, pp. 1689–1692, 2004.

[13] M. J. Samways, P. M. Caldwell, and R. Osborn, "Ground-living invertebrate assemblages in native, planted and invasive vegetation in South Africa," *Agriculture, Ecosystems and Environment*, vol. 59, no. 1-2, pp. 19–32, 1996.

[14] M. G. Cripps, M. Schwarzländer, J. L. McKenney, H. L. Hinz, and W. J. Price, "Biogeographical comparison of the arthropod herbivore communities associated with Lepidium draba in its native, expanded and introduced ranges," *Journal of Biogeography*, vol. 33, no. 12, pp. 2107–2119, 2006.

[15] Y. T. Wu, C. H. Wang, X. D. Zhang et al., "Effects of saltmarsh invasion by Spartina alterniflora on arthropod community structure and diets," *Biological Invasions*, vol. 11, no. 3, pp. 635–649, 2009.

[16] A. Yoshioka, T. Kadoya, S. I. Suda, and I. Washitani, "Impacts of weeping lovegrass (*Eragrostis curvula*) invasion on native grasshoppers: responses of habitat generalist and specialist species," *Biological Invasions*, vol. 12, no. 3, pp. 531–539, 2010.

[17] E. Gerber, C. Krebs, C. Murrell, M. Moretti, R. Rocklin, and U. Schaffner, "Exotic invasive knotweeds (Fallopia spp.) negatively affect native plant and invertebrate assemblages in European riparian habitats," *Biological Conservation*, vol. 141, no. 3, pp. 646–654, 2008.

[18] R. T. Paine, "Food webs: linkage, interaction strength and community infrastructure," *Journal of Animal Ecology*, vol. 49, no. 3, pp. 666–685, 1980.

[19] R. T. Mander, V. Kuusemets, K. Lohmus, and T. Mauring, "Efficiency and dimensioning of riparian buffer zones in agricultural catchments," *Ecological Engineering*, vol. 8, no. 4, pp. 299–324, 1997.

[20] C. R. Blinn and M. A. Kilgore, "Riparian management practices: a summary of state guidelines," *Journal of Forestry*, vol. 99, no. 8, pp. 11–17, 2001.

[21] O. Honnay, K. Verheyen, and M. Hermy, "Permeability of ancient forest edges for weedy plant species invasion," *Forest Ecology and Management*, vol. 161, no. 1–3, pp. 109–122, 2002.

[22] E. D. Yates, D. F. Levia Jr, and C. L. Williams, "Recruitment of three non-native invasive plants into a fragmented forest in southern Illinois," *Forest Ecology and Management*, vol. 190, no. 2-3, pp. 119–130, 2004.

[23] M. Corrozi, A. Homsey, G. Kauffman, E. Farris, and M. Seymour, "White Clay Creek state of the watershed report," Tech. Rep., Institute for Public Administration-Water Resources Agency University of Delaware, Newark, NJ, USA, 2008.

[24] R. H. Canfield, "Application of the line interception method in sampling range vegetation," *Journal of Forestry*, vol. 39, no. 4, pp. 388–394, 1941.

[25] K. Stinson, S. Kaufman, L. Durbin, and F. Lowenstein, "Impacts of garlic mustard invasion on a forest understory community," *Northeastern Naturalist*, vol. 14, no. 1, pp. 73–88, 2007.

[26] J. Miller, "Exotic invasive plants in southeastern forests," in *Proceedings of the USDA Forest Service and Tennessee Exotic

petitor Spartina pectinata," *Journal of Applied Ecology*, vol. 41, no. 5, pp. 888–896, 2004.

Plants Council, K. Britton, Ed., Exotic Pests of Eastern Forests, USDA, Nashville, Tenn, USA, 1997.

[27] C. Paris and T. Maney, "*Rosa multiflora* and its progeny," *Proceedings of the Iowa Academy of Sciences*, vol. 46, pp. 149–160, 1939.

[28] C. R. Rossell, S. Patch, and S. Salmons, "Effects of deer browsing on native and non-native vegetation in a mixed oak-beech forest on the atlantic coastal plain," *Northeastern Naturalist*, vol. 14, no. 1, pp. 61–72, 2007.

[29] T. M. Knight, J. L. Dunn, L. A. Smith, J. Davis, and S. Kalisz, "Deer facilitate invasive plant success in a pennsylvania forest understory," *Natural Areas Journal*, vol. 29, no. 2, pp. 110–116, 2009.

[30] C. D. Huebner, "Vulnerability of oak-dominated forests in West Virginia to invasive exotic plants: temporal and spatial patterns of nine exotic species using Herbarium records and land classification data," *Castanea*, vol. 68, no. 1, pp. 1–14, 2003.

[31] D. T. Patterson, *The ecology of oriental bittersweet, Celastrus orbiculatus, a weedy introduced Ornamental vine*, Ph.D. thesis, Duke University, Durham, NC, USA, 1974.

[32] J. Miller, "Nonnative invasive plants of Southern forests: a field guide for identification and control," USDA Forest Service General Technical Report SRS-62, 2003.

[33] J. P. Grime, J. G. Hodgson, and R. Hunt, *Comparative Plant Ecology*, Unwin Hyman, Boston, Mass, USA, 1988.

[34] J. F. Meekins and B. C. McCarthy, "Responses of the biennial forest herb *Alliaria petiolata* to variation in population density, nutrient addition and light availability," *Journal of Ecology*, vol. 88, no. 3, pp. 447–463, 2000.

[35] J. F. Meekins and B. C. McCarthy, "Effect of environmental variation on the invasive success of a nonindigenous forest herb," *Ecological Applications*, vol. 11, no. 5, pp. 1336–1348, 2001.

8

Mangrove Rehabilitation on Highly Eroded Coastal Shorelines at Samut Sakhon, Thailand

Matsui Naohiro,[1] Songsangjinda Putth,[2] and Morimune Keiyo[3]

[1] *Department of Environment, Kanso Technos Co., Ltd., Osaka 541-0052, Japan*
[2] *Fishery Department, Trang Coastal Aquaculture Station, Trang 92150, Thailand*
[3] *Power Engineering R&D Center, The Kansai Electric Power Co., Inc., Kyoto 609-0237, Japan*

Correspondence should be addressed to Matsui Naohiro, matui_naohiro@kanso.co.jp

Academic Editor: L. M. Chu

The study site is currently retreating at a rate of 20 m y^{-1} due to severe coastal erosion and found to be highly polluted as revealed from the water, sediment and biological analysis. In an attempt to prevent coastal erosion, 14,000 *Rhizophora mucronata* (RM) trees were planted across a heavily eroded shoreline at Samut Sakhon, Thailand. The survival rate of RM was high at the landward area and decreased at the offshore area. The most landward plot showed the highest survival rate when measured 4 years after planting (63.5%), while only 26.7% of trees survived at the most offshore plot. NPK and coconut fiber were shown to be significantly effective to enhance initial tree growths in heavily eroded area.

1. Introduction

Coastal erosion is one of the most severe environmental problems currently affecting Thailand, and the Thai Government has designated this problem as a high priority among national environmental problems to be solved urgently. Coastal erosion is accelerated with the destruction of mangrove forests that normally provide protection from erosion. Thousands of abandoned shrimp ponds in coastal areas of Thailand represent decades of mangrove destruction. It is therefore necessary to rehabilitate these sites in the interests of coastal protection.

Samut Sakhon is located approximately 50 km from Bangkok. Because of its proximity to the capital, coastal development began at this site in the 1970s. As there were no legislative restrictions on coastal development at that time, poorly planned coastal developments were carried out up to the very edge of the coast; this practice is now prohibited by law. Inland damming at the upper reaches also was actively conducted. Sediment is transported from the upper reach to lower lands and accumulates there. Reduced flux of sediment due to hydraulic dam had severe impacts on changes of

Samut Sakhon coastlines. Under the influence of direct wave/wind attack and without protection from mangrove forests, unprecedented coastal erosion has occurred in this region.

Mangrove plantations in vulnerable areas could be beneficial for long-term coastal protection both to continuous erosion and to severe hazard such as tsunami. The presence of mangrove as well as continuous riverine sediment flux is essential to maintain coastal stability. Over the years, a number of mangrove planting projects have been carried out to restore degraded mangrove forests [1–3], some of which aimed at protecting shoreline against storm and cyclone damage. Rehabilitation must be urgently implemented to prevent further coastal erosion; however, planting mangroves within eroded areas is far more difficult than planting at other sites because of the harsh physical conditions. Quite low survival rates of planted mangroves at the shoreline area were reported at the Visayas [4] and at other areas in the Philippines [5], as well as at Surat Thani [6] and at Samuth Songkram in Thailand [7]. Mangrove plantation in shoreline area leads to rapid accretion because the mangrove roots and pneumatophores effectively slow water movement and act as efficient sediment trappers [8]. Geomorphological process affects greatly this function of mangrove vegetation. However, mangroves explicitly promote sediment deposition, stimulate soil stability, and protect shoreline from erosion [3, 9–11]. The success of rehabilitation efforts largely depends on the early establishment of planted mangroves. After understanding the ecology of sites and adequate species as key issues for successful mangrove plantation, we shall head for the next step to enhance initial growth rates. Effects of nitrogen fertilizer on mangrove growths were identified in Florida, USA at the abandoned impoundment [12]. Also Matsui et al. (2010) revealed an importance of soil carbon in initial stage of mangrove growths [13]. Taking these into the account, soil amendment application will be effective in ensuring rapid initial growth. There is however still limited number of mangrove planting by the application of soil amendment, especially at shoreline area.

The objectives of this study are therefore: (1) to understand the growth characteristics of mangroves species planted upon highly eroded shorelines, (2) to assess the effectiveness of different soil amendments in promoting initial mangrove growth, and (3) to examine the pollution level of the study site from the viewpoints of water qualities and biological properties to determine the influence of the surrounding areas on the mangrove area.

2. Methods

2.1. Study Site. The study was conducted at an abandoned shrimp pond located close to the shoreline within the Samut Sakhon subdistrict/district, Samut Sakhon province, Thailand (13°28′N, 100°13′E) (Figure 1). Annual tidal range monitored at Tha Chin station (Samut Sakhon province) is about 0.36–3.46 m above mean sea level. The wave height rises up as high as 4 m at monsoon seasons.

Most of the mangroves disappeared from the study site but some species such as *Avicennia marina* and *Rhizophora mucronata* remained to grow in the edges along the road and of the former shrimp ponds. Exploitation of the area began with the construction of saltpans in the 1970s, followed by shrimp farming in the 1980s and associated clear-felling of the majority of mangrove forests. In addition to the destruction of mangroves in the area, a drastic decline was reported in fishery yields. Local fishermen reported that greater numbers of cockles and catfish were harvested prior to the large-scale conversion of mangroves to shrimp ponds. Today, few catfish are harvested in the area.

2.2. Outline of Coastal Erosion at Samut Sakhon. The location of the shoreline is generally determined by the dynamic balance between sedimentation and erosion. Accordingly, changes in sedimentation patterns affect the dynamic balance of the shoreline. At Nakhon Sawan (Figure 1), sediment loading was monitored for 33 years from 1960 to 1993 [14]. Figure 2 shows the changes in total sediment loading during this period. Sediment loading was variable during the period up to 1970, but subsequent to this time the total amount of sediment has gradually decreased, falling below 10^{11} metric tons per year for every year after 1976 except 1987. The

Bhumibol Dam was constructed in 1964 and the Sirikit Dam became operational in 1974 after a 9-year construction period from 1963 to 1972 (Figure 1). It is likely that dam construction resulted in reduced sediment loading into the Gulf of Thailand.

Sedimentation in this area is highest in September and October when peak rainfall occurs (Figure 3). The sediment patterns or the upstream area of the Chao Phraya River between 1942 and 1988 (Figure 3) resembles the pattern of rainfall distribution; however, there is likely to be an approximately 1-month time lag between rainfall and related sedimentation. For example, the highest rainfall is recorded in September, while the maximum sedimentation is recorded a month later in October.

Severe wave attack is concentrated in the period from March to the middle of May, although above-average wave action continues until the end of September [15]. As stronger wave attack leads to increased damage to planted mangroves, special attention should be given to these months to protect mangroves from wave damage.

The effects of wind and/or wave attack on the coast are more severe if mangrove forests have been cleared. The shoreline in the study region is retreating at a rate of 20 m y^{-1} [15], which is much higher than the figure of 3.6 m y^{-1} recorded in the eastern part of the Thai Gulf [16]. According to a feasibility study conducted by SEATEC and BEST [17] on a 14 km stretch of coastline in this area, the shoreline retreated by 127 m, equivalent to 1,105 rai (1 rai = 40 × 40 m), during the 7-year period from 1987 to 1994, and retreated by 351 m (3,056 rai) during the 25-year period from 1969 to 1994.

2.3. Mangrove Plantation and Tree Measurements. A total of approximately 14,000 *Rhizophora mucronata* (RM) seedlings were planted over an area of approximately 6 ha in September 2001. Subsequently, tree height was measured on five occasions: June 2002, January 2003, June 2003, January 2005, and October 2005, representing periods of 9 months, 1 year and 3 months, 1 year and 8 months, 3 years and 3 months, and 4 years after planting, respectively. To monitor the planted trees, we designated three plots within the plantation site (Figure 4). Each plot contains 8 lines of 25 trees with a total number of 200.

2.4. Soil Amendment Application. At sites of severe coastal erosion, it is important to induce the rapid growth of planted mangroves, especially during the initial growth stages. To this end, we applied three types of soil amendment: NPK, humic acid, and coconut fiber to 50 trees, respectively, at three plots. Coconut fiber is mixed into the seedling pot in Thai governmental mangrove nurseries to improve soil physical condition. Mangrove ecosystems were identified as either nitrogen (N) or phosphorus (P) limited [18]; hence NPK was applied in this study to enhance initial growth of mangroves. Contents of organic carbon influence initial growths of *Rhizophora apiculata* [13] so that humic acid as the fast-acting organic carbon was applied. Analysis of variance (ANOVA) was applied to examine the difference of three types of soil amendment as well as to examine the effects of locality on

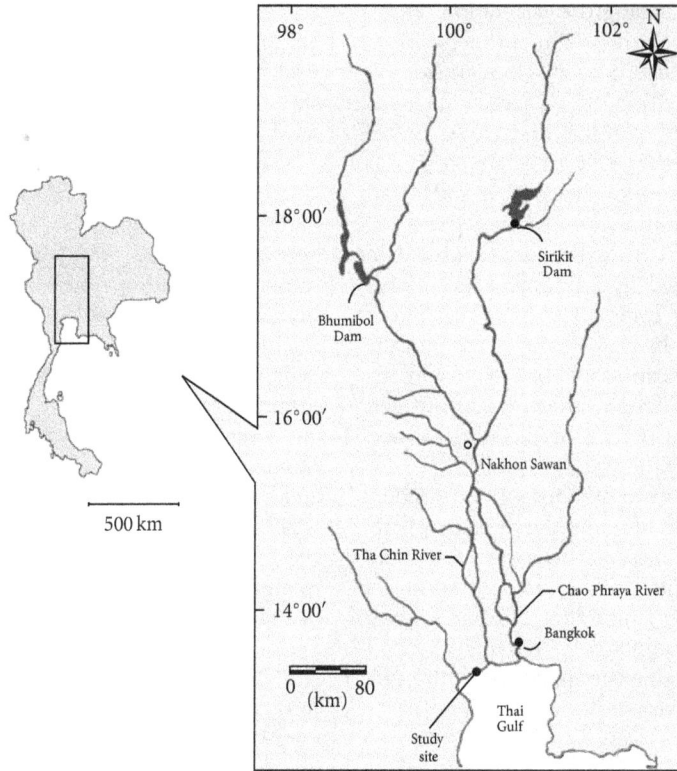

FIGURE 1: Location of the study site. Samut Sakhon province, Thailand.

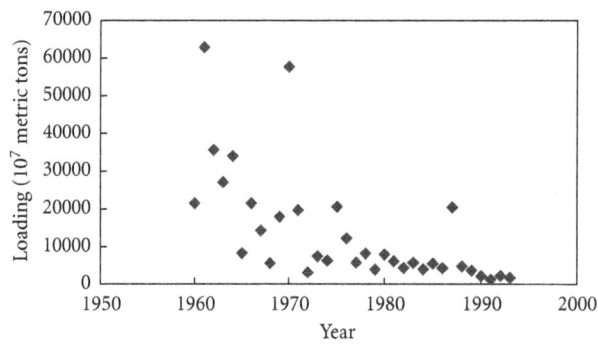

FIGURE 2: Total sediment loading at Nakhon Sawan from 1960 to 1993 (cited from [1]).

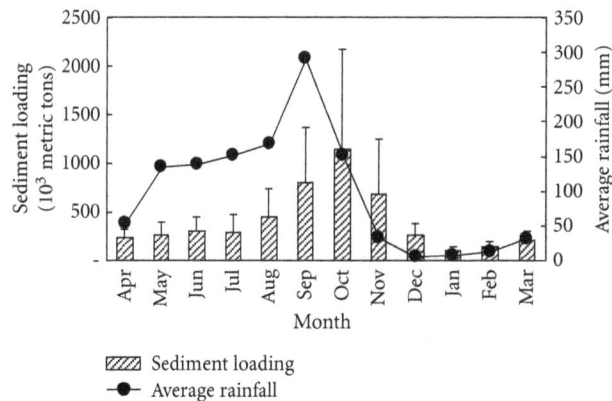

FIGURE 3: Average monthly sediment loading at Nakhon Sawan from 1976 to 1993 excluding the data of 1987 (Error bars denote standard deviation) and average rainfall (mm) within the Chao Phraya River basin between 1942 and 1988 (cited from [1]).

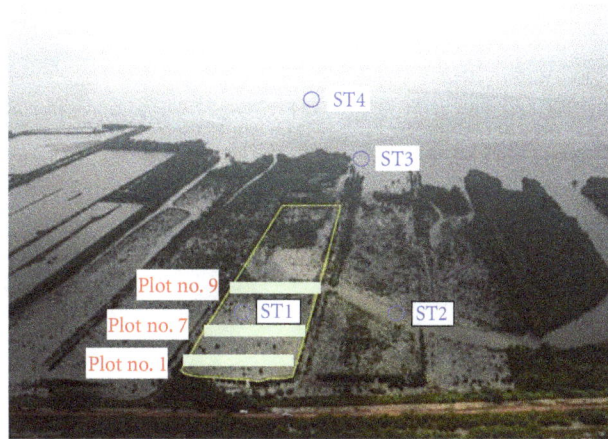

FIGURE 4: View of the study site taken by remote-controlled helicopter from 500 m above sea level. The enclosed area shows the location of the study site (approximately 6 ha). Blue circles show the stations where the samples were taken for measurements of water, sediment, and biological properties.

mangrove growth by the statistical software JMP 8.0.2. The statistical significance of differences was determined by ANOVA followed by a multiple comparison test (Tukey-Kramer test) at 95% significance level ($P < 0.05$).

2.5. Water/Sediment Sampling and Biological Measurements. Water was collected from the study site (ST1) and an offshore station (ST4) (Figure 4) and transported to the laboratory while kept chilled. The water samples were analyzed for temperature, dissolved oxygen, salinity, pH, ammoniacal-nitrogen (NH_4^+ plus NH_3), nitrite-nitrogen, nitrate-nitrogen, total dissolved nitrogen (TDN), dissolved organic nitrogen (DON), particulate organic nitrogen (PON), total nitrogen (TN), dissolved inorganic phosphorus (DIP), total dissolved phosphorus (TDP), dissolved organic phosphorus (DOP), particulate phosphorus (PP), total phosphorus (TP), particulate organic carbon (POC), chlorophyll *a*, and biological oxygen demand (BOD). All of the above water analyses were conducted according to the standard methods [19].

Sediment samples collected at the study site (ST1), a nearby study site (ST2), the edge of shoreline (ST3), and an offshore station (ST4) were analyzed for ammoniacal-nitrogen, nitrite-nitrogen, nitrate-nitrogen, phosphorus, total phosphorus (the sum of all forms of phosphorus), total nitrogen, organic carbon, and acid volatile sulfide (AVS-S). Ammoniacal-nitrogen was measured by the phenolhypochlorite method [20]. Nitrate and nitrite were determined together using the cadmium reduction method on a flow injection analyzer (Lachat QuikChem 8000). Phosphorus was analyzed using the spectrophotometric method [21] while total phosphorus was analyzed by ash/acid extraction [22]. Nitrogen and organic carbon were analyzed according to the high temperature combustion method using CHN elemental analyzer (LECO CHN-900) [23]. AVS-S was analyzed from the surface sediment using H_2S absorbent columns (GASTEC, Kanagawa, Japan) according to the procedure described in Tsutsumi and Kikuchi (1983) [24]. The texture of each sediment sample was determined using the hydrometer method.

The abundance of plankton was measured from samples collected in a 40 μm mesh plankton net and preserved in 5% formalin solution. The biomass of phytoplankton was studied according to the enumeration method and chlorophyll *a* content. Benthos abundance was calculated from the number of benthic animals retained after passing the sample through a 0.5 mm benthos sieve; collected animals were preserved in 10% formalin solution.

3. Results

3.1. Survival Rate and Tree Growth. The survival and growth rates of planted mangroves have varied since the mangroves were planted in September 2001. In October 2005, 4 years after planting, the survival rate decreased in the order of Plot 1 > Plot 7 > Plot 9, with survival rates of 63.5%, 50.0%, and 26.7%, respectively (Figure 5). The survival rate varied with distance from the shoreline and may correspond to the intensity of wave attack. As Plot 9 was located at the most offshore, it experienced the most severe erosion resulting from wind and wave action, thereby resulting in the lowest survival rate of the measured plots. Changes in the survival rates for all plots were relatively minor over the first 3 years after planting, but the survival rate for Plot 9 began to decrease markedly from January 2005, 3 years and 3 months after planting. The survival rate for Plot 1 oddly increased in Oct 2005 from that of Jan 2005, which is because the trees identified as dead in Jan 2005 had revived in Oct 2005.

Tree heights measured 4 years after planting (October 2005) were greatest in Plot 1, with an average height of 308 cm. Mangrove showed the highest growths where the survival rate was highest, indicating that growth and survival conditions were interrelated. Tree heights 4 years after planting were greater in Plot 9 (248 cm ± 61.6 cm) than in Plot 7 (214 cm ± 96.7 cm), possibly due to different degree of wave/wind attack between the two plots. It is thus speculated that physical conditions were critical for the survival and growth of mangroves in the study region.

We also established an experimental Rhizophora plantation within an abandoned shrimp pond at Nakorn Sri

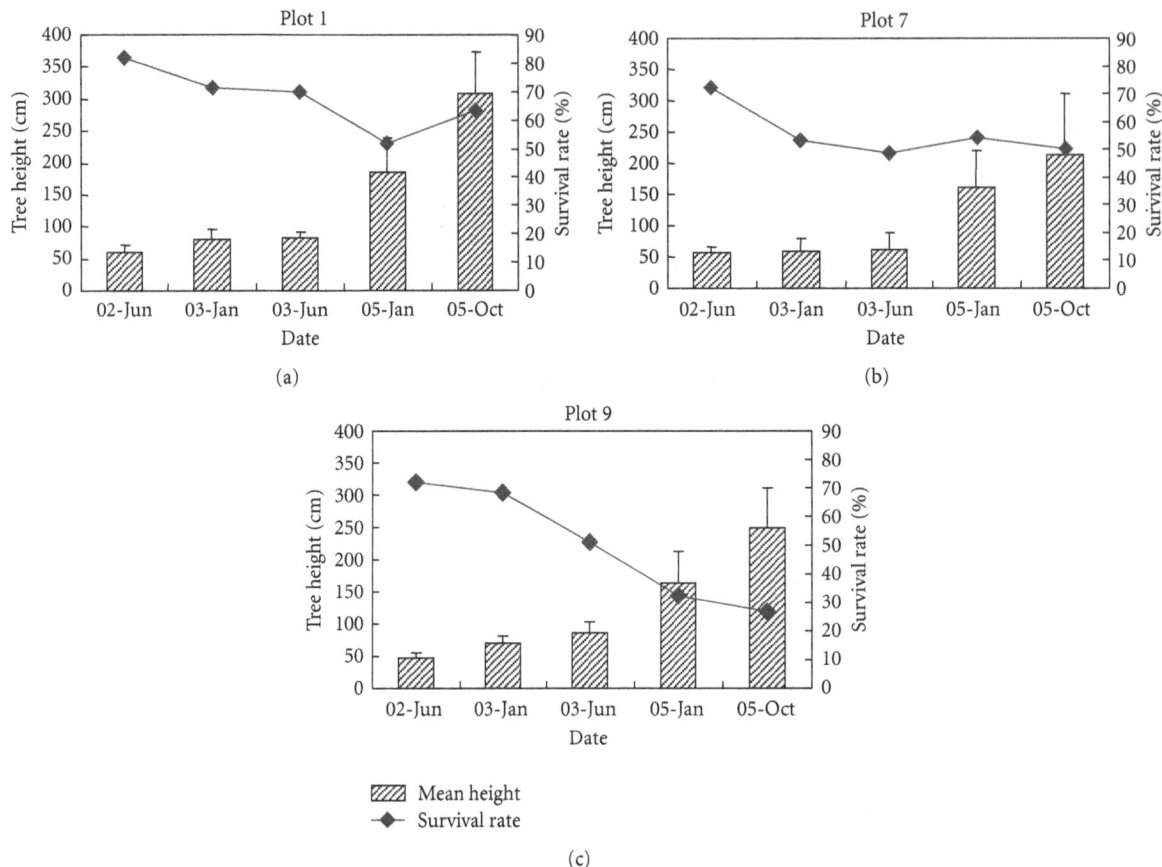

FIGURE 5: Survival and growth of planted mangroves measured in June 2002, January 2003, Jun 2003, January 2005, and October 2005. Locations of the plots are shown Figure 4. Error bars denote standard deviation.

Thammarat, 700 km south of the present study site. The southern site recorded a low frequency of tidal inundation, and tree heights measured 4 years after planting were around 200 cm [13]. Accordingly, the average height of 308 cm measured in the present study could indicate that growing conditions of the study site were relatively favorable in terms of nutrient supply or local hydrology.

3.2. Soil Amendment Application. Plots to which the three types of soil amendment were applied showed higher growth rates than the unfertilized control plot (Figure 6). In ANOVA we found significant difference in tree heights among the three types of soil amendment at each measurement from the first (June 2002) to the last (October 2005) despite no apparent difference in the means (Figure 6). Effect of soil amendment on tree height was most significant in NPK treatment at Plot 1 and Plot 9 while so in the coconut fiber treatment at Plot 7. Lower tree heights in Plot 7 than Plot 9 (Figure 5) assured that growing condition was harsher in Plot 7. Coconut fiber would have amended to stabilize erodible soils of Plot 7 which was more effective than NPK and humic acid.

If soil amendment effect with either NPK or coconut fiber was noticed in the first measurement (June 2002), its effect had continued to the subsequent measurements till the last

measurement (October 2005). This suggests that fertilization of mangrove plants in early stage is important when they are planted in eroded area. It appears that the effects of soil amendment are more pronounced in landward areas where erosion is less severe. In terms of cost-effectiveness, fertilization therefore would be recommendable to be done first in landward areas.

3.3. Water Quality. pH values were lowest at ST1, while salinity levels were almost similar at all sampling sites (Table 1). The low pH recorded at ST1 could have resulted from the high concentrations of POC at this site. Large amounts of organic matter were produced by mangroves which were then decomposed and consequently acted to lower the pH at ST1.

DON and PON were major nitrogen compounds as shown by higher concentrations than inorganic nitrogen compounds (ammoniacal-nitrogen, nitrite-nitrogen, and nitrate-nitrogen). The concentrations of chlorophyll a ranged from 11.3 to 28.1 in ST1 and from 8.4 to 16.3 $\mu g\,L^{-1}$ in ST4. Possible correlation between PON and chlorophyll a indicated that PON was originated from plankton [25]. Concentrations of chlorophyll a fluctuated seasonally with the high concentration in April 2001 and with the relatively low concentration at other times.

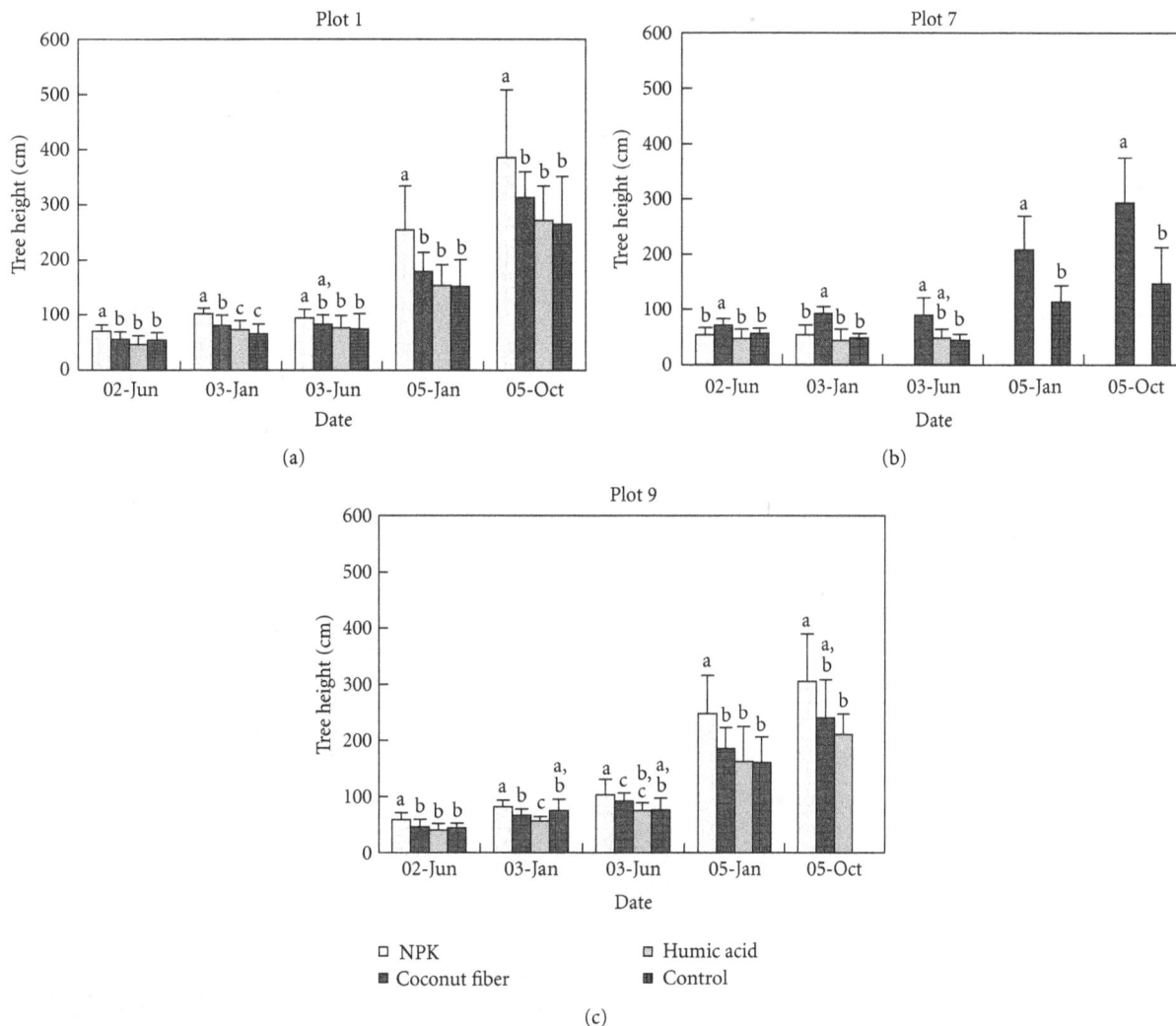

FIGURE 6: Effects of the fertilizer application on tree heights of June 2002, January 2003, Jun 2003, January 2005, and October 2005. Error bars denote standard deviation. Different letters are significantly different at $P < 0.05$ by the Tukey-Kramer test.

BOD was high in all measurements, indicating that the study site was heavily polluted (Table 1). Tookwinas reported that the average BOD of effluent from shrimp ponds was $8.47\,mg\,L^{-1}$ [26]. At the mangrove plantation site (ST1), BOD was $4.3\,mg\,L^{-1}$ when measured in August 2001, and $4.9\,mg\,L^{-1}$ in December 2001. These values, being approximately half of those for effluent from shrimp ponds, indicated that the mangrove plantation site has been subjected to the inflow of polluted water from adjacent sites or from the Tha Chin and Chao Phraya rivers (Figure 1).

The TN concentration at ST1 (1.9, $4.3\,mg\,N\,L^{-1}$) was higher than that at ST4 (1.8, $2.7\,mg\,N\,L^{-1}$), with these values being higher than the average value for Kung Krabaen Bay, Chantaburi province (0.4-$0.5\,mg\,N\,L^{-1}$) and lower than values recorded from effluent water sampled from an intensive shrimp pond ($4.9\,mg\,N\,L^{-1}$) [26]. Values of TP, PP, and TDP were also higher at ST1, suggesting that the loadings were derived from external sources. Similarly, as for nitrogenous compounds, all phosphorous compounds (TDP, DOP, PP, TP, POC) also increased in December 2001 from April

2001, probably reflecting the high loading of effluent from the surroundings of the study site.

The levels of chlorophyll a at the plantation site (ST1) were higher than those in the offshore area (ST4), indicating that a primary producer (phytoplankton) was generated at the plantation site. Levels of chlorophyll a showed temporal fluctuations, being especially high in April 2001 but relatively low at other times. Levels of BOD varied between 4.3 and $4.9\,mg\,L^{-1}$ at the plantation site, probably reflecting the influence of effluent from outside of the plantation site.

3.4. Sediment Quality. Nitrogen levels were relatively high in the mangrove plantation area. Ammoniacal-nitrogen was dominant over nitrate-nitrogen and nitrite-nitrogen, indicating reducing conditions in the sediment (Table 2). The sulfide content was relatively high in December 2001, reflecting the spontaneous input of organic matter from other sites and the resulting increase in reducing conditions. Levels of nitrate-nitrogen, which is the reduced form of nitrogen, also

TABLE 1: Results of water analyses conducted in April 2001, December 2001, and June 2002.

	April 2001		December 2001		June 2002
	ST1	ST4	ST1	ST4	ST4
Water temperature (°C)	30.4	31.1	27.0	28.0	32.6
Dissolved oxygen (mg L^{-1})	3.8	4.5	7.1	5.4	3.8
Salinity (ppt)	21.8	21.3	26.1	27.7	28.5
pH	8.00	8.40	7.85	7.96	8.29
Ammoniacal nitrogen (mg L^{-1})	0.012	0.021	1.005	0.014	0.010
Nitrite-nitrogen (mg L^{-1})	0.008	0.028	0.199	0.014	0.005
Nitrate-nitrogen (mg L^{-1})	0.009	0.012	0.024	0.005	0.028
Total dissolved nitrogen (mg L^{-1})	0.607	0.691	2.540	1.165	0.375
Dissolved organic nitrogen (mg L^{-1})	0.579	0.630	1.312	1.132	0.332
Particulate organic nitrogen (mg L^{-1})	1.320	1.139	1.774	1.577	0.847
Total nitrogen (mg L^{-1})	1.927	1.830	4.314	2.742	1.222
Dissolved inorganic phosphorus (mg L^{-1})	0.091	0.086	0.425	0.150	0.029
Total dissolved phosphorus (mg L^{-1})	0.107	0.086	0.852	0.819	0.078
Dissolved organic phosphorus (mg L^{-1})	0.017	0.000	0.427	0.669	0.049
Particulate phosphorus (mg L^{-1})	0.218	0.167	0.315	0.109	0.021
Total phosphorus (mg L^{-1})	0.325	0.253	1.167	0.928	0.099
Particulate organic carbon (mg L^{-1})	7.98	6.63	10.2	4.99	5.04
Chlorophyll a (μg L^{-1})	28.1	16.3	11.3	8.44	10.2
BOD (mg L^{-1}: 5 days at 20°C)	4.3	3.3	4.9	4.3	7.1

TABLE 2: Results of sediment analyses conducted in April 2001, December 2001, and June 2002.

Date	Station	Total nitrogen %	Ammoniacal-nitrogen mg kg^{-1}	Nitrite-nitrogen mg kg^{-1}	Nitrate-nitrogen mg kg^{-1}	Total phosphorus mg kg^{-1}	Available phosphorus mg kg^{-1}	Total organic carbon %	AVS-S mg kg^{-1}
April 2001	ST1	0.27	59.8	0.04	3.41	45.3	5.17	2.52	25.0
	ST2	0.30	112	0.06	4.20	44.6	18.1	2.70	77.3
	ST3	0.02	63.4	0.16	5.74	55.3	10.3	1.18	77.8
	ST4	0.04	47.0	0.04	3.25	86.9	8.79	0.97	214
December 2001	ST1	0.27	41.8	0.23	6.55	52.1	8.13	2.08	257
	ST2	0.31	91.9	0.33	8.36	80.4	27.2	2.36	213
	ST3	0.25	146	0.15	2.74	44.8	18.9	2.03	146
	ST4	0.26	42.1	0.15	2.46	88.7	10.4	2.02	92.1
June 2002	ST1	0.14	24.8	0.03	4.53	31.0	6.64	1.49	30.5
	ST2	0.20	25.9	0.00	5.11	35.7	9.57	2.06	51.4
	ST3	0.12	16.2	0.03	3.68	16.2	13.8	1.47	57.7
	ST4	0.15	20.6	0.11	2.77	30.2	2.17	1.97	134

increased at this time. Constantly high levels of sulfide at ST4 indicate large amounts of organic matter, even in the offshore areas.

In June 2002, reducing conditions were measured in the offshore area (ST4), probably due to effluent from major rivers. Considering that typical coastal sediment in Thailand contains 1.06% organic carbon and 0.05% total nitrogen [26], the mangrove plantation site (including the offshore area) experienced a high degree of eutrophication at this time. Available P was higher at ST2 than at ST1. Since ST2 is located in the middle of a small creek, we consider that phosphorus was transported along the creek from offshore areas.

3.5. Benthos. The abundance and species composition of benthos varied considerably at each measurement period;

however, values were generally higher during the monsoon season (April, June) than during the dry season (December) (Table 3). Environmental factors such as sediment composition [27] and salinity [28] affected the benthic community. Sediment texture coarsened in the order of ST1 < ST2 < ST3 < ST4 (Table 4); the number of species and abundance also increased in this order (Table 3). The diversity of *Polychaeta* has been reported to be related to the degree of salinity of the host sediment [29, 30]. In this study, *Polychaeta* was dominant at ST1 during the monsoon season when the influence of salinity was relatively minor while the diversity of *Crustacea* increased with lower salinity.

3.6. Phytoplankton. We identified 3 groups and 33 genera of plankton (Table 5). The blue-green algae group yielded

TABLE 3: Species of benthic fauna and abundance (individuals/m^2) within sediment and bottom of the replanted mangrove areas.

	No.	Group	April 2001 ST1	April 2001 ST4	December 2001 ST1	December 2001 ST2	December 2001 ST3	December 2001 ST4	June 2002 ST3	June 2002 ST4
Crustacea	1	Balanus		133						
	2	Leucosia		44						
	3	Nassarius					30			
	4	Uca spp.			15					
Gastopoda	1	Littorina		89			15			
	2	Paludinella			30					
Pelecypoda	1	Anadana		44						
	2	Mactra		178						134
	3	Musculus		222						
	4	Mytella		89						
	5	Pholas spp.					2,682			
	6	Tellina	89	533		15	30		134	
	7	Tellina foliacea						15		
Polychaeta	1	Aquilaspio	133							
	2	Cerithium	44							
	3	Glycera	267							
	4	Eunic	133	1,689					400	
	5	Nephytys	356							
	6	Nereis					504	445	1,067	1,600
	7	Notomastus	578							
	8	Polyodontes		222						
	9	Sabella		267						
	10	Sigambra	667							

TABLE 4: Results of texture analysis.

Station	Clay (%)	Silt (%)	Sand (%)	Type
ST1	49.7	19.3	31.1	Clay
ST2	35.1	27.6	37.3	Clay loam
ST3	28.8	31.2	40.1	Clay loam
ST4	14.2	29.3	56.5	Sandy loam

5 genera, green algae 3, and diatoms 25. Except for ST1 in April 2001, diatoms were the dominant group in terms of the number of species and the number of genera. Diatoms bloomed at the low rainfall periods (December 2001), with the increased number of genera to 12 genera and of the total cell number to 62,837 cells. At the most offshore site (ST4), the number of phytoplankton genera and of total phytoplankton cells increased in the order of April 2001 (9 genera, 3,164 cell L^{-1}), December 2001 (13 genera, 66,922 cell L^{-1}), and June 2002 (15 genera, 33,201 cell L^{-1}). This trend appears to correspond to the salinity level: a greater number of genera occurred during the periods of higher salinity. Similar trends have been reported for diatoms (the greater number of genera during the periods of low precipitation and at the sites with the higher salinity) in Songkhla Lake, Thailand [31].

The species composition of phytoplankton was related to the water quality. In December 2001, the water was eutrophicated as reflected by the higher contents of phosphorus and nitrogen (Table 1). Under these conditions, diatoms became dominant at the expense of other groups.

4. Discussion

Nitrogen fertilization is reported to increase growth rates of mangroves which were planted in an abandoned impoundment in Florida [32] while our study showed that fertilization was also effective in the open area with high-energy environments.

Establishment of mangroves in heavily eroded environments has been difficult. Total area losses accounted for 0.91 km^2 y^{-1} for the Thai Gulf coast. However the presence of mangroves reduces the erosion rates in areas where erosion prevailed [33, 34]. Enhancement of initial mangrove growth through soil amendment therefore should be an effective way to reduce erosion rates at heavily eroded places. The effects of soil amendment are limited if the plantation site is heavily eroded, as the case of Plot 7 in the present study. This suggests that the planting of mangroves along eroded shorelines should be initiated within the least erosion-prone areas.

Physical measures can be undertaken to protect seedlings from wave/wind attack; however, this approach is ineffective

TABLE 5: Results of phytoplankton (cell L^{-1}) measurement conducted in April 2001, December 2001, and June 2002.

| Phytoplankton group | No. | Genera | Density of phytoplankton (cell L^{-1}) | | | | |
| | | | April 2001 | | December 2001 | | June 2002 |
			ST1	ST4	ST1	ST4	ST4
Blue green algae	1	*Anabaenopsis*	397				
	2	*Oscillatoria*	1,008	336			225
	3	*Gloeocapsa*	14,463				
	4	*Lyngbya*		140			
	5	*Spirulina*	117				24
Green algae	1	*Gonatozygon*	332				
	2	*Schroederia*	91	14			
	3	*Staurastrum*	117				
Diatoms	1	*Skeletonema*	2,581	462			7,125
	2	*Coscinodiscus*	11,427	742	827	22	18
	3	*Dactyliosolen*				11	33
	4	*Rhizosolenia*	468	882	30,645	8,974	1,410
	5	*Bacteriastrum*				33	13,688
	6	*Chaetoceros*	273		7,425	22,000	8,250
	7	*Odontella*			281	28	51
	8	*Thalassionema*			6,210	72	1,185
	9	*Gyrosigma*			146		27
	10	*Navicula*	163	252	11	44	
	11	*Nitzschia*	345				
	12	*Pseudonitzschia*			17,145	35,132	
	13	*Campylodiscus*			11		
	14	*Dictyocha*					3
	15	*Noctiluca*					1,050
	16	*Ceratium*					105
	17	*Alexandrium*					6
	18	*Fragilaria*	137	238			
	19	*Eucampia*			11	6	
	20	*Melosira*	245				
	21	*Hemiaulus*			11	44	
	22	*Pleurosigma*	124	98	114		
	23	*Pseudosolenia*				550	
	24	*Thalassiosira*				6	
	25	*Stephanodiscus*	98				

in most cases. We erected bamboo fencelines at other study sites to protect mangrove seedlings from wave/wind attack, but the fences lasted no more than 1 year. The construction of physical barriers is also extremely expensive. According to the Japan International Cooperation Agency (JICA) study, the planting of mangroves across the 4,800 m shoreline of the present study area using civil engineering techniques to prevent coastal erosion will cost 983 million baht (approximately 31 million US dollars as of July 2011) [15]. In Malaysia, a 90 m long rubble mound structure (breakwater) was constructed along an eroding tropical shoreline at a cost of US$ 42,850 which is more than 10 times lower [35]. This kind of cost is quite variable according to the sort of construction method which is adopted. However civil engineering operation for coastal protection is generally rather expensive.

The disastrous tsunami that hit the Thailand coast in December 2004 revealed the important role mangroves play in terms of coastal protection. It appears that areas covered in mangroves suffered less damage than mangrove-free areas, and the protective effect of mangroves against tsunami damage differed among different species. Kamali and Hashim (2011) reported that *Rhizophora* spp. were found to have a greater protective effect than *Sonneratia* or *Avicennia* spp. [35]. These differences are attributed to root architecture, as the prop roots of *Rhizophora* spp. are stronger than the cable roots of *Sonneratia* and *Avicennia* spp.

As any reduction in sediment supply from the major rivers is likely to influence the retreat of the coastline in the study region, complete protection against coastal erosion is not possible; however, the planting of mangroves is a potentially effective measure in lowering the present erosion rate. Mangrove-planting techniques should therefore be developed to cope with coastal erosion, especially in terms of adequate species selection, site identification, and additional measures such as soil amendment application.

The study site was highly polluted as indicated by water, sediment, and biological measurements, which could be influenced by the surrounding areas. Mangrove soils have a large buffering capacity for pollutants [36]. However, its capacity has a limit. With excessive loadings, mangrove soil can no longer retain pollutants, and consequently marine ecology and fishery resources will be severely damaged. Considering the pollution level of the study area, it should be necessary to watch carefully a status of pollution level of mangrove area in order to prevent drastic marine pollution.

References

[1] JAM, "Development and dissemination of re-afforestation techniques of mangrove forests," in *Proceedings of the Workshop on ITTO Project*, p. 216, Japan Association for Mangroves, NATMANCOM, Bangkok, Thailand, 1994.

[2] "Ecology and management of mangrove restoration and regeneration in East and Southeast Asia," in *Proceedings of the 4th ECOTONE*, C. Khemnark, Ed., Surat Thani, Thailand, 1995.

[3] C. D. Field, Ed., *Restoration of Mangrove Ecosystems*, International Society for Mangrove Ecosystems, Okinawa, Japan, 1996.

[4] Silliman University, *Assessment of the Central Visayas Regional Project-1: Nearshore Fisheries Component*, Silliman University Marine Laboratory, Dumaguete City, Philippines, 1996.

[5] J. H. Primavera and J. M. A. Esteban, "A review of mangrove rehabilitation in the Philippines: successes, failures and future prospects," *Wetlands Ecology and Management*, vol. 16, no. 5, pp. 345–358, 2008.

[6] S. Angsupanich and S. Havanond, "Effects of barnacles on mangrove seedling transplantation at Ban Don Bay, Southern Thailand," in *Proceedings of the Tropical Forestry in the 21st Century, (FORTROP '96)*, pp. 72–81, Kasetsart University, Bangkok, Thailand, 1996.

[7] P. Sakunathawong, "Mangrove planting on new mudflats at Tambon Klong Khon, A. Muang, Samut Songkram," in *Proceedings of the 9th National Seminar on Mangrove Ecology*, p. 9, 1995.

[8] C. Woodroffe, "Mangrove sediments and geomorphology," in *Tropical Mangrove Ecosystems. Coastal and Estuarine Studies 41*, A. I. Robertson and D. M. Alongi, Eds., pp. 7–41, American Geophysical Union, Wash, USA, 1992.

[9] P. G. E. F. Augustinus, "Geomorphology and sedimentology of mangroves," in *Geomorphology and Sedimentology of Estuaries. Developments in Sedimentology 53*, G. M. E. Perillo, Ed., pp. 333–357, Elsevier Science, Amsterdam, The Netherlands, 1995.

[10] K. Furukawa and E. Wolanski, "Sedimentation in mangrove forests," *Mangroves and Salt Marshes*, vol. 1, no. 1, pp. 3–10, 1996.

[11] J. M. Smoak and S. R. Patchineelam, "Sediment mixing and accumulation in a mangrove ecosystem: evidence from ^{210}Pb, ^{234}Th and ^{7}Be," *Mangroves and Salt Marshes*, vol. 3, no. 1, pp. 17–27, 1999.

[12] I. C. Feller, D. F. Whigham, K. L. McKee, and C. E. Lovelock, "Nitrogen limitation of growth and nutrient dynamics in a disturbed mangrove forest, Indian River Lagoon, Florida," *Oecologia*, vol. 134, no. 3, pp. 405–414, 2003.

[13] N. Matsui, J. Suekuni, S. Havanond, A. Nishimiya, J. Yanai, and T. Kosaki, "Determination of soil-related factors controlling initial mangrove (*Rhizophora apiculata* BL.) growth in an abandoned shrimp pond," *Soil Science and Plant Nutrition*, vol. 54, no. 2, pp. 301–309, 2008.

[14] Royal Irrigation Department, Government of Thailand, *History of Water Resources Development in Thailand*, 2002.

[15] SANYU and PANTA, "The feasibility study on mangrove revival and extension project in the Kingdom of Thailand," JICA Report, 2001.

[16] U. Thampanya, J. E. Vermaat, S. Sinsakul, and N. Panapitukkul, "Coastal erosion and mangrove progradation of Southern Thailand," *Estuarine, Coastal and Shelf Science*, vol. 68, no. 1, pp. 75–85, 2006.

[17] SEATEC, BEST, "Feasibility study and preliminary design for alleviation of coastal erosion problem in the Upper Gulf of Thailand," Harbour Department of Thailand, 1994.

[18] C. E. Lovelock, I. C. Feller, M. C. Ball, B. M. J. Engelbrecht, and M. L. Ewe, "Differences in plant function in phosphorus- and nitrogen-limited mangrove ecosystems," *New Phytologist*, vol. 172, no. 3, pp. 514–522, 2006.

[19] APHA, AWWA, WPCF, *Standard Method for the Examination of Water and Wastewater*, American Public Health Association, Wash, USA, 17th edition, 1985.

[20] I. Solorzano, "Determination of ammonia in natural waters by the phenolhypochlorite method," *Limnology and Oceanography*, vol. 14, pp. 799–801, 1969.

[21] J. Murphy and J. P. Riley, "A modified single solution method for the determination of phosphate in natural waters," *Analytica Chimica Acta*, vol. 27, no. C, pp. 31–36, 1962.

[22] K. I. Aspila, H. Agemian, and A. S. Y. Chau, "A semi-automated method for the determination of inorganic, organic and total phosphate in sediments," *The Analyst*, vol. 101, no. 1200, pp. 187–197, 1976.

[23] J. I. Hedges and J. H. Stern, "Carbon and nitrogen determinations of carbonate-containing solids," *Limnology & Oceanography*, vol. 29, no. 3, pp. 657–663, 1984.

[24] H. Tsutsumi and T. Kikuchi, "Benthic ecological of a small cove with seasonal oxygen depletion caused by organic pollution," *Publications from the Amakusa Marine Biological Laboratory*, vol. 7, pp. 17–40, 1983.

[25] F. F. Perez, F. G. Figueiras, and A. F. Rios, "Nutrient depletion and particulate matter near the ice-edge in the Weddell Sea," *Marine Ecology Progress Series*, vol. 112, no. 1-2, pp. 143–154, 1994.

[26] S. Tookwinas, *The mitigation measures for the impacts of marine shrimp farming on coastal environment; A case study at Kung Krabaen Bay, Eastern Thailand*, Doctoral thesis, Graduate School of Biosphere Sciences, Hiroshima University, Japan, 2001.

[27] B. V. Bhat and T. R. C. Gupta, "Macrobenthos of Nethravati-Gurupur Estuary, Mangalore," in *Proceedings of the Symposium on Coastal Aquaculture*, p. 1465, 1986.

[28] S. Angsupanich and R. Kuwabara, "Macrobenthic fauna in Thale Sap Songkla, a brackish lake in southern Thailand," *Lakes and Reservoirs: Research and Management*, vol. 1, no. 2, pp. 115–125, 1995.

[29] M. Amino, "Macrobenthos and aquatic animals," in *Report on the Effect of Waste Water Effluent from Sewage Disposal Plant in Takamatsu City to Fishing Grounds*, pp. 59–86, 1979.

[30] R. Kuwabara, "Environment of sandy and muddy bottoms in shallow waters," *Fisheries Engineering*, vol. 21, pp. 53–60, 1985.

[31] Y. Yamaguchi, "Ecological characteristics and phytoplankton dynamics of lagoonal lake, Thale Sap Songkhla, Thailand," in *The Coastal Environment and Ecosystem in Southeast Asia, Studies on the Lake Songhla Lagoon System, Thailand*, pp. 40–71, Faculty of Bio-industry, Tokyo University of Agriculture, Japan, 1995.

[32] I. C. Feller, D. F. Whigham, K. L. McKee, and C. E. Lovelock, "Nitrogen limitation of growth and nutrient dynamics in a disturbed mangrove forest, Indian River Lagoon, Florida," *Oecologia*, vol. 134, no. 3, pp. 405–414, 2003.

[33] Y. Mazda, M. Magi, M. Kogo, and Phan Nguyen Hong, "Mangroves as a coastal protection from waves in the Tong King Delta, Vietnam," *Mangroves and Salt Marshes*, vol. 1, no. 2, pp. 127–135, 1997.

[34] U. Thampanya, J. E. Vermaat, S. Sinsakul, and N. Panapitukkul, "Coastal erosion and mangrove progradation of Southern Thailand," *Estuarine, Coastal and Shelf Science*, vol. 68, no. 1, pp. 75–85, 2006.

[35] B. Kamali and R. Hashim, "Mangrove restoration without planting," *Ecological Engineering*, vol. 37, no. 2, pp. 387–391, 2011.

[36] M. Ebato, N. Matsui, M. Nomura, and K. Yonebayashi, "Estimation of buffering capacity of mangrove soils by using hydrophobic organic compoiunds, Atrazine and Linuraon," *Soil Science and Plant Nutrition*, vol. 50, no. 4, pp. 477–484, 2004.

9

Changes in Herbaceous Species Composition in the Absence of Disturbance in a *Cenchrus biflorus* Roxb. Invaded Area in Central Kalahari Game Reserve, Botswana

Shimane W. Makhabu[1] and Balisana Marotsi[2]

[1] *Department of Basic Sciences, Botswana College of Agriculture, Private Bag 0027, Gaborone, Botswana*
[2] *Department of Wildlife and National Parks, Ministry of Environment, Wild life and Tourism, P.O. Box 4, Tsabong, Botswana*

Correspondence should be addressed to Shimane W. Makhabu, smakhabu@bca.bw

Academic Editor: Herman Shugart

A-nine year study was carried out to investigate changes in herbaceous species composition in an area invaded by *Cenchrus biflorus* Roxb, an exotic invader grass species. The study ensued termination of livestock and human activities in the area when residents of the area were relocated to another area. Vegetation characteristics from the disturbed sites (previous occupied areas) and undisturbed sites (previously unoccupied areas) were determined. The results show that *C. biflorus* has high tolerance to disturbance. It comprised the larger proportion of grasses in disturbed sites at the inception of the study. However, it decreased in abundance with time in disturbed areas and was absent in the undisturbed areas, suggesting that its ability to invade undisturbed sites is limited. Perennial species successfully reestablished on the third year after termination of disturbance. The study reveals that *C. biflorus* invasion in the Kalahari ecosystem can be controlled by termination of disturbances.

1. Introduction

The slow and uncertain rate of return of herbaceous species composition to the original status after severe disturbance is a major problem in wildlife management areas and on rangelands [1]. It is a function of the range condition at any given time, which is assessed as a function of grass composition, biomass, and cover [2, 3]. The dynamics of range condition is affected by environmental factors and land use such as animal grazing pressure. Heavy grazing pushes the range condition to degraded status dominated by weedy or unpalatable grass and forb species typical of disturbed areas [3, 4]. This has been supported by various studies done in semiarid Southern Africa [5–7]. The studies showed that basal cover, proportion of annuals and proportion of unpalatable species were higher in heavily grazed areas than in light grazed areas [5–7]. Vegetation outcrops on a disturbed area and surroundings are a function of availability of seeds or propagules of species or species groups.

Debates have ensued on the relative importance of both the biotic and abiotic factors on range conditions. Explanations for these are based on the equilibrium and nonequilibrium models [3, 8–11]. The equilibrium model stresses the importance of biotic feedbacks between herbivores and their resource, while the nonequilibrium model sees stochastic abiotic factors as the primary drivers of vegetation and livestock dynamics [3]. However, in most arid and semiarid rangelands both equilibrium and nonequilibrium models are at play, though at different scales.

It takes a considerable time for a species composition in a disturbed area to return to its previous condition. Samuel and Hart [1] reported that after 61 years of monitoring a disturbed area, its species composition did not reach the point of that of an undisturbed area it was compared with since there were visible differences between the two areas. The differences were attributed to various factors that included increase of invasive and alien species that often dominate in disturbed sites out competing native vegetation or disrupting native plant communities and nutrient cycling. On cessation of disturbance, expectations are that vegetation development follows a definite sequence of functional groups where early stages of recovery are dominated by species of light seeds, producing seeds in great numbers, fast growing and dispersing over vast areas [12]. These are hardened, annual plants that grow in very unfavorable conditions and improve the growth conditions resulting in plant communities that are adapted to the new, improved growth conditions, and replacing the existing plant communities. Species with heavy seeds, slower growth, and long life span dominate in later stages [13]. Studies by O'Connor and Roux [14] on the Kalahari landscape have shown that vegetation is able to reestablish after grazing pressure is lessened. Seitshiro [15] and Perkins [16] also observed that areas in the Kalahari sandveld vegetation communities, denuded of grass during dry years, had some rapid return of perennial grass cover in subsequent years of average or better rainfall or cessation of grazing. Indications are that, during prime growing conditions, it can take 8–10 years for disturbed vegetation to reestablish in the Kalahari landscape [17]. The Kalahari is a substantial part of Southern Africa interior covering countries among them Botswana, Namibia, and South Africa [18]. It is extensively elevated, flat and covered by Kalahari sand soils. Kalahari sand soils typically consist of over 95% fine sand-sized, aeolian-deposited sediment and are predominantly deep, structureless, and lacking in N, P, and organic matter [18].

Within the Kalahari ecosystem in Botswana lies Central Kalahari Game Reserve, the largest game reserve in the world that covers an area of some 52 800 km². Some pockets of the Central Kalahari Game Reserve (CKGR) are inhabited by people. When people inhabited the area in the 1960s, their main source of food was hunting and gathering. However, with the decrease of natural resources, people started keeping domestic animals such as horses, donkey, and goats. Over time, it was realized that the area had been invaded by a noxious weed, *Cenchrus biflorus* Roxb. It is common in the savannas of sub-Sahara Africa. *Cenchrus biflorus* Roxb. is an annual grass of the Poaceae family with synonyms: *C. annularis* Andersson, *C. barbatus* Schumach, *C. catharticus* Delile, *C. leptacanthus* A.Camus, and *C. perinvolucratus* Stapf & C.E.Hubb. Its culms are 4–90 cm high. The leaf: lamina is 2–25 cm long, 2–6 mm wide and the panicle 2–15 cm long. Its spikelets are 1–3 per bur and 3.5–6 mm long. Molecular analyses suggest that the species invading the Kalahari originates from India hence the common name "Indian sandbur". The grass was first observed in Botswana in the 1940 s growing around Nata. It is currently spreading rapidly in the western and central part of the country. The grass species is regarded as a noxious weed because of its objectionable burs, which adhere to animal skin, clothes, shoes and machinery. These burs are the likely vehicle to its rapid spread. The burs are harmful to grazers and may cause ulcers in the mouths of animals. The grass has also spread to arable fields making weeding difficult. As a result, some farmers in Botswana have opted to abandon their *C. biflorus* invaded arable fields. Intensive land use increases the potential to *C. biflorus* invasion as this increases the

area of disturbed soil, thus creating conditions under which the invasive grass thrives. Control of this grass species is achievable in arable fields through weeding, cultivation, and herbicides application before seed formation. However, controlling this grass species in range and pasture areas still remains a challenge since there is lack of a selective herbicide that will kill only this grass species. As such other control measures have to be explored. Environmental friendly control measures will be more appropriate to be used in Game Reserves were minimal interventions are allowed.

In 1997, Central Kalahari Game Reserve (CKGR) residents at Xade were relocated outside the reserve as their activities were no longer compatible with wildlife conservation. This abandoned Xade area has been under disturbance for four decades. The nature of the disturbance ranged from simple cumulative trampling on the herbaceous layer to large-scale clearing of trees and shrubs. Establishment of modern and traditional homesteads and institutions like the "Kgotla" (traditional gathering place headed by a Chief) contributed to land disturbance. The keeping of livestock (horses, donkeys, sheep, and goats) exerted more pressure on the soil and vegetation leading to the disappearance of nutritious grass species and the increase of annuals [19]. Concerns were then raised that this would negatively affect wildlife populations and eventually deteriorate the biodiversity of the area.

The relocation of Xade residents prompted this study as an opportunity to determine how herbaceous species composition and cover changes overtime in the absence of livestock and human disturbance. Wildlife activities that had previously been observed to be absent or minimal were left to go on as usual. Thus, this study aimed at determining whether a disturbed area invaded by an exotic invader grass species can recover if human and livestock activities were terminated.

2. Methods and Materials

2.1. Study Site. Xade is in Central Kalahari Game Reserve (CKGR) and lies at Latitude 23.01 degrees south, Longitude 22.33 degrees east. The vegetation is characterized by medium to open bush savannah with occasional isolated tress. The vegetation has mainly been kept in shape by frequent bush fires. The climatic conditions of the area can be summarized as semiarid. Generally, the CKGR lies between 350 mm and 400 mm isohyets [19]. The Xade rainy season is between November and April with long-term annual rainfall of 350 mm. There is however considerable variation in the amount and pattern of rainfall between years 1998 and 2006 (Figure 1). The area is covered mainly by deep Kalahari sands. Its topography is flat to undulating with low immobile sand dunes.

2.2. Experimental Design. The vegetation data was collected from seven (10) strategically located plots at Xade. Strategically here means that it was made sure that five of the plots were in disturbed sites while two were in undisturbed sites (reference sites). The disturbance arose from livestock

Changes in Herbaceous Species Composition in the Absence of Disturbance in a Cenchrus biflorus Roxb. Invaded Area in Central Kalahari Game Reserve, Botswana

85

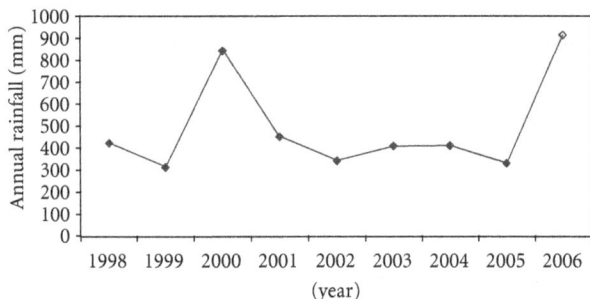

FIGURE 1: Annual rainfall amounts in Xade during the study period.

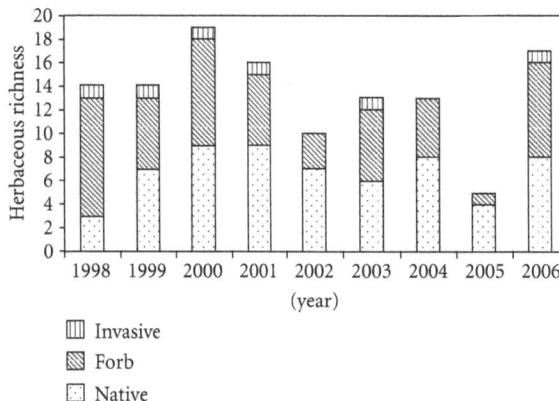

FIGURE 2: Mean species richness at disturbed sites (Xade) from 1998 to 2006. Native species refers to all indigenous species while invasive refers to Cenchrus biflorus.

and human activities in the area before relocation in 1997. The plots measured 30 m × 20 m in size. Sampling was done from April 1998 to April 2006. Sampling was carried out in April of each year, the peak plant growing period. During April, herbaceous species are mature and easy to identify since most are either flowering or in seed. Within each of the 30 m × 20 m plots, five transects each measuring 30 m long and being 5 m apart were sampled. The herbaceous species frequency was recorded every two (2) meters along each transect using the wheel point method [20]. Numbers of individuals of each species by life form (forb, native grass and exotic) were recorded. The data was used to calculate species richness, composition, and herbaceous cover. Grass species nomenclature is according to Field [21] and Van Wyk and Van Oudtshoor [22].

Rainfall data was obtained from Botswana Department of Meteorological Service, Ghanzi offices.

2.3. Statistical Analysis. STATISTICA program, version Kernel 5.5 A, was used to calculate species richness, composition, and herbaceous cover. Species richness here refers to the number of individual plants of each species in each life form group (forbs, native, and invasive). Composition refers to relative abundance of each group. Herbaceous cover means the cover contributed by each group excluding that covered by litter and bareground. Excel was implored to draw charts showing species richness, composition, and cover. Spearman's rank correlation coefficient (r_s) was used to calculate the correlation between herbaceous species richness and site age (time). Spearman's rank correlation was also applied to determine correlation between the native grass species, invasive species, and native forbs over time. The Spearman's rank correlation coefficient values were tested for significance using the t statistic. The student t-test was also used to test if there was some significant difference in species richness and cover and functionality composition between sites (disturbed and undisturbed). STATISTICA [23] was used to perform these calculations.

The Czekanowski coefficient [24] was used to assess the similarity or dissimilarity in herbaceous species composition between disturbed and undisturbed plots (reference sites). The coefficient values range from zero (0), complete dissimilarity, to one (1), total similarity.

3. Results

3.1. Herbaceous Species Richness. The herbaceous species richness varied during the study period (Figures 2 and 3). Mean species richness was higher in disturbed sites than in undisturbed sites. Initially an average of fourteen (14) herbaceous species was recorded in the disturbed sites and six (6) in undisturbed sites. These included forbs, native grass species, and the exotic grass species *Cenchrus biflorus*. The native grass species included *Stipagrostis uniplumis, Pogonarthria squarrosa, Schmidtia pappophoroides, Aristida meridionalis, Eragrostis lehmanniana, Aristida congesta, Melinis repens, Urochloa trichopus, Enneapogon cenchroides,* and *Schmidtia kalihariensis.* The forbs included *Tribulus terrestris, Amaranthus thunbergii, Heliotropium steudneri, Xenostegia tridentata, Harpagophytum procumbens, Hermbstaedtia odorata,* and *Cucumis myriocarpus.*

Cenchrus biflorus disappeared after four years of termination of disturbance but some reappeared during the sixth and ninth years after termination of disturbance (Figures 2 and 3). Species continued to recruit to reach a peak mean of nineteen (19) species and eighteen (18) for disturbed and undisturbed sites in 2000 and 2006, respectively (Figures 2 and 3). The lowest herbaceous species mean number of five (5) was recorded in 2005 for disturbed sites and of four (4) in 2003 for undisturbed sites. Species richness ranged from 5 to 18 (mean ± SE of 13 ± 1.26) plant species for disturbed sites and 4–18 (mean ± SE of 10 ± 1.26) for undisturbed sites. There was no significant difference in species richness between undisturbed and disturbed sites (t value = 1.71 P-value = .126).

3.2. Herb Layer Composition. The herbaceous layer composition differed substantially between sampling periods in both disturbed plots and undisturbed plots (Figure 4). The herbaceous layer in disturbed plots was composed of exotic grass species (*C. biflorus*), native grass species, and forbs and in undisturbed plots only native grass species and forbs were recorded. No exotic grass species were recorded in

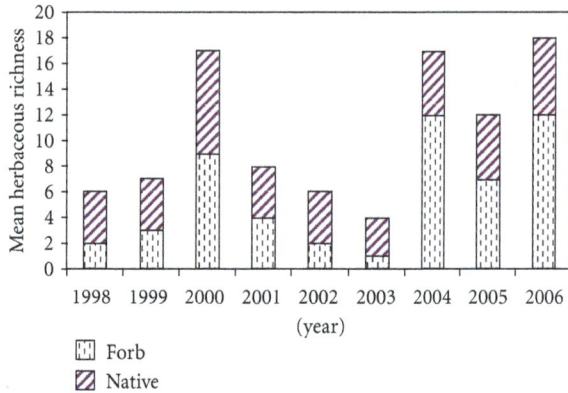

FIGURE 3: Mean species richness at undisturbed sites (Xade) from 1998 to 2006.

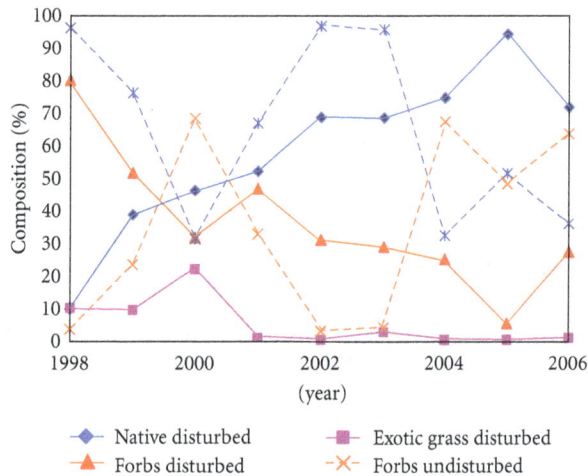

FIGURE 4: Herbaceous species composition at the disturbed (distbd) and undisturbed (undistbd) sites in CKGR: 1998 to 2006.

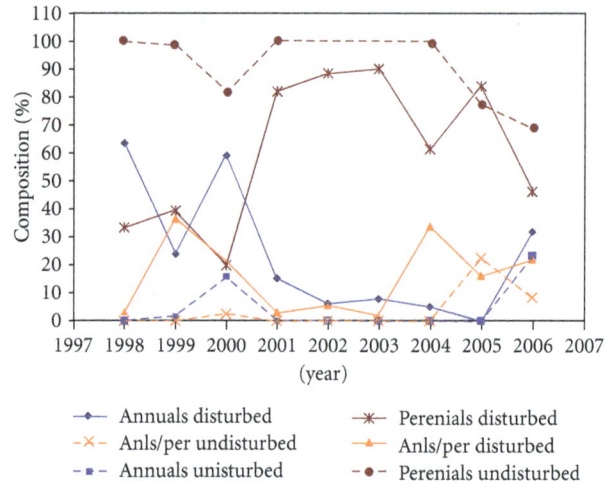

FIGURE 5: Composition of annuals and perennial grass species in disturbed (distbd) and undisturbed (undistbd) sites in CKGR:1998 to 2006.

The native species reached a peak average of 89% on the 7th year of sampling. In disturbed plots, both forbs and native grass species fluctuated between years but native grass species were much higher throughout the study period except in 2000, 2004, and 2006 (Figure 4). In comparison, forbs dominated in disturbed plots while native grass species dominated in undisturbed plots during the first two years of sampling. Native grass species recruited gradually in the disturbed plots, exceeding that of undisturbed plots in 2004.

3.3. Perenniality. Annual grass species were relatively high during the first three (3) years in disturbed sites but gradually declined (Figure 5). In contrast, from 2001 until 2003 perennial species recruited gradually to a species composition of about 80% exceeding annuals but subsequently declined during the last three years of sampling. Nonetheless, the decline was not significant. The fluctuations indicate that the vegetation had not yet fully recovered. In undisturbed sites, perennial species dominated throughout the study period.

3.4. Herbaceous Cover. Generally the disturbed sites showed some relatively higher percentage cover compared to undisturbed sites (Figure 6) but the difference was not significant (t value = -0.37, $P = .721$). The highest herbaceous cover was recorded in 2000 and the least in 2005 for both disturbed sites and undisturbed sites, respectively. The peak score was 94.46% while the least score was 41%.

3.5. Similarity. There were great dissimilarities in species composition during the first two years of sampling (Table 1). However, as the range rested following termination of disturbance, the similarity index increased gradually reaching 0.48 in 2001. Nonetheless, the similarity index continued to fluctuate indicating that the range condition had not stabilized.

undisturbed sites. The Spearman's rank correlation coefficient indicated that there were some negative relationships between percentage of forbs and time postdisturbance for disturbed sites ($R = -0.933$, $P = .0024$). A similar pattern was also observed for *C. biflorus* ($R = -0.746$, $P = .021$). In contrast, Spearman's correlation coefficient showed a strong positive correlation between native grass species and time postdisturbance for disturbed sites ($R = 0.933$, $P = .0002$).

During the first two years of sampling, forbs dominated in disturbed plots but gradually declined with time. In the first year, forbs contributed 80.56%, *C. biflorus* 9.88%, while native grass species contributed 9.56%. *Cenchrus biflorus* gradually recruited reaching an average score of 21.94% in 2000 coinciding with the year of high rainfall (Figure 1) but disappeared in the fourth year with its traces being noted on the 6th and 9th years postdisturbance. Similarly, forbs decreased gradually reaching a minimum of 5.10% on the 8th year of study but went up to 27.23% on the subsequent 9th year.

On the contrary, native grass species recruited gradually in disturbed plots dominating in the third year of sampling.

Changes in Herbaceous Species Composition in the Absence of Disturbance in a Cenchrus biflorus Roxb. Invaded Area
in Central Kalahari Game Reserve, Botswana

87

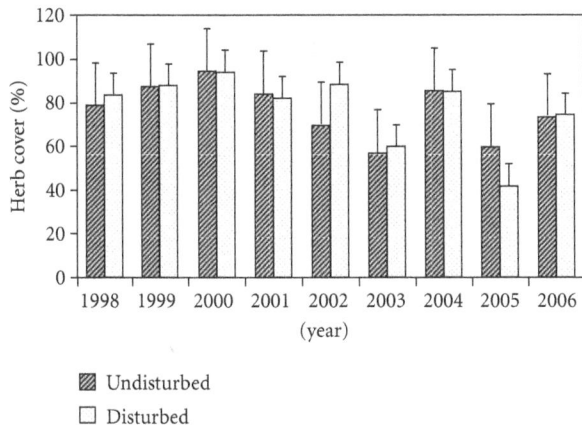

FIGURE 6: Herbaceous cover at disturbed and undisturbed sites in CKGR: 1998 to 2006. Error bars signify standard errors.

TABLE 1: Similarity Czekanowski coefficients of grass and forbs species between the disturbed and undisturbed sites.

Year	Coefficients
1998	0.07
1999	0.14
2000	0.29
2001	0.48
2002	0.33
2003	0.26
2004	0.26
2005	0.40
2006	0.32

4. Discussion

The study showed that forbs were the first to colonize the disturbed area and continued to dominate within the first two years posttermination disturbance. Dominance of forbs on disturbed area may be attributed to their hardness, ability to grow in very unfavorable conditions, adaptation to disturbed sites, and ability to grow vigorously to outcompete other plants [22].

Native grass species recruited gradually in disturbed plots dominating in the third year of sampling. Initially pioneer grass species that include Schmidtia kalihariensis, Enneapogon cenchroides, and Melinis repens dominated during the first five years of sampling but climax species which included Schmidtia pappophoroides and Eragrostis lehmanniana dominated during the last two years of sampling. This transition pattern of dominance by forbs and pioneer grass species in the early years of abandonment followed by an increase in climax species after some years postdisturbance are consistent with succession patterns observed by Holl [25].

Cenchrus biflorus recruited substantially as well in the first three years postdisturbance but declined significantly in the fourth year coinciding with the dominance of native grass species. This suggests that C. biflorus is outcompeted by native grass species if disturbance is reduced

or minimized. Of significance is that C. biflorus did not spread to undisturbed plots suggesting that the species is not competitive, and when factors promoting range deterioration are minimized, the species may not spread in the Central Kalahari Game Reserve. Also, the removal of livestock which degraded the area might have reduced the proportion of disturbance hence limiting the spread of C. biflorus to undisturbed areas. These results concur with previous studies that reported that reducing intensity of use of affected land is an efficient tool to control C. biflorus.

Annual grass species were relatively high during the first three (3) years in disturbed sites but gradually declined. In contrast, perennial species recruited gradually exceeding annuals from 2001 until 2003 but subsequently declined during the last three years of sampling. Nonetheless, the decline was not significant as the species continued to dominate. This pattern of vegetation succession growth was observed by Seitshiro [15] and Perkins [16] who concluded that perennial grass cover return rapidly to areas denuded of grass during dry years in subsequent years of average or better rainfall or cessation of grazing in the Kalahari sandveld vegetation communities. The presence of perennial grass species is vital since it protects the soil from erosion. It also changes the morphology of the landscape which would otherwise be bare during the dry season if it was dominated by annuals.

The results of this study concur with Wiggs [17] who concluded that it takes 8–10 years for a disturbed Kalahari biome to fully recover. However, studies by Samuel and Hart [1] have shown that it takes a longer time for the disturbed areas to fully recover if the sites were severely disturbed. This could be the situation here considering that the study area was under immense pressure of disturbance from human activities that included decades of small scale cultivation, livestock production, harvesting of grass and wood, as well as general infrastructure developments.

Generally the disturbed sites showed some relatively high percentage cover compared to undisturbed sites. The high cover in disturbed sites could be attributed to invasive species ability to vigorously regenerate and recruit rapidly in abandoned disturbed sites [22]. Cenchrus biflorus, though classified as a weed in Botswana, has potential of being a nutritious food plant for herbivores. The National Research Council [26] quotes it has "potential to improve nutrition, boost food security, foster rural development and support sustainable landcare".

In conclusion, the results show that Cenchrus biflorus has some high tolerance to disturbance as it comprised the larger proportion of grasses in disturbed sites at the inception of the study. However, it decreased in abundance and was absent in the undisturbed area, suggesting that its ability to invade undisturbed sites is limited. The results indicate that resting an area invaded by C. biflorus might rehabilitate it, and this takes several years depending on rainfall variability. It is thus important to mitigate and prevent factors which promote range deterioration such as human activities (ploughing, trampling, developments, harvesting of grasses for thatching, and firewood collection) and overgrazing in wildlife protected areas. This action will reduce the probabilities of range deterioration and intrusion of undesirable invasive exotic species in protected areas. This will promote viable wildlife populations and its habitat, thus conserving the biodiversity in such a protected area.

Acknowledgments

Many thanks to N. Tsopito, B. Batsile, B. Seretse, L. Maswena, Morake Moetsabatho, and B. Thibedi of Department of Wildlife and National Parks, Botswana who assisted in the study at its different stages and times. Thanks also go to Dr. D. P. Lebatha for reviewing an earlier draft of this paper. This work was supported by the Department of Wildlife and National Parks, Botswana.

References

[1] J. M. Samuel and R. H. Hart, "Sixty one years of secondary succession on rangelands of Wyoming high plains," *Journal of Range Management*, vol. 47, pp. 184–191, 1994.

[2] E. J. Dyksterhuis, "Condition and management based on quantitative ecology," *Journal of Range Management*, vol. 2, pp. 104–115, 1949.

[3] S. Vetter, "Rangelands at equilibrium and non-equilibrium: recent developments in the debate," *Journal of Arid Environments*, vol. 62, no. 2, pp. 321–341, 2005.

[4] S. W. Todd and M. T. Hoffman, "A fence-line contrast reveals effects of heavy grazing on plant diversity and community composition in Namaqualand, South Africa," *Plant Ecology*, vol. 142, no. 1-2, pp. 169–178, 1999.

[5] D. A. B. Parsons, C. M. Shackleton, and R. J. Scholes, "Changes in herbaceous layer condition under contrasting land use systems in the semi-arid lowveld, South Africa," *Journal of Arid Environments*, vol. 37, no. 2, pp. 319–329, 1997.

[6] A. Hoshino, Y. Yoshihara, T. Sasaki et al., "Comparison of vegetation changes along grazing gradients with different numbers of livestock," *Journal of Arid Environments*, vol. 73, no. 6-7, pp. 687–690, 2009.

[7] C. Skarpe, "Decertification, no-change or alternative states: can we trust simple models on livestock impact in dry rangelands?" *Applied Vegetation Science*, vol. 3, no. 2, pp. 261–268, 2000.

[8] Y. Pueyo and C. L. Alados, "Abiotic factors determining vegetation patterns in semi-arid mediterranean landscapes: different responses in gypsum and non-gypsum substrates," *Journal of Arid Environments*, vol. 69, pp. 490–505, 2006.

[9] J. E. Ellis and D. M. Swift, "Stability of African pastoral ecosystems: alternate paradigms and implications for development," *Journal of Range Management*, vol. 41, pp. 450–459, 1988.

[10] M. Westoby, B. Walker, and I. Noy-Meir, "Opportunistic management for rangelands not at equilibrium," *Journal of Range Management*, vol. 42, no. 4, pp. 266–274, 1989.

[11] M. E. Fernandez-Gimenez and B. Allen-Diaz, "Testing a non-equilibrium model of rangeland vegetation dynamics in Mongolia," *Journal of Applied Ecology*, vol. 36, no. 6, pp. 871–885, 1999.

[12] Z. Bernacki, "Secondary succession of the vegetation in the young shelterbelt (Turew area, Western Poland)," *Polish Journal of Ecology*, vol. 52, no. 4, pp. 391–404, 2004.

[13] M. Rees, S. Pacala, D. Tilman, R. Condit, and M. Crawley, "Long-term studies of vegetation dynamics," *Science*, vol. 293, no. 5530, pp. 650–655, 2001.

[14] T. G. O'Connor and P. W. Roux, "Vegetation changes (1949-71) in a semi-arid, grassy dwarf shrubland in the Karoo, South Africa: influence of rainfall variability and grazing by sheep," *Journal of Applied Ecology*, vol. 32, no. 3, pp. 612–626, 1995.

[15] G. Seitshiro, "Gradient analysis of five non-operational boreholes in the Kweneng District," Unpublished report, Land Utilisation Division, Ministry of Agriculture, Government of Botswana, Gaborone, Botswana, 1978.

[16] J. S. Perkins, *The impact of borehole dependant cattle grazing on the environment and society of the eastern Kalahari sandveld, Central District, Botswana*, Ph.D. thesis, University of Sheffield, Sheffield, UK, 1991.

[17] G. Wiggs, *Geomorphic Thresholds: Aeolian Dune Activity in the Southwest Kalahari*, Geographic Research 9, Centre for the Environment, University of Oxford, Oxford, UK, 2007.

[18] A. J. Dougill and A. D. Thomas, "Kalahari sand soils: Spatial heterogeneity, biological soil crusts and land degradation," *Land Degradation and Development*, vol. 15, no. 3, pp. 233–242, 2004.

[19] Bonifica, *Initial Measures for the Conservation of the Kalahari Ecosystem*, Technical Assistance to the Department of Wildlife and National Parks, Gaborone, Botswana, 1992, Project N.

[20] C. E. M. Tidmarsh and C. M. Havenga, "The wheel-point method of survey and measurement of semi-open grasslands and karoo vegetation in South Africa," *Memoirs of the Botanical Survey of South Africa*, vol. 29, p. iv-149, 1955.

[21] D. Field, *A Handbook of Common Grasses in Botswana*, Ministry of Agriculture, Gaborone, Botswana, 1976.

[22] E. Van Wyk and F. Van Oudtshoorn, *Guide to Grasses of Southern Africa*, Briza Publications, Pretoria, South Africa, 1st edition, 1999.

[23] Statistica, ""STATISTICA program, version Kernel 5.5 A (Stat Soft, Inc (2000)," STATISTICA for Windows [Computer program manual]," Tulsa, Okla, USA, 2000.

[24] M. Kent and P. Coker, *Vegetation Description and Analysis. A Practical Approach*, Belhaven Press, London, UK, 1992.

[25] K. D. Holl, "Long-term vegetation recovery on reclaimed coal surface mines in the eastern USA," *Journal of Applied Ecology*, vol. 39, pp. 960–970, 2002.

[26] National Research Council, *Wild Grains*, vol. 1, The National Academies Press, Washington, DC, USA, 1996.

10

Bone Accumulations of Spotted Hyaenas (*Crocuta crocuta*, Erxleben, 1777) as Indicators of Diet and Human Conflict; Mashatu, Botswana

Brian F. Kuhn

Institute for Human Evolution, School of Geosciences, University of the Witwatersrand (WITS), Johannesburg 2050, South Africa

Correspondence should be addressed to Brian F. Kuhn, kuhnbf@gmail.com

Academic Editor: Bruce Leopold

In a region where free ranging domestic species mix with wildlife, it is imperative to determine what, if any, predation may have occurred on domestic stock. As human settlements continuously encroach upon wild habitats, determining the types of predator-human conflicts that exist can be crucial to conserve numerous predator species. The partial diet of spotted hyaenas (*Crocuta crocuta*) of the Mashatu Game Reserve, Botswana, was established via analyses of faunal remains associated with four dens to determine predation/scavenging on wild or domestic species. Domestic species composed less than 3% of identified faunal remains. We acknowledge that this methodology is biased against small mammals, but, when combined with sociological studies, this methodology will aid in determining alleged predation on domestic stock by spotted hyaenas. Results indicated that the spotted hyaenas in question feed primarily on wild species.

1. Introduction

Spotted hyaenas (*Crocuta crocuta*) have long been the source for studies on feeding behaviour [1–12]. Once thought to be primarily a scavenger, more recent studies have shown spotted hyaenas to be consummate hunters, killing up to 95% of the animals consumed [11], and preferring the most abundant prey at a site weighing between 56 and 182 kg [13]. Having a reputation for killing domestic stock [14] but also thought to adjust behavioural patterns to avoid human presence [15, 16], it was uncertain if the resident spotted hyaenas of the Mashatu Game Reserve, Tuli Block, Botswana, were feeding (via predation or scavenging) on domestic animal populations. Reports from the Botswana government do not list hyaenas (they do not appear to differentiate between spotted hyaenas or brown hyaenas (*Parahyaena brunnea*, Thunberg, 1820) in the top five species of problem animals countrywide when it comes to conflicts with humans, but they are listed as potential problem animals. In the Central District (where the Tuli Block is located), hyaenas in general rank fourth, after lion (*Panthera leo*, Linnaeus, 1758), leopard (*Panthera pardus*, Linnaeus, 1758), and elephant (*Loxodonta*

africana, Blumenbach, 1797), amongst problem animals [17]. A study in East Africa has shown that lions have a much greater effect on livestock populations than spotted hyaenas, but the local pastoralists have a greater animosity towards hyaenas [18]. In contrast, a survey in Ethiopia indicated that spotted hyaenas are responsible for direct predation on domestic stock [19]. For the most part, studies determining which species has the most detrimental effect on livestock are conducted using scat analyses of said carnivore species and/or interviewing the local population and relying on their identifications and interpretations. It must also be acknowledged until recently that the Botswana government policy was to pay out for livestock killed by lions but not for livestock killed by hyaenas (L. Frank Pers Comm. 2011), which may have influenced the populations' identifications of conflict species.

In addition to the multitude of behavioural and ecological studies on spotted hyaenas, specific taphonomic studies involving extant spotted hyaenas have also been conducted to better understand their role in fossil deposits [20–31]. The taphonomic implications of this study have been previously published [28, 30, 31]; the same data and methodologies

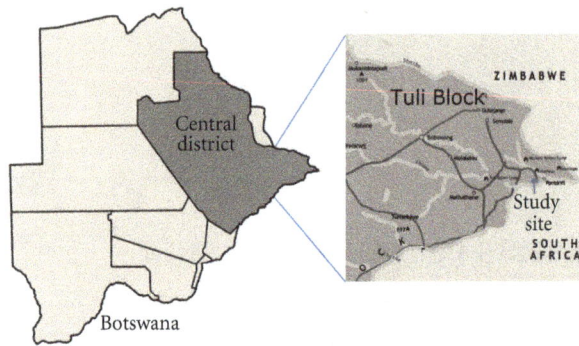

Figure 1: Map showing Botswana's Central District, the Tuli Block and study area.

are used to examine patterns of predation/scavenging as documented by skeletal accumulations at den sites. These methodologies, specifically the examination and identification to prey species, of faunal remains accumulated at hyaena dens. Here we demonstrate how methodologies used in taphonomic studies can be implemented in modern human-animal conflict issues to help gauge the amount of predation spotted hyaenas may inflict on domestic stock.

In Botswana, total livestock units (LUs) combining wild and domestic species in 2003 were 4.39 LU/Km^2. Domestic species composed 79% of the total LU biomass (the remaining 21% consisting of wildlife). The Central District, the largest district covering 147,730 km^2, of which the Tuli Block makes up a very small portion, 300,000 hectares, reported 55.3% of nation-wide problem-animal incidents. Three hundred twenty one of these incidents involved hyaena compared to lions involved in 1,857-recorded incidents [17]. Results indicated that, despite a high number of domestic species available, the spotted hyaenas of the Mashatu Game Reserve appear to be surviving primarily by feeding on wild species.

2. Study Site

The Mashatu Game Reserve lies in the eastern portion of the Central District of Botswana in a region known as the Tuli Block. The Tuli Block lies on the eastern central border region of Botswana, bordered by Zimbabwe to the north and northeast and South Africa to the south and southeast (Figure 1). Unfenced, the reserve covers approximately 30,000 ha of semi arid Bushvelt receiving approximately 177 mm of rainfall per annum (Mashatu.com 2006). The reserve supports numerous species including large herds of elephants (*Loxodonta africana*, Blumenbach, 1797), kudu (*Tragelaphus strepsiceros, Pallas, 1766*), wildebeest (*Connochaetes taurinus,* Burchell, 1823), zebra (*Equus quagga,* Boddaert, 1785), and impala (*Aepyceros melampus,* Lichtenstein, 1812). In addition to the wide variety of herbivores the area also supports numerous carnivore species such as the aforementioned spotted hyaenas and lions, leopards, caracals (*Caracal caracal,* Schreber, 1776), black-backed jackals (*Canis mesomelas,* Schreber, 1775), and wild dogs (*Lycaon pictus,* Temminck, 1820). The region has a diverse fauna (Table 1). While the region is well

within the natural range of brown hyaenas, there has only been a single confirmed sighting of brown hyaena in the past decade (J. Selier pers. com.). This, combined with observed occurrences of only spotted hyaenas associated with the four dens, enables us to conclude that spotted hyaenas primarily use the dens. All four dens collected from were caves, three of which were located at the base of the escarpment, and the fourth den/cave located just below the top edge of the escarpment. Den 2 had juvenile and adult hyaenas present prior to the study beginning and Den 4 had both adult and juvenile hyaenas present during the study. It has been shown that hyaenas collect bones in greater numbers when young are present thus, we infer that all four dens were used as maternity dens at some point in the recent past [2–4, 6–8, 24, 27, 31]. The condition of the bones collected suggested a time frame of a few months to over six years since the material was deposited at the dens.

The human population of the Tuli Block was recently estimated at 7,954 people [32] and domestic livestock units at approximately 3.4 LU/Km^2 [17]. The present study was based out of the Kgotla camp in the southwest part of the reserve and concentrated on regions adjacent to the Motloutse River and Motloutse archaeology site.

The human populations on the fringes of the reserve are surviving via the farming of livestock. Thus, even though the local populations take care with their livestock by the use of kraals and dogs, livestock is still available for the local carnivores to prey on. During the day it was common to see livestock roaming in the study area. While the hyaenas living deeper inside of the reserves would likely have less contact with domestic species, the clans living in the study area would have greater access to livestock.

3. Materials and Methods

In addition to long-term general observations made by staff of the reserve, which identified three dens, our study was conducted over three months (October–December of 2004). During this time dens of spotted hyaenas were identified and observed. Additional dens were located via random searches consisting of hikes up to 20 kilometres, all starting from the Motloutse archaeology site and radiating outward. No two searches covered the same area, and none could maintain a straight line due to the terrain. Most searches formed circular patterns skirting the base of the escarpments or covering the tops of said escarpments where spotted hyaenas were known to frequent. At the end of the study period faunal remains from four of the eight identified dens were collected and transported to lab facilities at the University of the Witwatersrand, Johannesburg, South Africa. The dens collected were chosen purely for logistical reasons as they were within 5–7 kilometres from where I was required to leave my vehicle. Specimens were identified to skeletal element and species or class size where possible, using reference collections housed at the Institute for Human Evolution and the Bernard Price Institute, University of the Witwatersrand. In addition bone identification manuals by Schmid [33], Walker [34], and Hillson [35] were used to

TABLE 1: Mammal species living in Mashatu and surrounds.

	Common name	Latin name
Bovid class 1	Klipspringer	*Oreotragus oreotragus*
	Steenbok	*Raphicerus campestris*
	Common duiker	*Sylvicapra grimmia*
Bovid class 2	Bushbuck	*Tragelaphus scriptus*
	Impala	*Aepyceros melampus*
	Goat (domestic)	*Capra hircus*
Bovid class 3	Kudu	*Tragelaphus strepsiceros*
	Waterbuck	*Kobus ellipsiprymnus*
	Blue wildebeest	*Connochaetes taurinus*
	Cow (domestic)	*Bos sp.*
Bovid cslass 4	Eland	*Taurotragus oryx*
Equid	Burchell's zebra	*Equus quagga*
	Donkey (domestic)	*Equus asinus*
Primate	Chacma Baboon	*Papio cynocephalus ursinus*
	Vervet monkey	*Cercopithecus aethiops*
	Thick-tailed bushbaby	*Galago crassicaudatus*
Carnivore	Lion	*Panthera leo*
	Leopard	*Panthera pardus*
	Cheetah	*Acinonyx jubatus*
	Caracal	*Caracal caracal*
	Serval	*Leptailurus serval*
	African wild cat	*Felis silvestris*
	Bat-eared fox	*Otocyon megalotis*
	Black-backed jackal	*Canis mesomelas*
	Cape fox	*Vulpes chama*
	Aardwolf	*Proteles cristatus*
	Spotted hyaena	*Crocuta crocuta*
	Brown hyaena	*Parahyaena brunnea*
	Cape clawless otter	*Aonyx capensis*
	Honey badger	*Mellivora capensis*
	Striped polecat	*Ictonyx striatus*
	African civet	*Civettictis civetta*
	Large-spotted genet	*Genetta tigrina*
	Small-spotted genet	*Genetta genetta*
	Banded mongoose	*Mungos mungo*
	Selous mongoose	*Paracynictis selousi*
	Slender mongoose	*Galerella sanguinea*
	White-tailed mongoose	*Ichneumia albicauda*
Suid	Bushpig	*Potamochoerus larvatus*
	Warthog	*Phacochoerus africanus*
Rodent	Porcupine	*Hystrix africaeaustralis*
	Springhare	*Pedetes capensis*
Hyrax	Rock dassie	*Procavia capensis*
	Yellow-spotted dassie	*Heterohyrax brucei*
Other	Hippopotamus	*Hippopotamus amphibius*
	Giraffe	*Giraffa camelopardalis*
	Elephant	*Loxodonta africana*
	Aardvark	*Orycteropus afer*
	Scrub hare	*Lepus saxatilis*
	Pangolin	*Manis temminckii*

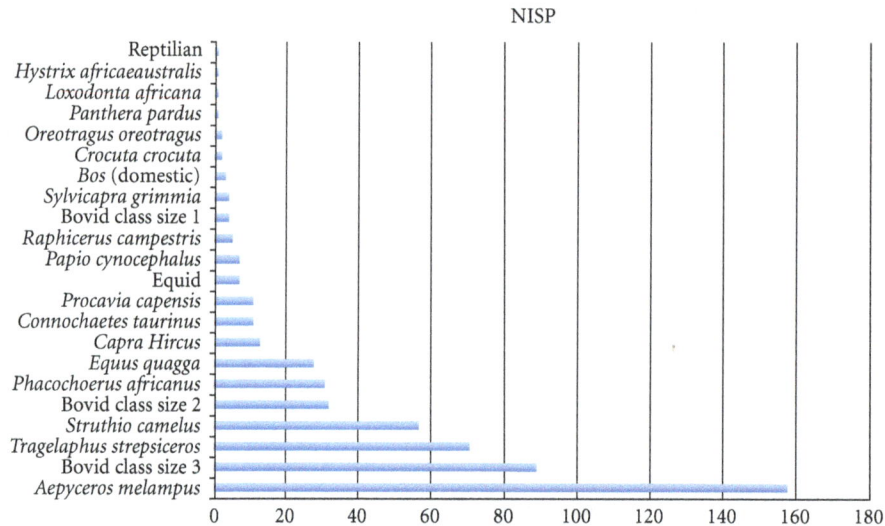

FIGURE 2: Abundance of identified species by NISP (Number of Identified Specimens).

TABLE 2: Identified species or class size, with (number of identified specimens) (NISP), (minimum number of individuals) (MNI) and which dens they were found. Bovid class size based on kilograms and follows Brain [36].

Species	NISP	MNI	Dens
Aepyceros melampus	158	13	1,2,3,4
Bovid class size 3	89	5	1,2,3,4
Tragelaphus strepsiceros	71	4	1,3,4
Struthio camelus	57	4	4
Bovid class size 2	32	2	1,2,3,4
Phacochoerus africanus	31	2	2,3,4
Equus quagga	28	4	1,2,3,4
Capra hircus	13	5	4
Connochaetes taurinus	11	3	2,3,4
Procavia capensis	11	2	1,2,3,4
Equid	7	2	3
Papio cynocephalus	7	2	2,3,4
Raphicerus campestris	5	2	4
Bovid class size 1	4	1	1,4
Sylvicapra grimmia	4	2	1,4
Bos (domestic)	3	1	4
Crocuta crocuta	2	1	2
Oreotragus oreotragus	2	1	4
Panthera pardus	1	1	4
Loxodonta africana	1	1	3
Hystrix africaeaustralis	1	1	4
Reptilian	1	1	1
Total	539	60	

aid in the identification of the collected remains. From this the number of identified specimens (NISP) and minimum number of individuals (MNI) for each den was calculated. All specimens are currently stored at the Palaeosciences Centre, University of the Witwatersrand, Johannesburg, South Africa. Bovid size classes are based on weight and follow Brain [36].

4. Results

We collected 976 specimens of faunal remains from four dens. Of this total, 69.4% of the specimens were identified to skeletal element and 55.2% to species or class size (Table 2, Figure 2). Examination of the faunal remains collected at the den sites indicated that impala (Aepyceros melampus) was the most common prey species and consisted of 29.3% of the identified remains. The second most common prey item identified was that of unknown bovid class size 3 (16.5%), followed by kudu (Tragelaphus strepsiceros) (13.2%) and ostrich (Struthio camelus, Linnaeus, 1758) (10.6%). The remaining identified species composed less than 10% (per species) of identified remains. Combined domestic species identified as cow (Bos sp.) and goat (Capra hircus, Linnaeus, 1758) composed less than 3% of the total identified remains.

Breakdown of skeletal elements is shown for taxonomically identified family groups, domestic groups, and class size in Table 3. Bovids of class size 2 and 3 show a much greater variety in skeletal elements, with class 2 having at least two samples of each skeletal element listed in Table 3. Domestic Bos yielded only three elements, while domestic Capra had 12 total identified elements, eight of which were mandibles. Carnivores were the group with the least number of elements, two.

5. Discussion

Considering numbers of domestic species available (79% of the total livestock unit (LU) biomass), results indicate that the spotted hyaenas in the study region are consuming primarily wild species that inhabit the region. Unknown

TABLE 3: Number of select skeletal elements identified for the domestic species, wild bovid class size 1, 2, and 3 [36], equid, carnivore, suid, and primate.

	Domestic *bos*	Domestic *Caprid*	Wild class 1	Wild class 2	Wild class 3	Equid	Carnivore	Suid	Primate
Skull	0	0	2	3	3	0	1	3	0
Maxilla	0	0	0	2	0	0	0	0	1
Mandible	0	8	0	5	1	1	0	6	0
Pelvic	0	0	1	6	3	3	0	0	0
Scapula	0	1	0	8	8	1	0	0	0
Humerus	1	1	1	20	17	5	0	0	1
Radius	0	0	2	12	8	3	0	2	0
Ulna	0	0	1	5	6	0	1	0	0
Metacarpal	1	0	0	11	7	7	0	2	0
Femur	0	0	2	19	7	3	0	1	2
Tibia	0	0	2	15	11	4	0	2	0
Calcaneus	0	1	0	4	4	0	0	0	0
Astragalus	0	0	0	2	5	0	0	0	0
Metatarsal	1	0	0	8	8	1	0	1	1
Phalanx 1	1	1	1	20	7	5	0	0	0
Phalanx 2	0	0	0	11	4	2	0	0	0
Phalanx 3	0	0	0	7	2	1	0	0	0
Total bones	4	12	12	158	101	36	2	17	5

bovid class 2 (which may include domestic goats and wild bovids, such as impala) and bovid class 3 (which could have domestic cattle and wild bovids) combined compose 22.5% of the identified assemblage. While it is acknowledged that some of this 22.5% may include domestic goat or cattle, it is argued here that the ratios of bovid species represented in the unidentifiable remains would likely follow the ratios of identified remains. Thus, the most unknown bovid class 2 would most likely be comprised of impala, and the bulk of unknown bovid class 3 kudu. Of note is that in addition to impala, only zebra (*Equus quagga*), hyrax, and unknown bovids of class sizes 2 and 3 are present in all four dens. This observation is suggestive that these prey types are in all probability the preferred prey of resident spotted hyaenas.

Examination of the identified skeletal elements indicates that bovid class sizes 2 and 3 (both identified to species as well as those identified only to class size) along with equids (zebra) show the greatest variation of identified elements. Unlike carnivores, domestic species and primates provided relatively few skeletal elements to the hyaena den assemblages. The bovids and equids illustrate that most body parts are being collected at the den sites. The difference between identified skeletal remains of the wild bovids and equids compared to the warthogs, domestic species, carnivores, and primates may very well be indicative of the difference between hunted prey and prey which has been scavenged. In particular the wild class 2 and class 3 bovids that have all major skeletal elements represented in the assemblages, which may infer predation. However, the reduced number of skeletal elements from domestic *Bos, Caprid,* carnivore, suid, and primate would suggest scavenging from human settlements or other predator kills.

This study concurs with reports published by the Botswana government [17] in that spotted hyaenas, while potentially a problem species as evidenced by a low percentage of identified domestic remains at den sites are not a major contributor to predation upon livestock in the Tuli Block. It must also be acknowledged that the remains studied here may or may not have come solely from hyaenid predation but could have been scavenged from the kills of leopard and or lion. Additionally, it must be emphasised that presence of so few domestic remains in the dens suggests that these particular elements were likely scavenged from human settlements.

Whether or not these results are typical of spotted hyaenas across their range is unknown. This study was conducted in a small area on the edge of the reserve. While this area was chosen for known hyaena activity within close proximity to human populations, it cannot be said whether these particular hyaenas are behaving the same as hyaenas from other regions. What has been demonstrated is a methodology from which hyaena-human conflict can be explored beyond and in addition to the previous use of interviewing the local populations. While there will always be conflict between wildlife and humans, especially when it comes to depredation of domestic stock, through bone-based ecological research and education, we should be able to reduce the unwarranted animosity many rural populations have towards spotted hyaenas. This study takes a critical, initial step to demonstrate that spotted hyaenas in the study area are not directly preying on large numbers of domestic stock. In addition we have illustrated another way in which to determine whether or not hyaenas in a given region are indeed problem animals or not.

Acknowledgments

This work was supported by the University of Pretoria and the Palaeontological Scientific Trust (PAST). Grateful appreciation goes to The Mashatu Game Reserve, especially Pete Le Roux, Jeanetta Selier, and Paul Grobler. Also the author is grateful to Ryan Franklin, Joost Segera, Michiel van Silfhout, and Jess Kuhn for aiding in collecting faunal remains. Additional appreciation to Bruce Leopold and six anonymous reviewers whose comments on earlier versions of this manuscript were most helpful.

References

[1] A. J. Sutcliffe, "Spotted hyaena: crusher, gnawer, digester and collector of bones," *Nature*, vol. 227, no. 5263, pp. 1110–1113, 1970.

[2] H. Kruuk, *The Spotted Hyena: A Study of Predation and Social Behavior*, University of Chicago Press, Chicago, Ill, USA, 1972.

[3] S. K. Bearder, "Feeding habits of spotted hyaenas in a woodland habitat," *East African Wildlife Journal*, vol. 15, pp. 163–290, 1977.

[4] J. R. Henschel, R. Tilson, and F. von Blottnitz, "Implications of a spotted hyaena bone assemblage in the Namib Desert," *South African Archaeological Bulletin*, vol. 3, pp. 127–131, 1979.

[5] J. D. Skinner and R. J. van Aarde, "The distribution and ecology of the brown hyaena *Hyaena brunnea* and spotted hyaena *Crocuta crocuta* in the central Namib Desert," *Madoqua*, vol. 12, no. 4, pp. 231–239, 1981.

[6] R. L. Tilson and W. J. Hamilton, "Social dominance and feeding patterns of spotted hyaenas," *Animal Behaviour*, vol. 32, no. 3, pp. 715–724, 1984.

[7] J. R. Henschel and J. D. Skinner, "The diet of the spotted hyaenas *Crocuta crocuta* in Kruger National Park," *African Journal of Ecology*, vol. 28, no. 1, pp. 69–82, 1990.

[8] M. G. L. Mills, *Kalahari Hyaenas: The Comparative Ecology of Two Species*, Unwin-Hyman, London, UK, 1990.

[9] S. M. Cooper, "The hunting behaviour of spotted hyaenas (*Crocuta crocuta*) in a region containing both sedentary and migratory populations of herbivores," *African Journal of Ecology*, vol. 28, no. 2, pp. 131–141, 1990.

[10] K. E. Holekamp, L. Smale, R. Berg, and S. M. Cooper, "Hunting rates and hunting success in the spotted hyena (*Crocuta crocuta*)," *Journal of Zoology*, vol. 242, no. 1, pp. 1–15, 1997.

[11] S. M. Cooper, K. E. Holekamp, and L. Smale, "A seasonal feast: long-term analysis of feeding behaviour in the spotted hyaena (*Crocuta crocuta*)," *African Journal of Ecology*, vol. 37, no. 2, pp. 149–160, 1999.

[12] K. E. Holekamp, S. T. Sakai, and B. L. Lundrigan, "The spotted hyena (*Crocuta crocuta*) as a model system for study of the evolution of intelligence," *Journal of Mammalogy*, vol. 88, no. 3, pp. 545–554, 2007.

[13] M. W. Hayward, "Prey preferences of the spotted hyaena (*Crocuta crocuta*) and degree of dietary overlap with the lion (*Panthera leo*)," *Journal of Zoology*, vol. 270, no. 4, pp. 606–614, 2006.

[14] IUCN-Hyaena specialist group, 2009, http://www.hyaenidae .org/the-hyaenidae/spotted-hyena-crocuta-crocuta/crocuta-diet-and-foraging.html.

[15] E. E. Boydston, K. M. Kapheim, H. E. Watts, M. Szykman, and K. E. Holekamp, "Altered behaviour in spotted hyenas associated with increased human activity," *Animal Conservation*, vol. 6, no. 3, pp. 207–219, 2003.

[16] J. M. Kolowski and K. E. Holekamp, "Ecological and anthropogenic influences on space use by spotted hyaenas," *Journal of Zoology*, vol. 277, no. 1, pp. 23–36, 2009.

[17] *Wildlife Statistics*, Central Statistics Office, Gaborone, Botswana, 2004.

[18] L. G. Frank, "Spotted hyenas and livestock in Laikipia, Kenya," *Hyaena Specialist Group Newsletter*, vol. 7, pp. 12–17, 2000.

[19] G. Yirga and H. Bauer, "Livestock depredation of the spotted hyena (*Crocuta crocuta*) in Southern Tigray, Northern Ethiopia," *International Journal of Ecology and Environmental Sciences*, vol. 36, no. 1, pp. 67–73, 2010.

[20] L. R. Binford, M. G. L. Mills, and N. M. Stone, "Hyena scavenging behavior and its implications for the interpretation of faunal assemblages from FLK 22 (the Zinj floor) at olduvai gorge," *Journal of Anthropological Archaeology*, vol. 7, no. 2, pp. 99–135, 1988.

[21] K. Cruz-Uribe, "Distinguishing hyaena from hominid bone accumulations," *Journal of Field Archaeology*, vol. 18, pp. 467–486, 1991.

[22] Y. M. Lam, "Variability in the behaviour of spotted hyaenas as taphonomic agents," *Journal of Archaeological Science*, vol. 19, no. 4, pp. 389–406, 1992.

[23] T. R. Pickering, "Reconsideration of criteria for differentiating faunal assemblages accumulated by hyenas and hominids," *International Journal of Osteoarchaeology*, vol. 12, no. 2, pp. 127–141, 2002.

[24] J. D. Skinner, "Bone collecting by hyaenas: a review," *Transactions of the Royal Society of South Africa*, vol. 61, no. 1, pp. 4–7, 2006.

[25] J. T. Faith, "Sources of variation in carnivore tooth-mark frequencies in a modern spotted hyena (*Crocuta crocuta*) den assemblage, Amboseli Park, Kenya," *Journal of Archaeological Science*, vol. 34, no. 10, pp. 1601–1609, 2007.

[26] J. T. Faith, C. W. Marean, and A. K. Behrensmeyer, "Carnivore competition, bone destruction, and bone density," *Journal of Archaeological Science*, vol. 34, no. 12, pp. 2025–2034, 2007.

[27] J. T. Pokines and J. C. Kerbis Peterhans, "Spotted hyena (*Crocuta crocuta*) den use and taphonomy in the Masai Mara National Reserve, Kenya," *Journal of Archaeological Science*, vol. 34, no. 11, pp. 1914–1931, 2007.

[28] B. F. Kuhn, L. R. Berger, and J. D. Skinner, "Variation in tooth mark frequencies on long bones from the assemblages of all three extant bone-collecting hyaenids," *Journal of Archaeological Science*, vol. 36, no. 2, pp. 297–307, 2009.

[29] S. W. Lansing, S. M. Cooper, E. E. Boydston, and K. E. Holekamp, "Taphonomic and zooarchaeological implications of spotted hyena (*Crocuta crocuta*) bone accumulations in kenya: a modern behavioral ecological approach," *Paleobiology*, vol. 35, no. 2, pp. 289–309, 2009.

[30] B. F. Kuhn, L. R. Berger, and J. D. Skinner, "Examining criteria for identifying and differentiating fossil faunal assemblages accumulated by hyenas and hominins using extant hyenid accumulations," *International Journal of Osteoarchaeology*, vol. 20, no. 1, pp. 15–35, 2010.

[31] B. F. Kuhn, *Hyaenids: Taphonomy and Implications for the Palaeoenvironment*, Cambridge Scholars Publishing, London, UK, 2011.

[32] 2011, http://www.traveljournals.net/explore/botswana/map/ m2006523/tuli_block.html.

[33] E. Schmid, *Atlas of Animal Bones*, Elsevier, Amsterdam, The Netherlands, 1972.

[34] R. Walker, *A Guide to Post-Cranial Bones of East African Animals: Mrs Walker's Bone Book*, Hylochoerus Press, Norwich, UK, 1985.

[35] S. Hillson, *Mammal Bones and Teeth: An Introductory Guide to Methods of Identification*, Dorset Press, Dorchester, UK, 1992.

[36] C. K. Brain, *The Hunters or the Hunted? An Introduction to African Cave Taphonomy*, The University of Chicago Press, Chicago, Ill, USA, 1981.

11

Modeling Critical Forest Habitat in the Southern Coal Fields of West Virginia

Aaron E. Maxwell,[1] Michael P. Strager,[2] Charles B. Yuill,[2] and J. Todd Petty[3]

[1] Natural Resource Analysis Center, West Virginia University, Morgantown, WV 26506, USA
[2] Division of Resource Management, West Virginia University, Morgantown, WV 26506, USA
[3] Division of Forestry and Natural Resources, West Virginia University, Morgantown, WV 26506, USA

Correspondence should be addressed to Michael P. Strager, mstrager@wvu.edu

Academic Editor: Bradford Hawkins

Throughout the Central Appalachians of the United States resource extraction primarily from coal mining has contributed to the majority of the forest conversion to barren and reclaimed pasture and grass. The loss of forests in this ecoregion is significantly impacting biodiversity at a regional scale. Since not all forest stands provide equal levels of ecological functions, it is critical to identify and map existing forested resources by the benefits that accrue from their unique spatial patterns, watershed drainage, and landscape positions. We utilized spatial analysis and remote sensing techniques to define critical forest characteristics. The characteristics were defined by applying a forest fragmentation model utilizing morphological image analysis, defining headwater catchments at a 1 : 24,000 scale, and deriving ecological land units (ELUs) from elevation data. Once critical forest values were calculated, it was possible to identify clusters of critical stands using spatial statistics. This spatially explicit method for modeling forest habitat could be implemented as a tool for assessing the impact of resource extraction and aid in the conservation of critical forest habitat throughout a landscape.

1. Introduction

Forests provide many benefits to society and the natural environment that include providing wildlife habitat [1], maintaining biodiversity [2], regulating climate and providing carbon storage [3, 4], nitrogen cycling [5, 6], and altering and moderating hydrologic function, including evapotranspiration rates and surface runoff volumes [7]. Society land use needs and resource extraction often result in the loss of forested land cover, leading to an impact on biodiversity from habitat conversion [8–10]. In reference to forest and biodiversity conservation, there has been a shift in focus from rare or endangered species management to ecosystem and landscape scale management in which the health and function of the ecosystem as a whole is considered [11]. This focus requires that a diversity of ecological processes be considered, such as the health and structure of the forest [11]. The need is to focus on conservation of habitat richness and biodiversity at the landscape-level.

Conservation efforts should be directed at spatially explicit, landscape-level attributes, such as minimizing structural contrast between adjacent landscape units [12]. The goal of this paper was to implement such a methodology in analyzing the conservation of biodiversity and forest habitat in the Southern Coal Fields of West Virginia.

We have expanded on previous work by suggesting and implementing a methodology that allows for the indexing of critical forest stands on a pixel-by-pixel basis relative to fragmentation patterns, watershed drainage, and landscape position. We implemented a methodology for extracting high-resolution, spatially explicit forest cover and disturbance data from National Agricultural Imagery Program (NAIP) orthophotography using object-based image classification and ancillary datasets. A fragmentation model utilizing morphological image analysis was then applied to the land cover data after Vogt et al. [13] to obtain high-resolution forest fragmentation data. These data were then related to watershed and landscape position in order to index forest pixels relative to importance of conservation for terrestrial habitat and biodiversity conservation within the region. An objective was to implement a spatial analysis method that utilized high-resolution data to describe the distribution and patterns of forest stands critical to biodiversity as a potential tool for evaluating and mitigating the effect of resource extraction in this unique and biodiverse landscape.

2. Background

Not all forest stands are equal in terms of crucial habitat. Wickham et al. [14] documented the importance of core or interior forests in Southern Appalachia. Core and fragmented forests differ in terms of structure, composition, and ecological processes. Edge forests have higher rates of atmospheric decomposition [15] and higher proportions of exotic species [16]. In addition, it has been determined that the amount of forest cover is positively correlated with forest breeding bird populations [17]. Habitat edges have an influence on ecological function, including biogeochemical nutrient transport [18] and species interactions [19, 20]. Edges impact ecological mechanisms, patterns, and dynamics at variable spatial scales [21], including the landscape scale [22]. Fragmentation is of critical concern when modeling processes at the landscape scale [16].

Meyer et al. [23] documented the importance of headwater reaches. Such areas within a watershed contribute to overall biodiversity in a watershed, provide refuge from predators, competitors and alien species, and serve as links to adjacent watersheds. The unique conditions of an individual headwater reach contribute to the heterogeneity of the entire watershed and landscape and enhance biodiversity [24]. Appalachia supports 10% of the global salamander and freshwater mussel diversity [25, 26], and their diversity is high in headwater streams [27].

Strausbaugh and Core [28] documented the unique cove hardwoods or mixed mesophytic forests that support a wide range of species in Appalachia, such as a host of arachnid species [29, 30]. These communities are highly productive due to high moisture and fertile soil [31]. Cove hardwood forests are species rich [32, 33], and loss and fragmentation of these communities influence habitat suitability and potentially the persistence of the species on the landscape [31].

Weakland and Wood [34] and Wood et. al. [35] documented the importance of upland, core forest for preservation and protection of the Cerulean Warbler, a species of concern that is threatened by habitat perturbations [34, 36, 37]. As a result, we argue that forest fragmentation patterns, watershed position, and landscape position have an influence on how critical a forest stand is in terms of habitat conservation.

The Southern Coal Fields of West Virginia provides large expanses of temperate deciduous forest that are rare worldwide and provide for biodiversity [14]. For example, the eastern deciduous forest supports a wide variety of avian and large carnivore species; however, the extent and fragmentation of the forest have a direct impact on species abundance and composition [38]. In the temperate zone, the southern

Appalachian region is among the highest in biodiversity and endemism with more than 2,000 documented vascular plant species and the highest freshwater biodiversity in North America [25, 39]. In the Southern Coal Fields of Appalachia, it has been suggested that increased seasonality of climate during the transition from glacial to interglacial conditions in the late Pleistocene and Holocene induced landscape instability that resulted in a heterogeneous mosaic of habitats [40]. This resulted in the rich and diverse biota observed today, and environmental changes can have a significant impact on the survival of species that have adapted to such specific conditions [40]. Southern Appalachian forests can be characterized as spatially extensive interior forests, and the extent of forest and its spatial patterns and distribution have been shown to have a direct influence on habitat suitability [41–43].

Land use and land disturbance impact forest extent, composition, and spatial patterns, which have consequences for biodiversity and other environmental services [44, 45]. Surface mining, and specifically mountaintop removal with valley fills (MTR/VF), is currently the leading cause of land cover and land use change in the Southern Coal Fields of West Virginia [46–49]. MTR/VF consists of clearing upland forests, removing top soil, and blasting away overburden rock material to extract coal seams. Overburden material is often placed in adjacent valleys, filling headwater steam segments. Land cover, landscape contours, and surface flowpaths are altered [24]. Multiple watersheds in West Virginia have more than 10% of their surface area disturbed by surface mining [50]. The mining processes often results in a conversion of forested land cover to barren land cover. Later reclamation produces grasslands on the topographically altered mountaintops; however, productivity of the grasslands is often low due to poor soil conditions [51]. It has been estimated that all surface mining in Appalachia has resulted in a net loss of 420,000 ha of forest [48].

Wickham et al. [14] suggest that the amount of interior forest lost is greater than the amount of forest lost resulting from direct conversion of forest to surface mines. Forests loose interior character by the introduction of nonforest edges. Thus, the influence of surface mining extends into adjacent forest due to edge effects, and the spatially extensive nature of the forest is interrupted. Weakland and Wood [34] and Wood et al. [35] found that MTR/VF mining alters the spatial configuration of forested habitats, creating edge and area effects that negatively affect Cerulean Warbler abundance and occurrence on the landscape.

3. Study Area

The study area for this project is a 2,280,423 ha area that encompasses the Southern Coal Fields of West Virginia (Figure 1). This area was defined relative to hydrologic unit code (HUC) 8 boundaries that intersect the MTR/VF region in West Virginia.

The study area exists within the Appalachian Plateau physiographic province, a dissected, westward-tilted plateau dominated by Pennsylvanian bedrock. Pennsylvanian stratigraphy is characterized as cyclic sequences of sandstone, shale, clay, coal, and limestone [52]. The terrain is dissected by a dendritic stream network and shows fine texture with moderate-to-strong local relief. In comparison to the northern Appalachian Plateau, the Southern Coal Fields is generally more rugged due to resistant strata. Precipitation levels, temperature, and terrain allow for a variety of forest communities to flourish including oak-pine, oak-chestnut, cove hardwoods or mixed mesophytic forests, and floodplain communities [28]. Although surface mining has altered this landscape, it is still predominantly forested, in contrast to other regions of the eastern United States that have experienced development [46, 47].

4. Methodology

Indexing critical forest habitat incorporated several remote sensing image classification and spatial analysis processes. First, high-resolution land cover was extracted using object-based feature extraction and GIS decision rules. Object-based image classification allowed for the creation of high-resolution land cover/land use data from available aerial imagery. Second, a high-resolution forest fragmentation model was created for the region using a model developed by Vogt et al. [13]. Third, small catchment areas relative to 1 : 24,000 scale stream segments were extracted and headwater catchments were identified. And lastly, we incorporated digital elevation model (DEM) data at 1 : 4,800 scale to produce ecological land units (ELUs) for a thematic map representation of landforms. These products were then combined to describe critical forest habitat relative to fragmentation, watershed position, and landscape position.

4.1. Extraction of Land Cover/Land Use Data. Land cover/land use was produced utilizing an object-based feature extraction process along with decision rules. NAIP imagery was collected during leaf-on conditions in 2009 and made available in October 2009. The imagery has a 1 m nominal pixel spacing, a scale of 1 : 10,000, and is rectified to a horizontal accuracy of within ±5 meters of reference digital ortho quarter quads from the National Digital Ortho Program. This imagery was classified and results were merged to produce a continuous thematic dataset for the region.

Feature Analyst 4 by Visual Learning Systems (VLSs) within Erdas Imagine 9.3 was utilized to extract the following land cover types: barren/developed, grasslands, and forest. This software tool examines both the spectral and textural information within the imagery to produce thematic map data. As a result, unlike supervised classification techniques, feature extraction classifies an image using more than just the digital number (DN) or spectral information contained by each pixel. Spatial context information such as spatial association, size, shape, texture, pattern, and shadow are considered [53]. Studies have shown that feature extraction or objected-based algorithms are more effective and accurate at extracting information from high-resolution imagery than traditional image classification methods, such as supervised classification, because additional image characteristics are considered such as spatial context [54–57]. Feature extraction takes into account the spatial context that is available

FIGURE 1: Study area. Southern Coal Fields of West Virginia.

in high-resolution imagery, such as NAIP orthophotography. Due to the high spatial resolution and reduced spectral resolution of NAIP orthophotography in comparison to satellite imagery traditionally used to obtain thematic map data, we used feature extraction as the classification approach in order to make use of the spatial context and textural information contained in the imagery.

User-assisted feature extraction requires user-defined knowledge, as training data, to recognize and classify target features in an image [58]. As a result, we collected training samples as polygons by manually interpreting the imagery. For example, for a single dataset, we collected 101 examples of barren/developed land cover, 165 examples of grasslands land cover, and 147 examples of forested land cover. Our goal was to provide training data that accurately described the spectral and textural signature and variability of the land cover of interest.

An error assessment was performed by comparing the thematic map results to manual interpretation of the imagery at randomly selected locations. This assessment predicted 88% producer's accuracy and 99% user's accuracy for the forest class. As a result, this method allows for an accurate differentiation of forested land cover utilizing NAIP imagery to provide a recent and high spatial resolution representation of land cover and forest extent. It should be noted that datasets can be updated as new imagery is collected; for example, NAIP imagery has been collected in West Virginia in 2007, 2009, and 2011, which is a true benefit of utilizing NAIP. It is publically available, high-resolution data and is commonly flown every two years.

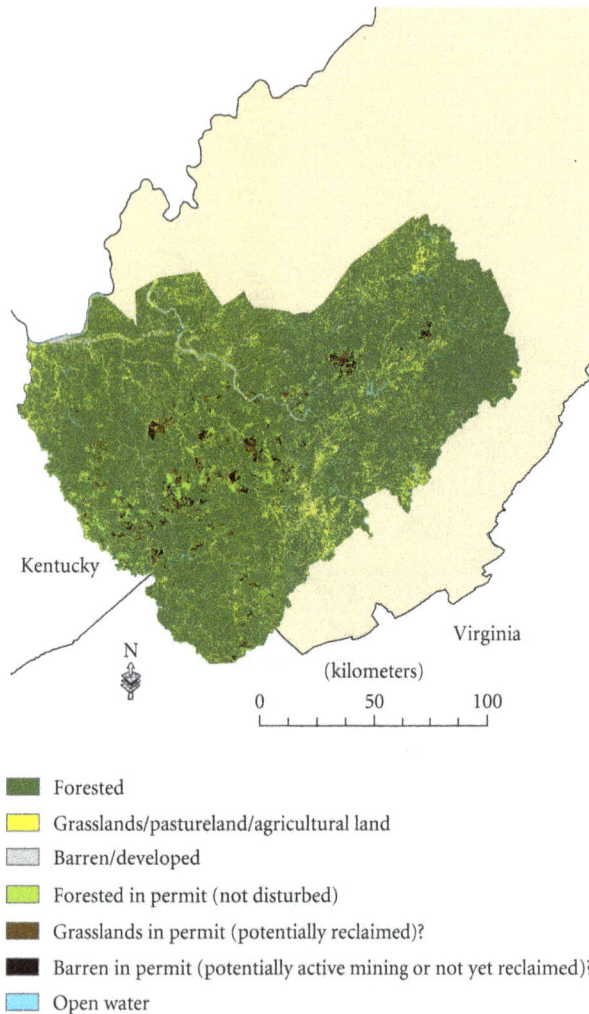

Forested

Grasslands/pastureland/agricultural land

Barren/developed

Forested in permit (not disturbed)

Grasslands in permit (potentially reclaimed)?

Barren in permit (potentially active mining or not yet reclaimed)?

Open water

FIGURE 2: Land cover/land use data from object-based image classification of NAIP orthophotography.

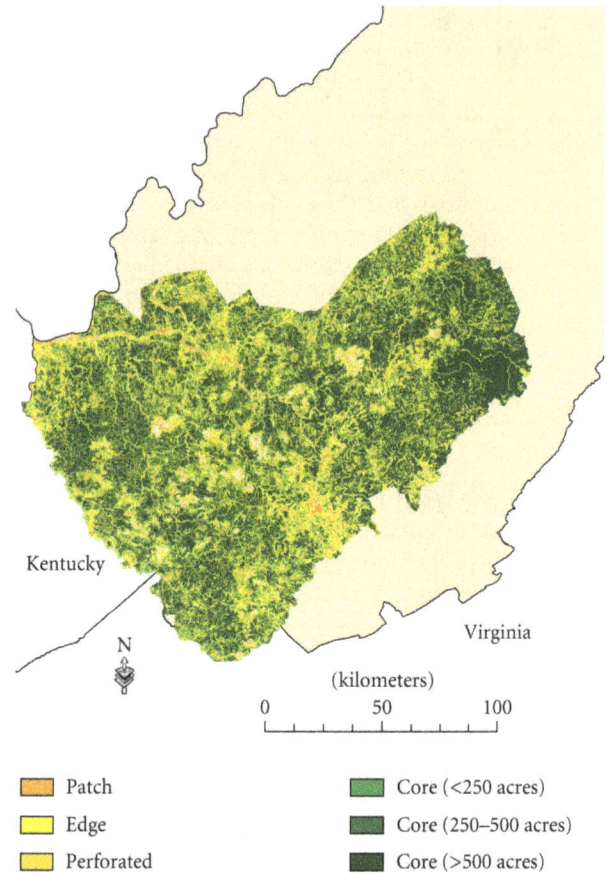

Patch

Edge

Perforated

Core (<250 acres)

Core (250–500 acres)

Core (>500 acres)

FIGURE 3: Fragmentation model after Vogt et al. [13].

To further differentiate the land cover, we incorporated ancillary GIS data. Surface mine permit data from the West Virginia Department of Environmental Protection (WVDEP) were utilized to extract barren and grassland land cover that resulted from surface mining. For example, if barren land cover existed within a permit, it was determined to be a result of mining. This method allowed us to obtain an estimate of the area that is currently permitted but is still forested, and this classification represents areas on the landscape that have a higher potential for future disturbance relative to areas outside of surface mine permits.

This feature extraction and decision rule method allowed for the development of a high-resolution land cover dataset for the region relative to leaf-on 2009 conditions. Figure 2 describes the resulting land cover.

4.2. Forest Fragmentation Model. The land cover data were reclassified as a binary forest/nonforest raster. The data were resampled using bilinear interpolation to a 9 m cell size to match the topographic data. Morphological image processing was applied to extract forest fragmentation data. The

methodology is described in detail by Vogt et al. [13]. Morphological image processing uses mathematical morphology to analyze shape and form of objects [13]. Forest pixels are classified as patch, edge, perforated, or core using this model. This method has been shown to provide a more accurate representation of fragmentation at the single pixel or landscape level when compared to image convolution [13]. Figure 3 shows the fragmentation model.

4.3. Defining Watershed Position. Headwater watersheds were produced at a 1:24,000 scale as described by Strager et al. [59]. The high-resolution National Hydrologic Dataset (NHD) stream segments were extracted and catchments delineated for all stream segments using DEM data. Photogrammetrical elevation data from 1:4,800 scale were interpolated to a 3 m and then 9 m scale. Stream segment pour points were compared to a flow direction grid to obtain the catchment areas.

The high-resolution NHD data have tabulated flow networks and connectivity information for each segment, to delineate headwater reaches. These are reaches that do not collect drainage from upstream. As a result, the headwater catchments in this project are 1:24,000 scale catchments derived by comparing high-resolution NHD data to DEMs and were defined as catchments collecting no upstream drainage.

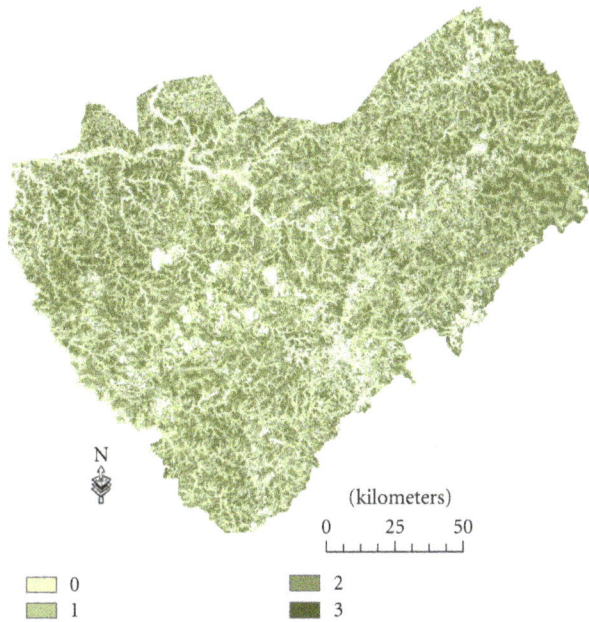

FIGURE 4: Critical forest model for Southern Coal Fields of West Virginia.

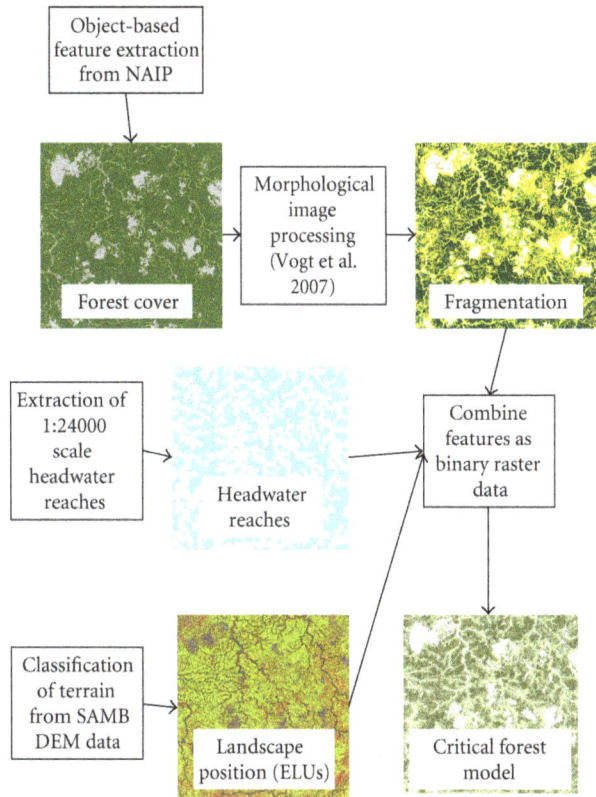

FIGURE 5: Flow chart for critical forest model creation.

4.4. Modeling Landscape Position. The 9 m DEM data for the region, resampled from 3 m, was used to classify the terrain into the following ecological land units (ELUs): cliff, steep slope, slope crest, upper slope, flat summit, sideslope, cove, dry flat, moist flat, wet flat, and slope bottom. The methodology described by Anderson et al. [60] provides a means to utilize DEM data to classify the landscape into different units of ecological significance. DEM derivatives were created including slope in degrees, a hydrologically filled DEM, flow direction, and flow accumulation within GIS. The slope and flow accumulation grids were utilized to calculate a moisture index. The moisture index is a relative measure of the moisture of a specific cell, and it assumes that the moisture level is a function of how much water flows into the cell, predicted by flow accumulation, and how fast the water can flow out, described by slope [61–66].

Landscape position was calculated following Fels and Zobel [67], by dividing the landscape into the following categories: ridge, wide ridge, and slope/flat, slope/cove. The approach is based on a local neighborhood analysis in which a cell is compared to its neighbors. This method is based on distance-weighted assessment of elevation values in which a cell is compared to the mean elevation of all values within search windows of different sizes and extents. Landscape position was combined with slope and moisture index data to derive the ELU classes.

Once the ELU data were created, certain classes were selected as critical for our model. This included upland areas (slope crest and flat summit) and coves. This provided two mutually exclusive positions upon the landscape that were considered critical for biodiversity.

4.5. Critical Forest Model. The critical forest model was created by combining the core forest, forest in headwater

catchments, cove forests, and upper slope forests. These layers were combined as binary raster layers to produce an index grid. Possible values ranged from 0 (not critical) to 3 (most critical). As the model was defined, cove and upper slope forests could not overlap, so the maximum index possible was 3. As a result, the final result of the analysis was an index grid that combines forest fragmentation, watershed position, and landscape position to describe forest critical for habitat and biodiversity at a 9 m pixel scale. The final output is shown in Figure 4. Figure 5 describes the overall process for the creation of the critical forest model.

Once the critical forest model was produced, we summarized the data relative to HUC (Hydrologic Unit Code) 8 and 11 catchments. We also assessed the data for spatial clustering using cluster and outlier analysis (Anselin Local Morans I) in GIS. Areas of surface mining, both active mining and reclamation, were converted to forested land cover, and the analysis was repeated using the potential premining land cover to analyze the distribution of premining critical forest habitat.

5. Results

5.1. Land Cover/Land Use and Forest Fragmentation. The land cover/use for the entire study area extent is summarized in Table 1. Although this landscape has been heavily surface mined (contributing 3% of the land cover/land use), forest is still the dominant cover type (86.1% of the land cover); however, 3.2% of the landscape has been permitted for mining,

TABLE 1: Land cover/land use data for study area.

Land cover/land use	Hectares	Percentage
Forested	1,892,774	82.9%
Grasslands/pastureland/agricultural land	172,316	7.5%
Barren/developed	54,770	2.4%
Forested in permit (not disturbed)	72,505	3.2%
Grasslands in permit (potentially reclaimed)	44,314	1.9%
Barren in permit (potentially active mining or not yet reclaimed)	25,621	1.1%
Open water	20,318	0.9%

TABLE 2: Land cover/land use data for the Coal River Watershed.

Land cover/land use	Hectares	Percentage of watershed
Forested	168,420	73.0%
Grasslands/pastureland/agricultural land	1,6420	7.1%
Barren/developed	4,140	1.8%
Forested in permit (not disturbed)	20,070	8.7%
Grasslands in permit (potentially reclaimed)	12,760	5.5%
Barren in permit (potentially active mining or not yet reclaimed)	7,640	3.3%
Open water	1,310	0.6%

TABLE 3: Critical forest index distributions.

Index	Study area		Forest in Permit	
	Hectares	Percentage	Hectares	Percentage
0	351,924	17.7%	14,406	21.5%
1	819,975	41.3%	35,768	53.3%
2	670,289	33.8%	15,305	22.8%
3	143,474	7.2%	1,596	2.4%

TABLE 4: Critical forest index distribution in mined areas.

Value	Hectares	Percentage
0	61,949.7	9.3%
1	222,677.1	33.4%
2	296,801.1	44.5%
3	84,903.3	12.7%

within large core extents, greater than 500 acres. Forested areas currently in permits were predicted as dominantly being edge and perforated forest explained possibly by the adjacency to active mine sites and reclamation or small forest stands surrounded by reclamation.

5.2. *Critical Forest Model.* Table 3 shows the distribution of critical forest index values throughout the region and only within the permitted forested areas. Based on visual interpretation of the results, the most critical areas as defined by the model are either cove or upper slope core forests within headwater reaches. As can be seen in Table 1, it was very common for a pixel to satisfy at least one of the criteria.

The data were summarized relative to HUC 8 and HUC 11 watershed that intersect the MTR/VF region and exist entirely within West Virginia. This data suggests that The Elk River HUC 8 watershed has the highest percentage of critical forest value pixels (values 1 through 3) within the MTR/VF region of West Virginia. At the HUC 11 scale, we found that local, spatial clusters of watersheds exist when the percentage of critical forest is considered.

Perhaps a more useful representation of spatial clustering for land management and regulation purposes would be the stand scale throughout the landscape as opposed to large catchment areas. We smoothed the raster critical forest model using majority filter operations within GIS then converted contiguous areas of a single critical forest value to polygon features, and a cluster and outlier analysis of these areas was performed relative to area of the polygons and critical forest value. This showed that clusters of critical forest areas exist across the landscape, especially when value 3 areas are of interest. Clusters of critical forest stands could represent areas that should be managed or protected in order to preserve biodiversity throughout the landscape and region.

Table 4 describes the distribution of critical forest values in areas of mine disturbance and it offers an estimate of the critical nature of mine disturbed areas based on the assumption that these areas were forested on the premining landscape, which we argue is a valid assumption based on comparisons to premining satellite data and the landscape position of the mines. Figure 6 offers a comparison of pre- and postmining conditions. This data suggests that critical forest areas overlap with areas of resource extraction. It has been documented that mining impacts headwater catchments and upland terrains, which are critical for biodiversity [14, 24, 35]. This data suggests that there is a need to delineate stands and clusters of critical forests habitats throughout the region because they are being deforested and impacted by resource extraction.

but is currently not disturbed (forested). As a result, mine expansion is likely to induce further land cover change and loss of forest over large extents.

The landscape disturbance is not evenly distributed throughout the region. Table 2 describes the land cover distribution for the Coal River HUC 8 watershed. According to our land cover calculations, 8.8% of the watershed is disturbed by active mining or reclamation while an additional 8.7% has been permitted but has not yet been disturbed or is still forested. The dominant land cover type in this watershed is forest; however, landscape disturbance due to surface mining is a dominant land use feature.

Of the forested land cover within the region, 51% was predicted as core forest using the Vogt et al. [13] model, 38.1% was classified as perforated, 9.2% as edge, and 1.2% as patch. Of the core forest, the majority of the pixels were

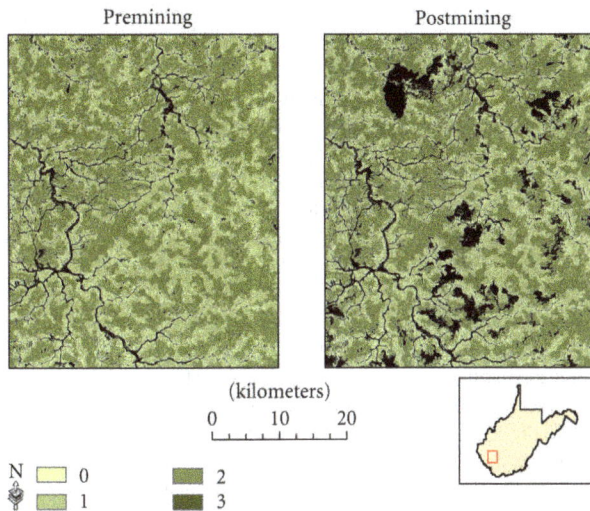

FIGURE 6: Pre- and postmining critical forest habitat.

6. Discussion

Our critical forest modeling is based on that all forest stands are not equal in terms of habitat suitability and supporting biodiversity. By taking into account high-resolution forest fragmentation patterns, watershed position, and landscape position, forest pixels were indexed relative to these factors that have an impact on habitat. Continued surface mining in the region will likely continue to alter forests, habitats, and the ecosystem and landscape as a whole, and this practice has been shown to produce both aquatic and terrestrial habitat degradation [24, 50]. Perhaps considering not just the extent of forest to be disturbed, but the ecological function of the forest, is necessary when continued mining and forest disturbance are proposed. Perhaps certain forests, such as core, headwater, cove forests, should be protected over other forested areas that are not as ecologically valuable. As resource extraction continues, perhaps regulators should take into account such abiotic factors of the landscape and adjust practices to minimize disturbance of critical areas. Perhaps spatial clusters of critical forest habitat should be considered during the permitting process.

We argue that a spatially explicit and high-resolution means of classifying or indexing critical forest stands is valuable as a regulatory tool for assessing the impact of continued mining and resource extraction in this region. McDermid et al. [68] suggest that remote sensing and GIS data that consider multiple attributes for modeling wildlife habitat are valuable, especially for large study areas. They also highlighted the use of land cover derived from object-based image classification, as was performed here. Their study focused on explaining patterns of grizzly bears habitat use within a study area in Alberta, Canada, and they found that incorporating data beyond categorical land cover was superior for their modeling purposes. Currently, there is a lack of guidelines for the selection and analysis of spatial data for exploring environmental and ecological questions [68]. We suggest that considering forest fragmentation (a derivative of land cover)

along with catchment and terrain-derived variables provided a better understanding of critical habitat in the Southern Coal Fields of West Virginia than could be obtained utilizing land cover alone.

Optimal spatial resolution of thematic map data for a region of this extent and for this mapping purpose is debatable. As an example, Landsat MSS data, offering a 60 m pixel, has been shown to be inadequate for urban mapping; also, as a general rule a pixel should be at least one-half the size of the shortest dimension of the smallest feature that is being mapped [69]. Our aim was to describe the forest cover at a stand level and to map disturbance features with small spatial extents that interrupt the canopy. As a result, a finer resolution dataset was required than what Landsat Multispectral Scanner (MSS) or Thematic Mapper (TM) data could offer; as a result, NAIP imagery and object-based classifiers were utilized to obtain the base thematic map data from which the forest cover and forest fragmentation were derived. Now that object-based image classification tools are available and high-resolution land cover and forest cover can be extracted from high-resolution and low spectral resolution imagery, forest cover data at the stand scale may become more widely available. We note that there is value in including such data in environmental modeling and assessment practices and combining that data with other datasets, such as topographic data, to model multiple attributes of the landscape.

Forest alteration resulting from mountaintop mining may not be characteristic of other human land use practices that result in deforestation, such as agriculture. Generally, forest cover change is most common in areas of low elevation and gentle terrain [70]. Mountaintop mining often induces land cover change in upland areas, headwater reaches, and rugged terrains [24, 35]. As a result, there is a need to understand the critical nature of these upland forests in order to assess the environmental impact of mountaintop mining, which produces unique landscape alteration.

Critical forest habitat modeling may also be useful in other landscapes impacted by resource extraction. For example, extraction of natural gas from the Marcellus Shale Formation throughout parts of New York, Pennsylvania, West Virginia, and Maryland in the eastern United States could contribute to habitat fragmentation and alteration of forest extents as access roads and well pads are established [71]. High-resolution land cover, fragmentation, watershed, and terrain data could be used to model the impact and predict critical habitats using similar methodology as proposed here.

7. Conclusion

We have presented a method for indexing of critical forest stands on a pixel-by-pixel basis relative to fragmentation patterns, watershed drainage, and landscape position. This methodology was implemented throughout the Southern Coal Fields of West Virginia and could be applied in other regions impacted by resource extraction. It relies on high-resolution spatially explicit forest cover from NAIP imagery using object-based feature extraction, creating a forest fragmentation model after Vogt et al. [13] utilizing morphological image analysis, defining headwater catchments at a 1 : 24,000 scale, and extracting ecological land units (ELUs) from DEM data. Once critical forest values are calculated, it is possible to identify spatial clusters of critical stands. This spatially explicit method for modeling forest habitat could be implemented as a tool for assessing the impact of resource extraction and aid in the conservation of critical habitat.

Future research should focus on incorporating additional landscape variables to model and differentiate forest stands. For example, canopy height data derived from LiDAR (light detection and ranging) could provide additional forest metrics for modeling. Additional pattern and connectivity analysis may provide for a better understanding of the network of critical forest stands throughout the region.

Our goal was to present a remote sensing and GIS method for indexing forest pixels throughout a disturbed landscape and was based on the assumption that all forests are not equal in terms of habitat. As resource extraction continues throughout the region, considering the nature of the forest to be disturbed may aid in mitigating environmental impacts of human land use.

References

[1] D. D. Chiras, J. P. Reganold, and O. S. Owen, *Natural Resource Conservation*, Prentice Hall, Upper Saddle River, NJ, USA, 8th edition, 2002.

[2] T. E. Lovejoy, R. O. Bierregaard, A. B. Rylands et al., "Edge and other effects of isolation on Amazonian forest fragmentation," in *Conservation Biology: The Science of Scarcity and Diversity*, M. Soule, Ed., pp. 257–285, Sinauer Associates, Sunderland, Mass, USA, 1986.

[3] R. R. Nemani and S. W. Running, "Satellite monitoring of global land cover changes and their impact on climate," *Climatic Change*, vol. 31, no. 2–4, pp. 395–413, 1995.

[4] R. A. Houghton, J. L. Hackler, and K. T. Lawrence, "The U.S. carbon budget: contributions from land-use change," *Science*, vol. 285, no. 5427, pp. 574–578, 1999.

[5] P. M. Vitousek, J. D. Aber, R. W. Howarth et al., "Human alteration of the global nitrogen cycle: sources and consequences," *Ecological Applications*, vol. 7, no. 3, pp. 737–750, 1997.

[6] S. R. Carpenter, N. F. Caraco, D. L. Correll, R. W. Howarth, A. N. Sharpley, and V. H. Smith, "Nonpoint pollution of surface waters with phosphorus and nitrogen," *Ecological Applications*, vol. 8, no. 3, pp. 559–568, 1998.

[7] R. E. Dickinson, "Global change and terrestrial hydrology—a review," *Tellus*, vol. 43, no. 4, pp. 176–181, 1991.

[8] J. Chow, R. J. Kopp, and P. R. Portney, "Energy resources and global development," *Science*, vol. 302, no. 5650, pp. 1528–1531, 2003.

[9] A. P. Dobson, A. D. Bradshaw, and A. J. M. Baker, "Hopes for the future: restoration ecology and conservation biology," *Science*, vol. 277, no. 5325, pp. 515–522, 1997.

[10] W. E. Westman, "Managing for biodiversity: unresolved science and policy questions," *BioScience*, vol. 40, no. 1, pp. 26–33, 1990.

[11] K. A. Poiani, B. D. Richter, M. G. Anderson, and H. E. Richter, "Biodiversity conservation at multiple scales: functional sites, landscapes, and networks," *BioScience*, vol. 50, no. 2, pp. 133–146, 2000.

[12] J. F. Franklin, "Preserving biodiversity: species, ecosystems, or landscape?" *Ecological Applications*, vol. 3, no. 2, pp. 202–205, 1993.

[13] P. Vogt, K. H. Riitters, C. Estreguil, J. Kozak, T. G. Wade, and J. D. Wickham, "Mapping spatial patterns with morphological image processing," *Landscape Ecology*, vol. 22, no. 2, pp. 171–177, 2007.

[14] J. D. Wickham, K. H. Riitters, T. G. Wade, M. Coan, and C. Homer, "The effect of Appalachian mountaintop mining on interior forest," *Landscape Ecology*, vol. 22, no. 2, pp. 179–187, 2007.

[15] K. C. Weathers, M. L. Cadenasso, and S. T. A. Pickett, "Forest edges as nutrient and pollutant concentrators: potential synergisms between fragmentation, forest canopies, and the atmosphere," *Conservation Biology*, vol. 15, no. 6, pp. 1506–1514, 2001.

[16] K. A. Harper, S. E. MacDonald, P. J. Burton et al., "Edge influence on forest structure and composition in fragmented landscapes," *Conservation Biology*, vol. 19, no. 3, pp. 768–782, 2005.

[17] M. K. Trzcinski, L. Fahrig, and G. Merriam, "Independent effects of forest cover and fragmentation on the distribution of forest breeding birds," *Ecological Applications*, vol. 9, no. 2, pp. 586–593, 1999.

[18] J. F. Kitchell, R. V. O'Neil, D. Webb et al., "Consumer regulation of nutrient cycling," *BioScience*, vol. 29, no. 1, pp. 28–34, 1979.

[19] P. Kareiva and G. Odell, "Swarms of predators exhibit "preytaxis" if individual predators use area-restricted search," *American Naturalist*, vol. 130, no. 2, pp. 233–270, 1987.

[20] J. Roland and P. D. Taylor, "Insect parasitoid species respond to forest structure at different spatial scales," *Nature*, vol. 386, no. 6626, pp. 710–713, 1997.

[21] T. B. Smith, R. K. Wayne, D. J. Girman, and M. W. Bruford, "A role for ecotones in generating rainforest biodiversity," *Science*, vol. 276, no. 5320, pp. 1855–1857, 1997.

[22] T. M. Donovan, P. W. Jones, E. M. Annand, and F. R. Thompson III, "Variation in local-scale edge effects: mechanisms and landscape context," *Ecology*, vol. 78, no. 7, pp. 2064–2075, 1997.

[23] J. L. Meyer, D. L. Strayer, J. B. Wallace, S. L. Eggert, G. S. Helfman, and N. E. Leonard, "The contribution of headwater streams to biodiversity in river networks," *Journal of the American Water Resources Association*, vol. 43, no. 1, pp. 86–103, 2007.

[24] E. S. Bernhardt and M. A. Palmer, "The environmental costs of mountaintop mining valley fill operations for aquatic ecosystems of the Central Appalachians," *Annals of the New York Academy of Sciences*, vol. 1223, no. 1, pp. 39–57, 2011.

[25] L. L. Master, S. R. Flack, and B. A. Stein, *Rivers of Life: Critical Watersheds for Protecting Freshwater Biodiversity*, The Nature Conservancy, Arlington, Va, USA, 1998.

[26] N. B. Green and T. K. Pauley, *Amphibians and Reptiles in West Virginia*, University of Pittsburgh Press, Pittsburgh, Pa, USA, 1987.

[27] A. Clarke, R. Mac Nally, N. Bond, and P. S. Lake, "Macroinvertebrate diversity in headwater streams: a review," *Freshwater Biology*, vol. 53, no. 9, pp. 1707–1721, 2008.

[28] P. D. Strausbaugh and E. L. Core, *Flora of West Virginia*, Seneca Books, Morgantown, WVa, USA, 1977.

[29] J. A. Coddington, L. H. Young, and F. A. Coyle, "Estimating spider species richness in a Southern Appalachian cove hardwood forest," *The Journal of Arachnology*, vol. 24, no. 2, pp. 111–128, 1996.

[30] F. A. Coyle, "Effects of clearcutting on the spider community of a Southern Appalachian forest," *The Journal of Arachnology*, vol. 9, pp. 285–298, 1981.

[31] M. G. Turner, S. M. Pearson, P. Bolstad, and D. N. Wear, "Effects of land-cover change on spatial pattern of forest communities in the Southern Appalachian Mountains (USA)," *Landscape Ecology*, vol. 18, no. 5, pp. 449–464, 2003.

[32] S. M. Pearson, A. B. Smith, and M. G. Turner, "Forest patch size, land use, and mesic forest herbs in the French Broad River Basin, North Carolina," *Castanea*, vol. 63, no. 3, pp. 382–395, 1998.

[33] C. E. Mitchell, M. G. Turner, and S. M. Pearson, "Effects of historical land use and forest patch size on myrmechocores and ant communities in the Southern Appalachian Highlands (USA)," *Ecological Applications*, vol. 12, pp. 1364–1377, 2002.

[34] C. A. Weakland and P. B. Wood, "Cerulean warbler (*Dendroica cerulea*) microhabitat and landscape-level habitat characteristics in Southern West Virginia," *The Auk*, vol. 122, no. 2, pp. 497–508, 2005.

[35] P. B. Wood, S. B. Bosworth, and R. Dettmers, "Cerulean warbler abundance and occurrence relative to large-scale edge and habitat characteristics," *The Condor*, vol. 108, no. 1, pp. 154–165, 2006.

[36] P. B. Hamel, D. K. Dawson, and P. D. Keyser, "How we can learn more about the cerulean warbler (*Dendroica cerulea*)," *The Auk*, vol. 121, no. 1, pp. 7–14, 2004.

[37] J. Jones, R. D. DeBruyn, J. J. Barg, and R. J. Robertson, "Assessing the effects of natural disturbance on a neotropical migrant songbird," *Ecology*, vol. 82, no. 9, pp. 2628–2635, 2001.

[38] P. A. Keddy and C. G. Drummond, "Ecological properties for the evaluation, management, and restoration of temperate deciduous forest ecosystems," *Ecological Applications*, vol. 6, no. 3, pp. 748–762, 1996.

[39] B. A. Stein, L. S. Kutner, and J. S. Adams, *Precious Heritage: The Status of Biodiversity in the United States*, Oxford University Press, New York, NY, USA, 2000.

[40] P. A. Delcourt and H. R. Delcourt, "Paleoecological insights on conservation of biodiversity: a focus on species, ecosystems, and landscapes," *Ecological Applications*, vol. 8, no. 4, pp. 921–934, 1998.

[41] SAMAB (Southern Appalachian Man and the Biosphere), "The Southern Appalachian assessment terrestrial," Tech. Rep. Report 5 of 5, US Department of Agriculture, Forest Service, Atlanta, Ga, USA, 2008, http://www.samab.org/saa/saa_reports.html.

[42] S. K. Robinson, F. R. Thompson III, T. M. Donovan, D. R. Whitehead, and J. Faaborg, "Regional forest fragmentation and the nesting success of migratory birds," *Science*, vol. 267, no. 5206, pp. 1987–1990, 1995.

[43] L. Fahrig, "Effect of habitat fragmentation on the extinction threshold: a synthesis," *Ecological Applications*, vol. 12, no. 2, pp. 346–353, 2002.

[44] D. J. Mladenoff, M. A. White, J. Pastor, and T. R. Crow, "Comparing spatial pattern in unaltered old-growth and disturbed forest landscapes," *Ecological Applications*, vol. 3, no. 2, pp. 294–306, 1993.

[45] K. H. Riitters, J. D. Wickham, R. V. O'Neill et al., "Fragmentation of continental United States forests," *Ecosystems*, vol. 5, no. 8, pp. 815–822, 2002.

[46] R. L. Hooke, "Spatial distribution of human geomorphic activity in the United States: comparison with rivers," *Earth Surface Processes and Landforms*, vol. 24, no. 8, pp. 687–692, 1999.

[47] K. L. Saylor, "Land Cover Trends Project: Central Appalachians," U.S. Department of the Interior, U.S. Geological Survey, Washington, DC, USA, 2008, http://landcovertrends.usgs.gov/east/eco69Report.html.

[48] M. A. Drummond and T. R. Loveland, "Land-use pressure and a transition to forest-cover loss in the Eastern United States," *BioScience*, vol. 60, no. 4, pp. 286–298, 2010.

[49] P. A. Townsend, D. P. Helmers, C. C. Kingdon, B. E. McNeil, K. M. de Beurs, and K. N. Eshleman, "Changes in the extent of surface mining and reclamation in the Central Appalachians detected using a 1976–2006 Landsat time series," *Remote Sensing of Environment*, vol. 113, no. 1, pp. 62–72, 2009.

[50] M. A. Palmer, E. S. Bernhardt, W. H. Schlesinger et al., "Mountaintop mining consequences," *Science*, vol. 327, no. 5962, pp. 148–149, 2010.

[51] J. A. Simmons, W. S. Currie, K. N. Eshleman et al., "Forest to reclaimed mine land use change leads to altered ecosystem structure and function," *Ecological Applications*, vol. 18, no. 1, pp. 104–118, 2008.

[52] WVGES (West Virginia Geologic and Economic Survey), "Physiographic provinces of West Virginia," 2005, http://www.wvgs.wvnet.edu/www/maps/pprovinces.htm.

[53] D. Opitz, "An automated change detection system for specific features," in *Proceedings of the International ESRI User Conference*, Environmental Systems Research Institute, San Diego, Calif, USA, July 2003.

[54] M. A. Friedl and C. E. Brodley, "Decision tree classification of land cover from remotely sensed data," *Remote Sensing of Environment*, vol. 61, no. 3, pp. 399–409, 1997.

[55] N. R. Harvey, J. Theiler, S. P. Brumby et al., "Comparison of GENIE and conventional supervised classifiers for multispectral image feature extraction," *IEEE Transactions on Geoscience and Remote Sensing*, vol. 40, no. 2, pp. 393–404, 2002.

[56] G. Hong, Y. Zhang, and D. A. Lavigne, "Object-based change detection in high resolution image," in *Proceedings of the American Society for Photogrammetry and Remote Sensing Annual Conference (ASPRS '06)*, American Society for Photogrammetry and Remote Sensing, Reno, Nev, USA, May 2006.

[57] J. V. Kaiser Jr., D. A. Stow, and L. Cao, "Evaluation of remote sensing techniques for mapping transborder trails," *Photogrammetric Engineering and Remote Sensing*, vol. 70, no. 12, pp. 1441–1447, 2004.

[58] Visual Learning Systems, "User Manual, Feature Analyst 4.2 for ERDAS Imagine," Visual Learning Systems, Missoula, Mont, USA, p. 354, 2002, http://www.featureanalyst.com/feature_analyst/publications/manuals/FA_4.2_RELEASE/FA_4.2_Reference_arc_040908_RELEASE.pdf.

[59] M. P. Strager, J. T. Petty, J. M. Strager, and J. Barker-Fulton, "A spatially explicit framework for quantifying downstream hydrologic conditions," *Journal of Environmental Management*, vol. 90, no. 5, pp. 1854–1861, 2009.

[60] M. G. Anderson, M. D. Merrill, and F. B. Biasi, *Connecticut River Watershed Analysis: Ecological Communities and Neo-Tropical Migratory Birds (Final Report)*, Eastern Conservation Science, The Nature Conservancy, 1998.

[61] R. B. Grayson, I. D. Moore, and T. A. McMahon, "Physically based hydrologic modeling 1. A terrain-based model for investigative purposes," *Water Resources Research*, vol. 28, no. 10, pp. 2639–2658, 1992.

[62] H. Mitasova, J. Hofierka, M. Zlocha, and L. R. Iverson, "Modelling topographic potential for erosion and deposition using GIS," *International Journal of Geographical Information Systems*, vol. 10, no. 5, pp. 629–641, 1996.

[63] D. M. Moore, B. G. Lees, and S. M. Davey, "A new method for predicting vegetation distributions using decision tree analysis in a geographic information system," *Environmental Management*, vol. 15, no. 1, pp. 59–71, 1991.

[64] M. Boer, G. Del Barrio, and J. Puigdefabregas, "Mapping soil depth classes in dry Mediterranean areas using terrain attributes derived from a digital elevation model," *Geoderma*, vol. 72, no. 1-2, pp. 99–118, 1996.

[65] E. M. O'Loughlin, "Prediction of surface saturation zones in natural catchments by topographic analysis," *Water Resources Research*, vol. 22, no. 5, pp. 794–804, 1986.

[66] A. J. Parker, "The topographic relative moisture index: an approach to soil- moisture assessment in mountain terrain," *Physical Geography*, vol. 3, no. 2, pp. 160–168, 1982.

[67] J. Fels and R. Zobel, "Landscape position and classified land-type mapping for the statewide DRASTIC mapping project," Tech. Rep. VEL.95.1, North Carolina State University, North Carolina Department of Environmental, Health, and Natural Resources, Division of Environmental Management, 1995.

[68] G. J. McDermid, R. J. Hall, G. A. Sanchez-Azofeifa et al., "Remote sensing and forest inventory for wildlife habitat assessment," *Forest Ecology and Management*, vol. 257, no. 11, pp. 2262–2269, 2009.

[69] J. R. Jensen, *Introductory Digital Image Processing: A Remote Sensing Perspective*, Pearson Prentice Hall, Upper Saddle River, NJ, USA, 3rd edition, 2005.

[70] D. N. Wear and R. O. Flamm, "Public and private forest disturbance regimes in the Southern Appalachians," *Natural Resource Modeling*, vol. 7, no. 4, pp. 379–397, 1993.

[71] D. J. Soeder, "The marcellus shale: resources and reservations," *EOS Transactions American Geophysical Union*, vol. 91, no. 32, pp. 277–288, 2010.

12

Use of Host-Plant Trait Space by Phytophagous Insects during Host-Associated Differentiation: The Gape-and-Pinch Model

Stephen B. Heard

Department of Biology, University of New Brunswick, P.O. Box 4400, Fredericton, NB, Canada E3B 5A3

Correspondence should be addressed to Stephen B. Heard, sheard@unb.ca

Academic Editor: Andrew Hendry

Ecological speciation via host shifting has contributed to the astonishing diversity of phytophagous insects. The importance for host shifting of trait differences between alternative host plants is well established, but much less is known about trait variation *within* hosts. I outline a conceptual model, the "gape-and-pinch" (GAP) model, of insect response to host-plant trait variation during host shifting and host-associated differentiation. I offer four hypotheses about insect use of plant trait variation on two alternative hosts, for insects at different stages of host-associated differentiation. Collectively, these hypotheses suggest that insect responses to plant trait variation can favour or oppose critical steps in herbivore diversification. I provide statistical tools for analysing herbivore trait-space use, demonstrate their application for four herbivores of the goldenrods *Solidago altissima* and *S. gigantea*, and discuss their broader potential to advance our understanding of diet breadth and ecological speciation in phytophagous insects.

1. Introduction

The insects have long been held up as providing spectacular examples of rapid diversification and high standing diversity (e.g., [1–3]). Among insects, phytophagous clades often undergo dramatic radiations [4], and phytophagous lineages tend to be more diverse than their nonphytophagous sisters [5, 6]. One likely driver of diversification among phytophagous insects is their tendency to specialize on host-plant species or organs [7–10] and to diversify via host or organ shifts followed by host-associated differentiation (HAD), the evolution of new specialist races or species [9, 11–14]. Because many cases of HAD appear to have proceeded in sympatry [15], a great deal of theoretical and empirical work has focused on understanding ways in which adaptation to different host plants can impose disruptive selection on nascent specialist forms and also reduce gene flow (or permit differentiation in the face of gene flow) between those forms [14–16]. Phytophagous insects, along with parasitoids [17], freshwater fishes [18], seed-eating birds [19], and habitat-specialist plants [20] and lizards [21] have therefore been central to the development of ideas about ecological speciation [22].

A common theme among case studies of ecological speciation is the existence of two alternative niches—microhabitats, resources, reproductive strategies, and so forth—that can be exploited by individuals of a single species, with the potential for disruptive selection to operate between the alternative niches. For phytophagous insects, the alternative niches are a pair of host plant species (or organs). One commonly imagines an evolutionary sequence beginning with an insect exploiting only one of the two alternative hosts. Perhaps via host-choice errors, some individuals occasionally attack individuals of the second host, and if fitness penalties for doing so are not too severe, a host shift occurs and the insect begins to exploit both alternative hosts. (Description of these events as "errors" is standard in the plant-insect literature, but of course this usage is teleological shorthand and can conceal interesting biology. For instance, it might be that genotypes with strong enough host preferences to avoid "errors" would also show costly rejection of some suitable hosts; in this case, the occurrence of host-choice "errors" is simply an adaptive compromise. Nonetheless, for simplicity I retain the standard usage here). Disruptive selection can now begin to favour genotypes better adapted to each alternative host. If reproductive isolation arises between nascent forms, then ecological speciation can proceed, and a single (perhaps polymorphic) generalist is replaced with a pair of host-specialist races or species. Because reproductive isolation is expected to take some time to evolve, if it can evolve at all, different insects exploiting a pair of alternative host plants are expected to fall on a continuum from generalists to nascent, poorly differentiated host forms to distinct host-specialist sister species [13, 23]. There will be analogous continua for ecological speciation across other kinds of alternative niches, for instance, in parasitoids speciating across hosts or fish across depth niches (e.g., [17, 24]).

The process of HAD in phytophagous insects has been widely discussed, both in general [14] and in the context of a few well-studied model systems (e.g., apple maggot fly [25, 26], goldenrod ball-gall fly [27, 28]). Perhaps unsurprisingly, nearly all studies of HAD have emphasized insect responses to differences in plant traits between the alternative hosts, while downplaying variation in plant traits among individuals within each host. Such an interspecific perspective is obviously appropriate for studies using population-genetic tools to detect host-associated forms and reconstruct their history (e.g., [13, 29–31]), but it is also near universal in studies discussing ecological mechanisms by which host shifts and HAD proceed (e.g., [25, 26, 28, 32–38]). An alternative approach would explicitly recognize within-species variation in host-plant traits and consider possible roles for such variation in favouring or impeding host shifting and HAD. This approach has yet to be applied in earnest to any system, but intriguing hints at its usefulness appear in the literature for the goldenrod ball-gall fly, *Eurosta solidaginis*, and its races on the goldenrods *Solidago altissima* and *S. gigantea*. For example, *Eurosta* of the *S. altissima* race prefer the largest ramets of their host [39], and since *S. gigantea* plants tend to be shorter when *Eurosta* oviposits [40], this preference might discourage host-choice errors by *altissima* flies. In contrast, if *gigantea* flies similarly prefer taller ramets, they could be susceptible to host-choice errors (although *gigantea* flies' preferences have not been assessed, and neither hypothesis raised here appears to have been tested). Work on the phenology of insect emergence and host-plant growth has similar implications. *Eurosta* adults emerge from *S. gigantea* earlier than from *S. altissima* [41], and this pattern is correlated with availability of rapidly growing ramets of each host to be attacked [40]. Thus, individual *S. altissima* ramets with earlier phenology, or *S. gigantea* ramets with later phenology, might be more likely to be attacked by the "wrong" host race. There is geographic variation in the abundance of such intermediate-phenology ramets, and How et al. [40] suggested that host shifts might be more easily initiated where host phenology overlaps more extensively.

Even for *Eurosta*, however, there are few plant traits for which insect responses have been studied on *both* alternative hosts, and so we know little about how insect responses to plant trait variation might relate to the ecology of host shifting and HAD. Furthermore, there is no system for which we can compare insect responses to plant trait variation for a set of insects attacking the same plants but differing in stage of host shifting and HAD. I outline here a conceptual model of host trait-space use during host shifting and HAD, along with a statistical approach for investigating trait-space use in phytophagous insects. I suggest hypotheses for temporal changes in host trait-space use over evolutionary time, from initial host-choice errors through to the independent evolution of a pair of well-isolated host-specialist sibling species. I call the overall model the "gape-and-pinch," or "GAP," model of trait-space use (the reason for this name will be apparent after the model is described). While I outline the model for plants and phytophagous insects, it will apply to many other systems with some straightforward vocabulary substitutions.

2. Conceptual "GAP" Model of Host Trait-Space Use

All plant species vary intraspecifically for numerous morphological, phenological, and chemical traits, with variation having genetic, epigenetic, and/or environmental causes (e.g., [42–45]). This variation defines a set of phenotypes that are available for attack by a phytophagous insect searching its environment for suitable hosts. This set of phenotypes can be depicted as a cloud of points in a multidimensional trait space, with each point representing an individual plant (or a ramet, for clonal plants; or even a module, when important variation occurs within individuals [46]). It is convenient to consider the two-dimensional case (Figure 1), which can represent either a system in which two plant traits show variation relevant to insect attack or a two-dimensional summary of a higher-dimensional trait space (using principal components to extract two dominant axes of trait variation). I use the term "available trait space" to describe this cloud of points, as combinations of plant traits falling inside it are available to attacking herbivores, while combinations outside are not available (i.e., they do not correspond to real plants which might be attacked). This trait space can be characterized by calculating its centroid (the point whose coordinate on each axis is the mean value of that coordinate for all individuals), its size (average distance from individual plants to the centroid), and its shape.

Consider first an insect interacting with a single host species. Some plant individuals will be attacked, but others will likely escape. The attacked individuals define the "attacked trait space" (filled circles in Figure 1), which must be a subset of the available trait space (and might be expected to be a smaller subset for more specialist herbivores [47, 48]). The relationship between available and attacked trait spaces will depend on active host-selection behavior by insects (insect preference), and also on whether insects can survive on an individual plant after initiating attack (insect performance). Both preference and performance will depend on plant traits—sometimes the same traits, but sometimes not. For simplicity, I use "herbivore attack" to denote the occurrence of feeding herbivores on plants, whether patterns in occurrence arise from preference or performance, and terms like "selective" attack should similarly be taken to include both preference and performance effects. The attacked trait space may represent a common preference by all herbivore individuals, or the sum of herbivore individuals' distinct preferences in species with strong individual specialization [49, 50].

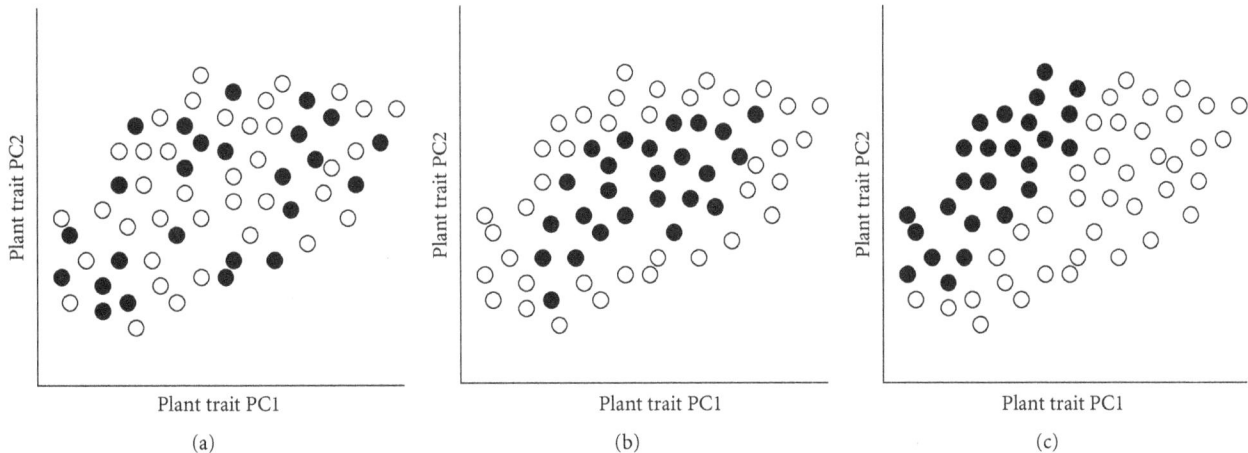

FIGURE 1: Possible patterns of host trait-space use by herbivorous insects. Available host-plant individuals form a cloud in trait space (which is likely multidimensional, but summarized here by the first two principal component axes). Open circles denote unattacked plants, and filled circles attacked ones.

When herbivore attack is random with respect to plant traits (Figure 1(a)), the attacked trait space will resemble the available trait space in centroid location, shape, and (once corrected for the smaller number of attacked plants) size. For a selective herbivore (one that rejects some available plants), in contrast, the attacked and available trait spaces will differ. Many patterns are possible, but two are particularly likely. First, the herbivore might attack typical plants (those with trait values near the population means) and reject extreme ones, leading to an attacked trait space that is central with respect to the available trait space: the attacked space is smaller than the available space, but the two spaces have similar centroids (Figure 1(b)). Such a pattern might be favoured by selection because (for example) typical plants are most common, and insects preferring them pay lower search costs and experience greater resource availability. Alternatively, herbivore attack might be associated with extreme trait values (e.g., herbivores might perform best on the largest or least-defended individuals), leading to an attacked trait space that is marginal with respect to the available trait space: the attacked trait space is again restricted in size, but in this case the two spaces have different centroids (Figure 1(c)).

Now consider a pair of plant species available for attack (Extension to larger numbers of hosts is possibly but considerably complicating.) Given a common set of measured traits, we should see two clouds of points in trait space (Figure 2) defining a pair of available trait spaces. The distance between available trait spaces defined by different plant species might be large compared to the size of each available trait space [38], but, for closely related pairs of phenotypically variable plants, this need not be so (e.g., for *Solidago altissima* and *S. gigantea*, see Figure 3). Two available trait spaces could even be touching or interdigitated, especially for hybrid swarms [51]. One can again consider attacked trait spaces in comparison to available trait spaces on each host, but now there are many more possibilities, as each attacked trait space could be nonselective, central, or marginal (and if marginal, toward or away from the other host). Among possible patterns, I emphasize here a set of

trait-space relationships predicted for an insect herbivore moving through a four-step evolutionary sequence: from original specialization on one of the two hosts, through a host shift, to early and late stages of HAD.

2.1. Stage 1: Single-Host Specialists and the Importance of Host-Choice Errors.

An insect attacking a single host could show virtually any pattern in the relationship between attacked and available trait spaces, but some patterns are of special interest in the context of possible host-shifting to an evolutionarily novel host. (By a "host shift" I mean the addition of a novel plant to the herbivore species' diet, which will normally occur without immediate abandonment of the old). Such host shifts are likely to begin when a few individuals attack the "wrong" (novel) host, making it possible for selection to favour the incorporation of the novel host into the insect's host range. Importantly, the likelihood of host-choice errors is likely to depend on the insect's use of plant trait space. In particular, imagine an insect showing a marginal attacked trait space on the ancestral host. That marginal attacked trait space could be adjacent to the available trait space defined by the novel host (Figure 2(a)), or could be distant from it. When it is adjacent, host-choice errors are more likely and insects making those errors are more likely to survive on the novel host [52]. In contrast, when the ancestrally attacked trait space is distant from the novel host, host-choice errors (and thus host shifting) should be less likely. I call this the "adjacent errors hypothesis." The logic mirrors the widespread expectation that host-choice errors and host shifts are more likely between species that resemble each other morphologically, chemically, or phylogenetically [38] but stresses that the distance in trait space that needs to be crossed for a host-choice error depends not only on the distance between available trait spaces but also on how insect preference and performance define the attacked trait space.

2.2. Stage 2: Oligophagous Feeding Following Diet Expansion.

Following a host shift that expands diet, our focal insect

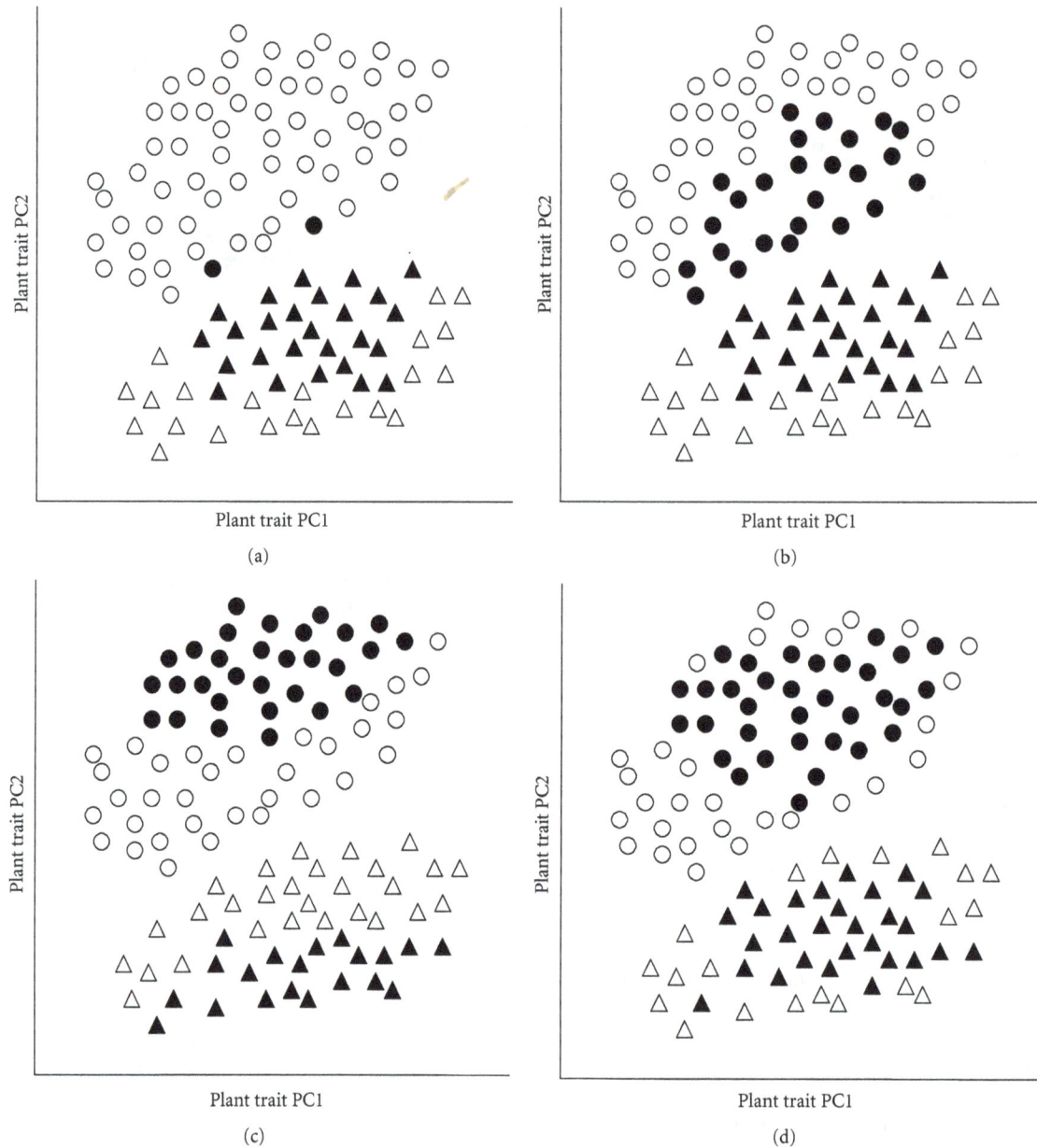

FIGURE 2: Some possible relationships between attacked trait spaces on two alternative host plants. Illustrated are hypothetical relationships for four stages during a host shift and subsequent host-associated differentiation. (a) Specialist on host 1, predisposed to host-choice errors due to marginal attack on trait space adjacent to the alternative host. (b) Following diet expansion, herbivore is oligophagous accepting both hosts; selection favours central attack on the combined trait space. (c) Early evolution of host-associated differentiation: use of marginal and distant trait spaces on the two hosts reduces host-choice errors and hence gene flow. (d) Reproductively isolated monophagous specialists on each host: selection favours central attack by each specialist on the trait space of its host. See text for further discussion of these scenarios.

species will be oligophagous, feeding on two hosts (ancestral plus novel) rather than one. Because this stage should follow from patterns in attack allowing host-choice errors (adjacent errors hypothesis), attacked trait space is likely to remain marginal on the ancestral host (Figure 2(a)). As host preference and performance evolve to include attack on the novel host, the attacked trait space on that host is likely to be marginal as well, but with the two attacked trait spaces adjacent (Figure 2(b)) because novel plants closer to

the ancestrally attacked trait space are more easily colonized. I call this the "adjacent oligophagy hypothesis."

Note that the adjacency pattern is equivalent to restricted but central use of an available trait space defined by the two hosts in combination (compare Figures 1(b) and 2(b)). If disruptive selection between alternative hosts does not act or is not powerful, oligophagy and the adjacency pattern could be evolutionarily persistent. Alternatively, this stage might be transient, persisting only until disruptive selection has time to drive HAD of insect subpopulations exploiting the two hosts. The contrast between these possibilities highlights an important fork in the evolutionary road [53], in which disruptive selection favouring HAD is or is not sufficient to overcome gene flow working to homogenize the herbivore population and to maintain an oligophagous diet.

2.3. Stage 3: Nascent Host-Specialist Forms and the Selection-Gene Flow Tension.

How might insect trait-space use favour or oppose the ability of disruptive selection to achieve HAD? Craig et al. [52] argued that persistent oligophagy is likely when the two available trait-spaces are very close, with HAD likely when they are more distant. However, their conceptual model assumes that attacked trait spaces on the two hosts remain indefinitely adjacent (their Figure 1). I suggest that there is another important possibility; a critical step in HAD may be the separation of the attacked trait spaces (Figure 2(c)) such that insect subpopulations on the two hosts (now appropriately thought of as nascent host forms) come to attack dissimilar individuals rather than similar ones. I call this the *"trait distance-divergence hypothesis"*. Separation of the two attacked trait spaces could arise in two different but complementary ways.

First, distance between attacked trait spaces could arise simply because disruptive selection for adaptation to the alternative hosts overpowers the homogenizing effect of gene flow between nascent host forms. Under this scenario, the attacked trait spaces could move to opposite ends of the available trait spaces, as in Figure 1(c), or merely further apart than adjacency, depending on the shape of the fitness landscape—that is, fitness optima on the alternative hosts might favour central or marginal trait spaces. Under this scenario, distance between attacked trait spaces is just a symptom by which the progress of HAD can be recognized.

Second, distance could be a product of selection to minimize host-choice errors by each nascent host form or hybridization between them [22]. Host-choice errors could be opposed by selection because they lead to preference-performance mismatches, or because they put larvae in competition with members of the other host form (encouraging divergence by character displacement). Alternatively, host-choice errors coupled with the tendency for phytophagous insects to mate on their host plants could lead to hybrid matings. Hybrid disadvantage is possible given tradeoffs in ability to exploit the alternative hosts, or if hybrids prefer or are best suited for trait-value combinations falling in the gap between the two available trait spaces (this gap is shown narrow in Figure 2 but will often be wider [38]). Selection to reduce hybridization by widening the distance between attacked trait spaces (Figure 1(c)) would be a form of reinforcement [54]. Under the reinforcement scenario, distance between attacked trait spaces is more than a symptom of HAD; once achieved, it serves to reduce gene flow between nascent host forms and permit HAD to progress. (In passing, I note that if selection simply overpowers gene flow, we would expect HAD to involve genetic divergence in genomic islands, whereas if selection opposes hybridization, genome-wide genetic divergence should result via "isolation by adaptation") [55–57].

In summary, the trait distance-divergence hypothesis holds that attainment of distance between attacked trait spaces (Figure 1(c)) can be both a symptom of HAD and also a factor permitting HAD. The larger the distance between attacked trait spaces, the more likely is the evolution of genetic differentiation between insects on the two hosts. Since genetic differentiation can ease the evolution of distance between attacked trait spaces, this stage of evolution can involve positive feedback [58–60]. In contrast, an insect for which attacked trait spaces remain adjacent (Figure 1(b)) is likely to remain an oligophagous insect with no host-associated structure to its gene pool.

2.4. Stage 4: Pair of Established Host Specialists.

As HAD proceeds and gene flow between nascent host forms declines, we would expect the gradual accumulation of more, and more effective, reproductive isolating mechanisms [24, 61, 62]. This should continue until ecological speciation is complete, and the two host-specialist forms attain the status of full biological species. As reproductive isolation becomes enforced by multiple, redundant mechanisms, the importance of separation between attacked trait spaces should decline. Selection will then be free to mould trait-space use independently for each species, and if reinforcement earlier in HAD pushed the attacked trait spaces apart (trait distance-divergence hypothesis), this force can now relax. If selection favours use of central trait space on each host, for example (Figure 2(d); or if it favours nonselective use of trait space on each host), the distance between the attacked trait spaces should decrease. I call this the *"distance relaxation hypothesis."*

Note that through the temporal sequence (Figures 2(a)–2(d)), the overall conceptual model suggests a pair of attacked trait spaces that begin close together, move apart, and then move back together like pincers. This movement underlies the terminology "gape-and-pinch" model of trait-space use.

2.5. What about Generalists?

The foregoing considered insects that begin as monophagous on one of the two hosts and remain narrowly oligophagous or monophagous at all stages of HAD. However, many herbivores are broader generalists [63] for which we would not expect any of the trait-space patterns shown in Figure 2. In particular, it would be very surprising if a broad generalist showed nonrandom separation between attacked trait spaces (Figures 2(c) and 2(d)). Instead, attacked trait space might be nonselective on both hosts [47, 48], or restricted but marginal along a trait axis orthogonal to the difference between the two hosts (e.g., insects might prefer larger individuals of each host and also attack other, larger species). Such broad generalists are much less likely than host specialists to undergo HAD because (being already adapted to multiple hosts) they are less likely to experience strongly disruptive selection for performance on one host versus another [64].

2.6. Testing the Hypotheses.

The four hypotheses that make up the GAP model are logically distinct; finding that one

TABLE 1: Relationships between patterns in trait-space use (Figures 2(a)–2(d)), the GAP model, and statistical tests implemented for analysis of attacked trait spaces.

Pattern in trait-space use	Hypothesis	Attacked trait spaces marginal?[1]	Attacked trait spaces restricted?[1]	Attacked trait spaces distant?
Monophagous, attacked trait space marginal and adjacent to alternative host (Figure 2(a))	Adjacency favours host shifting (adjacent errors hypothesis)	Ancestral host: marginal Novel host: marginal but rare[2]	—	Attacked spaces close[2]
Oligophagous, attacked trait spaces marginal and adjacent (Figure 2(b))	Adjacency persists after host shifting (adjacent oligophagy hypothesis)	Ancestral host: marginal Novel host: marginal	—	Attacked spaces close
Nascent host races, attacked trait spaces marginal and distant (Figure 2(c))	Distance permits, and is also symptomatic of, genetic isolation (trait distance-divergence hypothesis)	Ancestral host: marginal Novel host: marginal	—	Attacked spaces distant
Pair of monophagous species, attacked trait spaces central on each host (Figure 2(d))	Other isolating mechanisms reduce importance of trait-space distance (distance relaxation hypothesis)	Ancestral host: not marginal Novel host: not marginal	Ancestral host: restricted (central) or nonselective Novel host: restricted (central) or nonselective	Attacked spaces neither close nor distant

[1] More strictly, only restricted or marginal trait-space use along the PC axis (or axes) defining the difference between available hosts is directly relevant to the GAP model.
[2] But statistical detection is difficult because attack on the novel host is rare.

hypothesis holds (or fails) implies nothing about the others. For example, for a given herbivore, the adjacent errors and adjacent oligophagy hypotheses could hold, but the distance-divergence and distance relaxation hypotheses fail, if HAD proceeds to ecological speciation without any movement of attacked trait spaces away from each other following the host shift.

Each of the four hypotheses can also be posed, and tested, at two levels. First, we can test each hypothesis for a single herbivore. For instance, are attacked trait spaces adjacent on *Solidago altissima* and *S. gigantea* for the narrowly oligophagous [13] gallmaker *Epiblema scudderiana* (adjacent oligophagy hypothesis)? Of course, such tests focus on patterns, and confirmation of a pattern need not constitute a strong test of underlying mechanism. Second, and more powerfully, we can test each hypothesis for herbivorous insects as a class. For instance, are attacked trait spaces (statistically) further apart for recently divergent and incompletely isolated pairs of host races than they are for more ancient specialist species pairs (distance relaxation hypothesis)? At this level, the hypotheses can hold strongly or weakly (or not at all); that is, the empirical relationship between trait-space distance and extent of reproductive isolation could be stronger or weaker (or non-significant).

Testing the adjacent errors, adjacent oligophagy, trait distance-divergence, and/or distance relaxation hypotheses for individual herbivores will require trait-space use data for large numbers of individuals on the alternative hosts. Some tests will be difficult at the individual-herbivore level (e.g., testing the distance relaxation hypothesis for an individual herbivore would require historical data on past trait-space use, which will only rarely be available). Ultimately, though,

assessment of the GAP model depends more on comparative tests of the hypotheses for herbivores as a class, and this will require data on host trait-space use for multiple insect herbivores differing in host range. A particularly revealing approach will involve sets of herbivores that differ in the extent of their progression through HAD (e.g., [13]), because of the expectation that the trait-space relationships shown in Figures 2(a)–2(d) form a temporal evolutionary sequence. The cleanest comparative tests will involve herbivores sharing a common pair of alternative hosts, so that attacked trait spaces can be contrasted among herbivores (e.g., recent host forms and ancient specialist pairs, for the distance relaxation hypothesis) while seen against the simple backdrop of a common pair of available trait spaces.

Progress towards understanding the influence of trait-space use on the evolutionary trajectory of herbivore specialization, then, can be made by measuring for multiple insect herbivores the relationships between available and attacked trait spaces on the alternative hosts, for comparison with those suggested by the four hypotheses (summarized in Table 1). In particular, we will be interested in whether attacked trait space on each host is marginal (and in what direction), and whether the two attacked trait spaces are adjacent or distant.

Unfortunately, we do not yet have sufficient comparative data to test the GAP model. Members of my laboratory are beginning to gather such data for insect herbivores attacking the goldenrods *Solidago altissima* and *S. gigantea*. In following sections, I provide formal statistical tools for analysis of such data and apply them to an illustrative data set demonstrating a path towards the comparative hypothesis tests that are our ultimate goal.

3. Statistical Methods

I developed statistical methods to test for three patterns in host trait-space use by insect herbivores. These patterns are predicted, in different combinations, by the adjacent errors, adjacent oligophagy, distance-divergence, and distance relaxation hypotheses (Table 1) and thus provide windows on the overall GAP model. The tests share a common framework in that they are based on relationships between attacked and available trait spaces for host plants of two species (Figures 1 and 2). Two tests pertain to the pattern of attack on a single host, and the third to the pattern of attack on each host relative to the other.

First, I test for central versus marginal location of the attacked trait space on each host ("*Marginal trait-space test*"). I calculate the centroid of the available trait space (mean PC1 and PC2 scores for all available plants, attacked and unattacked) and that of the attacked trait space (mean PC scores for attacked plants only). I then calculate the distance between available and attacked centroids and compare this to a null distribution of 10,000 such distances calculated following random shuffling of attack status across all plant individuals. The fraction of randomization distances larger than the actual attacked-available distance is a P value, and when it is small we reject the null hypothesis that attacked and available plants have a common centroid (central or nonselective attack, Figures 1(a) and 1(b)) in favour of the alternative of marginal attack (Figure 1(c)).

When we are unable to detect marginal use of available trait space, we might seek to distinguish between nonselective (Figure 1(a)) and restricted but central (Figure 1(b)) alternatives. To do so, I use the "*restricted trait-space test*." I calculate the Euclidean distance from each attacked plant to the centroid of attacked trait space. The size of the attacked trait space is given by the sum of these distances. I then compare this trait-space size to a null distribution the sizes of 10,000 attacked trait spaces generated by randomly shuffling attack status across all plant individuals. Note that shuffling attack status maintains the number of attacked plants, which is critical when calculating the size of a trait space. The fraction of randomization attacked trait spaces smaller than the actual one is a P-value, and when it is small we reject the null hypothesis that attack is nonselective. Since we are using the restricted trait-space test following a nonsignificant marginal trait-space test, the alternative is that herbivores exploit a restricted but central subset of available trait space.

Finally, I test whether the distance between attacked centroids on the two host plants is smaller or larger than expected at random ("*Distant trait-spaces test*," Figure 2(b) versus 2(c)). I first calculate the distance between attacked trait-space centroids on the two alternative hosts. This distance is compared, in a two-tailed test, to a null distribution of 10,000 such distances calculated following random shuffling of attack status across individuals of each plant species (separately). When the actual centroids are farther apart than the mean distance from randomizations, then twice the fraction of randomization distances that are larger than the actual distance is a P-value, which when small supports rejection of the null hypothesis in favour of the alternative

that the two attacked trait spaces are significantly distant. On the other hand, when the actual centroids are closer than the mean distance from randomizations, then twice the fraction of randomization distances that are smaller than the actual distance is a P-value, which when small supports rejection of the null hypothesis in favour of the alternative that the two attacked trait spaces are significantly adjacent.

The marginal trait-space, restricted trait-space, and distant trait-space tests are implemented in TraitSpaces 1.20, a program written in Microsoft Visual Basic.NET for Windows. The software takes as input a datafile with a row for each individual host plant, and columns for host species identity, presence/absence of each herbivore, and first and second principal components calculated from the host trait matrix. (Principal components may be output from any standard statistical package.) Extension to trait spaces of higher dimensionality, if desired, is straightforward; one could even use an unreduced trait matrix at the cost of some complexity in displaying results. The analytical framework easily accommodates data for other host/attacker systems and could even be applied to cases where consumers use variable microhabitats or food resources. The current version of the TraitSpaces package is available from the author on request.

4. Field Methods

4.1. Study System. The goldenrods *Solidago altissima* L. and *S. gigantea* Ait. are clonal perennials codistributed over much of eastern and central North America. Intermixed stands of the two species are common in open habitats such as prairies, old fields, roadsides, and forest edges. Individual ramets grow in spring from underground rhizomes, flower in late summer and fall, and die back to ground level before winter. The two species differ most obviously in pubescence [65]: *S. altissima* stems are sparsely to densely short-hairy, especially basally, while *S. gigantea* stems are typically glabrous. Both species display extensive intraspecific variation (genetic and plastic) in most traits, including ramet size, pubescence, leaf shape, size, and toothiness, and chemical profiles ([27, 66, 67], S. B. Heard, unpubl. data).

S. altissima and *S. gigantea* are attacked by a diverse fauna of insect herbivores [68–70], which vary in diet specialization. Some are broad generalists that accept *Solidago* as part of a taxonomically diverse diet (e.g., the exotic spittlebug *Philaenus spumarius* [71]), and some are broadly oligophagous, feeding on *Solidago* among other members of the Asteraceae (e.g., the chrysomelid *Exema canadensis* [72]). Others are more narrowly oligophagous, attacking only *Solidago* spp. (e.g., the tortricid stem-galler *Epiblema scudderiana* [13, 73]). Finally, at least four herbivores have evolved monophagous host races or cryptic species on *S. altissima* and *S. gigantea* [13, 30, 74, 75], with divergence ranging from quite recent for the ball-gall fly *Eurosta solidaginis* (at most 200,000 years, but likely much less) to >2 × 10⁶ years old for the bunch-gall flies *Rhopalomyia solidaginis/R. capitata*.

Especially for the better-studied *S. altissima*, attack by various herbivores is known to vary among clones [68, 76, 77], and with plant traits including ramet size [39, 78],

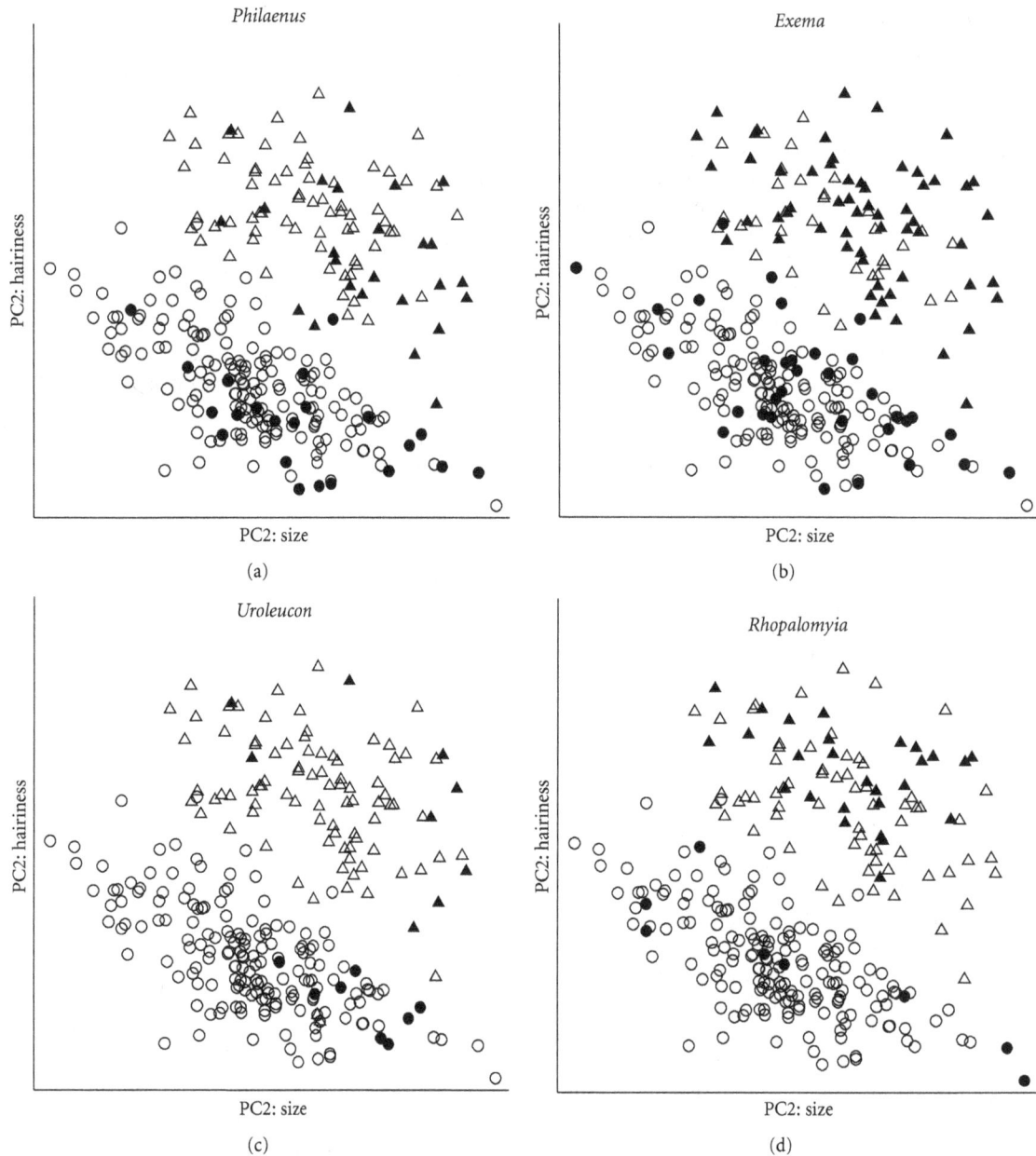

FIGURE 3: Attacked trait spaces for four goldenrod herbivores on *S. altissima* (triangles) and *S. gigantea* (circles). Filled symbols denote attacked plants, and open symbols unattacked ones. Axis labels are shorthand for the first two principal components from a 7-variable morphological dataset; full factor loadings are provided in Table 2.

growth rate [79], nutritional status [80, 81] and ploidy where this varies locally [82]. These trait-attack relationships involve both plant resistance and insect preference [27] and may be concordant or discordant among different herbivore species [68, 82].

4.2. Field Data. I and my field team gathered data on plant traits and herbivore attack in old-field and trailside *Solidago* populations in Fredericton, NB, Canada (45° 57′ 30″ N, 66° 37′ 1–20″ W). Here both *S. gigantea* and *S. altissima* are abundant along with *S. rugosa, S. juncea, S. canadensis, Euthamia graminifolia, Symphyotrichum* spp., and other Asteraceae. *S. altissima* is exclusively hexaploid in the east, and

S. gigantea exclusively diploid, so effects of ploidy on herbivore attack [83] need not be considered here.

It is important to assess the available and attacked trait spaces using traits measured before herbivore attack; otherwise, herbivore responses to plant traits could be confounded with herbivore-driven changes in the same traits. In early June 2004, we marked 104 *S. altissima* ramets and 186 *S. gigantea* ramets by setting line transects through well-mixed patches of the two species and marking each ramet touched by the line. At the time of marking, a few ramets had already been attacked by the stem-galler *Gnorimoschema gallaesolidaginis* (Lepidoptera: Gelechiidae; galls on 4 *S. altissima* and 3 *S. gigantea* ramets), but other herbivores

TABLE 2: Correlations among morphological variables measured for *S. altissima* and *S. gigantea*.

	Ramet height	Leaf length	Leaf width	Teeth	Water content	Trichomes
Stem width	0.88	0.72	0.18	0.40	−0.42	0.36
Ramet height		0.75	0.17	0.43	−0.44	0.36
Leaf length			0.47	0.56	−0.23	0.12
Leaf width				0.32	0.24	−0.57
Teeth					−0.18	−0.01
Water content						−0.48

had yet to attack. We measured 7 morphological traits of our marked ramets, focusing on easily measured traits that were likely to influence herbivore attack, that help distinguish the two study species, or both. We measured stem trichome density by counting, in the field with a hand lens, all trichomes in silhouette along a 10 cm length of stem just below the terminal bud. We measured stem width 5–10 cm above ground using a caliper, and stem height (from ground to the base of the terminal bud) using a measuring tape. For the largest leaf from each ramet, we measured leaf length, leaf width at the widest point, and the number of teeth along one leaf edge. Finally, we weighed each largest leaf before and after drying to constant mass at 45-55°C and calculated percent water content.

We surveyed marked ramets twice weekly until the end of August, identifying herbivores present as specifically as possible without disturbing them on the plant (for some groups, like larval *Trirhabda* beetles, species-level identifications require the removal of the insects to the laboratory, and we wanted to leave plants to experience natural levels of herbivory). When herbivores of the same species were present on consecutive surveys, we were usually unable to determine whether they were the same individuals, so rather than count individuals we classified each ramet as attacked or unattacked by each herbivore over the course of the entire season.

Some marked ramets were lost or damaged during the season, leaving 92 *S. altissima* ramets and 175 *S. gigantea* ramets with comprehensive herbivory and plant-trait data. Four herbivores were identifiable to species and abundant enough to give our analyses reasonable power: the xylem-sucking spittlebug *Philaenus spumarius*, which is broadly polyphagous [71]; the folivorous chrysomelid beetle *Exema canadensis*, which is oligophagous with many hosts in the tribe Astereae [72]; the phloem-sucking aphid *Uroleucon nigrotuberculatum*, which is narrowly oligophagous on *Solidago* spp. [84]; and the gall-making cecidomyiid fly *Rhopalomyia solidaginis/R. capitata*, which is a pair of monophagous specialists (*R. solidaginis* on *S. altissima* and *R. capitata* on *S. gigantea* [13]). All further analyses use this reduced set of 267 ramets and 4 herbivores.

5. Field Results and Discussion

5.1. Plant Traits. Among the 7 measured traits stem width, ramet height, and leaf length were strongly intercorrelated $(0.72 < r < 0.88)$, suggesting that all three reflect overall ramet size. The other 18 correlations were weak to moderate

TABLE 3: Factor loadings for the first two principal components from the morphological data matrix for *S. altissima* and *S. gigantea*.

Trait	Loading on PC1	Loading on PC2
Stem width	0.50	0.04
Ramet height	0.51	0.04
Leaf length	0.47	−0.22
Leaf width	0.14	−0.63
Teeth	0.34	−0.24
Water content	−0.29	−0.39
Trichomes	0.34	0.58

(Table 2). The first two principal component axes explained 47% and 28% of the morphological variance (75% total), while no other axis explained more than 9.3%. PC1 largely reflects ramet size (heavy loadings for stem width, ramet height, and leaf length; Table 3), but also leaf toothiness (positively) and water content (negatively). PC2 contrasts trichome counts (strong positive loading; Table 3) with leaf width (strong negative loading) but also includes leaf water content, leaf length, and toothiness (all negative). These two principal components do a good job of capturing both intraspecific and interspecific variation (Figure 3), with *S. altissima* and *S. gigantea* separated primarily along PC2 (the pubescent *S. altissima* with high scores, and the glabrous *S. gigantea* with low scores).

5.2. Herbivore Use of Phenotype Space. Attack rates by *Solidago* herbivores are generally low (often 1–10% or even less), with the exception of some diet generalists and outbreaking species in high-density years (S. Heard, unpubl. data). In our dataset, even though we worked with some of the most common herbivores, only one herbivore on one host had an incidence above 30% (*Exema canadensis* on *S. altissima*, 64% of ramets attacked). Other herbivore/host combinations had lower incidences, with several less than 10% (*Uroleucon nigrotuberculatum* on both hosts and *Rhopalomyia capitata* on *S. gigantea*; Table 4).

The patterns I document in trait-space use could have arisen via herbivore preference, or via performance if poor herbivore growth leads to death or departure of herbivores before surveys can detect them. For most herbivores, repeated surveys allow herbivore detection shortly after attack begins, and so preference is the most likely driver of patterns in attack. However, for gallmakers like *Rhopalomyia* performance at the stage of gall induction could be important.

TABLE 4: Tests of trait-space use on *Solidago altissima* and *S. gigantea* for four *Solidago* herbivores. *P*-values in bold are significant at $\alpha = 0.05$.

Herbivore	Host	# Attacked plants[1]	Distance from available centroid[2]		Marginal trait-space test *P*	Restricted trait-space test *P*	Distance between attacked trait-spaces[3]	Distant trait-spaces test *P*
			PC1	PC2				
Philaenus	*S. altissima*	25	1.00	−0.50	**<0.001**	—[4]	Small	0.71
	S. gigantea	22	1.11	−0.52	**0.001**	—		
Exema	*S. altissima*	59	0.30	0.05	**0.023**	—	Small	0.14
	S. gigantea	33	0.65	0.04	**0.023**	—		
Uroleucon	*S. altissima*	9	1.39	−0.10	**0.007**	—	Large	0.86
	S. gigantea	8	2.32	−0.62	**<0.001**	—		
Rhopalomyia	*S. altissima*	27	0.09	0.35	0.21	0.18	Large	0.80
	S. gigantea	8	0.66	−0.05	0.31	1.0		

[1] Of 92 available *S. altissima* and 175 available *S. gigantea* ramets.

[2] Attacked centroid minus available centroid (PC1 and PC2 components). A positive entry means that ramets with a large PC score are more likely to be attacked.

[3] "Small" if the two attacked trait spaces are adjacent (Figure 2(b)), and "large" if the two attacked trait spaces are distant (Figure 2(c)).

[4] This test is informative only when the marginal trait-space test is not significant.

Plant genotype effects on gall induction, mismatched with herbivore preference, are known (for example) for *Eurosta solidaginis* on *S. altissima* [85].

The two most generalist herbivores (*Philaenus* and *Exema*) showed similar patterns in trait-space use (Figures 3(a) and 3(b); Table 4). Both showed significant evidence for nonrandom use of available hosts (marginal trait-space test). For *Philaenus* on both hosts, attack was concentrated on larger but less pubescent ramets (higher PC1 and lower PC2), while for *Exema* on both hosts attack was concentrated on larger ramets but did not depend on pubescence (higher PC1). For both species, the distance between attacked trait-spaces on *S. altissima* and *S. gigantea* was slightly but not significantly smaller than expected under the null (distant trait-spaces test).

For the oligophagous *Uroleucon* (Figure 3(c)), attack on both hosts was significantly marginal, being concentrated on larger and less pubescent ramets. The distance between the two attacked trait spaces was slightly, but not significantly, larger than expected under the null. Because this herbivore had the smallest sample size (just 17 attacked ramets total), these tests have much less power than for the more common generalists.

For the monophagous *Rhopalomyia* species pair (Figure 3(d)), there was no evidence on either host for marginal use of trait space, and the restricted trait space test suggests nonselective rather than central use of available trait space (Table 4). The distance between the two attacked trait-spaces was slightly, but not significantly, larger than expected under the null. Sample size, however, was very small for *R. capitata* on *S. gigantea* (8 attacked ramets), so the tests for that species and for the distance between attacked centroids are likely not very powerful.

5.3. Interpretation and Prospects. The clearest pattern in the illustrative dataset is that for three of four herbivores, attack is significantly concentrated on larger ramets (large PC1). The stem gallers *Eurosta* [39] and *Gnorimoschema*

[78] also have well-documented associations with larger ramets, something that is common but not universal among phytophagous insects [86, 87]. Such concordance across herbivore species in the use of trait space increases the likelihood of multiple herbivores cooccuring on a single ramet—something very unlikely under the null hypothesis of independent occurrence, since most attack rates are low. Herbivores cooccurring on a plant may compete directly (via resource consumption) or indirectly (via induced resistance) or may even show facilitation [88] although we know little about potential interactions among goldenrod herbivores [89–91]. However, concordance among goldenrod herbivores in use of plant trait space is far from universal [68, 82].

How do the illustrative data fit with the GAP model of trait space use during host shifting and HAD laid out above? The tendency for most herbivores to attack larger ramets (larger PC1) generates pattern in trait-space use. However, this shared tendency means that both attacked trait spaces are offset from the available spaces, in parallel and orthogonally to the contrast between alternative host plants (PC2). The distance between attacked trait spaces is unaffected, and so this ramet-size pattern is not directly relevant to the GAP model. Three of the four herbivores analyzed (*Philaenus*, *Exema*, and *Uroleucon*) have host ranges broader than just the *S. altissima*-*S. gigantea* pair and might therefore be expected to be rather unselective about traits distinguishing the two hosts. Indeed, *Exema* showed no offsets between attacked and available trait spaces along the principal components axis contrasting *S. altissima* and *S. gigantea* (PC2; Table 4). *Philaenus* and *Uroleucon* did show offsets along this axis, but because they were in the same direction and roughly equal on the two hosts, separation of attacked trait-spaces was not significantly large for either herbivore (Table 4, distant trait spaces test). The fourth herbivore, *Rhopalomyia*, is a pair of relatively old monophagous species [13]. For such a species pair, the distance relaxation hypothesis suggests that the use of distant trait spaces may no longer be an important barrier to gene flow (Figure 2(d)). *Rhopalomyia*'s use of trait space

(Table 4: no evidence for marginal attack, and separation between the two attacked trait spaces no larger than expected at random) is consistent with this (Table 1).

Overall, none of the illustrative data are inconsistent with the GAP model, but none of the four herbivores analyzed provides a strong test of its hypotheses. I did not find any examples of the patterns hypothesized for a single-host specialist making host-choice errors, for a narrowly oligophagous species immediately following a host shift, or for a pair of nascent forms early in HAD (Table 1, Figures 2(a)–2(c)). This is not surprising, though, because species known to be in early stages of HAD on *S. altissima*/*S. gigantea*, such as the ball-gall fly *Eurosta solidaginis* and the spindle-gall moth *Gnorimoschema gallaesolidaginis*, were insufficiently abundant for analysis. At the broader level of hypothesis testing, four herbivores constitute just a small step towards assessing general patterns in trait-space use through HAD and ecological speciation. It will take many studies like mine, with herbivores on *Solidago* and other plants, before we can assess the generality of patterns in trait-space use.

Because attack rates for most *Solidago* herbivores are low, achieving powerful hypothesis tests for any herbivore will entail marking very large numbers of ramets—especially since ramet selection must be done before attack begins to avoid distortions of attack-space measurements if trait values change under herbivore attack. In *Solidago*, for instance, ramet biomass and height are often reduced by herbivory [91–93]. I am currently expanding on the illustrative study with the goal of securing larger sample sizes for the herbivores studied here and acceptable sample sizes for many more herbivores.

Another obvious limitation of the illustrative dataset is that it includes measurements of only seven plant traits, and conspicuously omits leaf-chemistry traits (and ploidy [83], which varies elsewhere but not in New Brunswick). *S. altissima* and *S. gigantea* have complex secondary chemistry, and variation in leaf chemistry is known to influence herbivore attack [66, 67]. Unmeasured morphological traits may also be relevant to insect attack; for instance, *Philaenus* prefers plant species and individuals with wider leaf axils [94], and this trait varies among *S. altissima* genotypes (Maddox unpubl. in [77]). Expanding the list of measured traits, and especially incorporating leaf chemistry, is a high priority for future work.

Despite the small numbers of attacked ramets and measured traits that earn the "illustrative" dataset its descriptor, my analysis of trait-space use for four herbivores establishes that the field and analytical approach outlined here is feasible and can detect nonrandom trait-space use. Because the goldenrod system includes such a diverse herbivore community attacking syntopic ramets of the alternative hosts, it offers the potential for great progress in testing hypotheses about host trait-space use during hostshifts and HAD.

6. General Discussion

The literature on how insect preference and performance vary with intraspecific variation in host-plant genotype, morphology, chemistry, and phenology is immense [95, 96].

Similarly, interspecific variation in the same kinds of traits has been widely held up as the key to the macroevolutionary fate of herbivore lineages (host shifting, diversification, specialization, and so forth [38, 63, 64, 97]). What is surprising is that the intersection of these perspectives is so little developed: we know almost nothing about trait-space use in systems where host shifting and HAD are suspected. This gap is clearly illustrated by the two best-studied cases of HAD in phytophagous insects: *Eurosta solidaginis* on *Solidago altissima* and *S. gigantea* and *Rhagoletis pomonella* on apple and hawthorn. For *Eurosta*, despite a wealth of information about how preference and performance relate to genetic and trait variation within *S. altissima* [27], few comparable data are available for flies attacking *S. gigantea* (except see [40]). For *Rhagoletis*, much has been written about the importance for HAD of apple-hawthorn differences in ripening phenology [26, 98, 99] and fruit size [100]. However, data on local intraspecific variation in phenology appear to be unavailable (although latitudinal clines have been documented [26]), and data on intraspecific fruit size variation appear limited to confirming significance of interspecific differences in average fruit size [100]. This is not to criticize work on these two model systems, which has pioneered the study of HAD, but rather to draw attention to a significant opportunity for progress.

Of course, the GAP model likely falls short of recognizing the full complexity of trait-space use in nature. While I have focused on snapshots of trait variation and insect use of trait space at a single site and in a single year, both available trait space and its use are likely to vary in space and time. This variation could have interesting and important consequences for HAD. For example, intraspecific variation in *Solidago* phenology and the difference in average phenology between *S. altissima* and *S. gigantea* change in space and time, and phenological differences are involved in host choice for at least two *Solidago* herbivores undergoing HAD (*Eurosta* [40]; *Gnorimoschema gallaesolidaginis*, S. B. Heard, unpubl. data). Hawthorn phenologies show latitudinal gradients across space favouring local adaptation rather like that required during *Rhagoletis'* host shift to apple [26]. There are thus likely to be places or times that are more conducive to host shifts and HAD than others [40, 52, 53]. Superimposed over this variation in available trait space can be strong geographic variation in insect preference (e.g., [82, 101]) and thus trait-space use. As a consequence, the places or times conducive to host shifting for one insect herbivore might not be so conducive for shifts by another. This is consistent with the evolutionary pattern seen in the *Solidago* system, in which three gallmakers have made host shifts from *S. altissima* to *S. gigantea* but have done so at different times [13].

Thinking about intraspecific variation in plant traits, and patterns of insect attack with respect to that variation, can expand and enhance our view of ecological speciation by phytophagous insects. Testing the hypotheses I frame about trait-space use for herbivores differing in diet breadth and in progress along the evolutionary sequence of HAD (Table 1) could take us a long way towards a predictive understanding of diet evolution and specialization in phytophagous insects. Ultimately, we would like to know for which taxa host-shifting and HAD are likely, and for which taxa they are not—and why [13, 38, 64, 102]. While much data collection and analysis lies ahead, the trait-space perspective promises a new and powerful window on the fascinating complexity of insect-plant interactions and herbivore diversification.

Acknowledgments

The author thanks the City of Fredericton for permission to conduct field work; William Godsoe and Susan Timmons collected the field data, and William Godsoe conducted preliminary statistical analysis. He thanks Andrew Hendry and two anonymous reviewers for comments on the manuscript. John Nason and John Stireman helped shape his view of the *Solidago* system through collaborations, and Warren Abrahamson and John Semple have provided advice and encouragement over the years. This work was supported by Discovery Grants from the Natural Sciences and Engineering Research Council (Canada).

References

[1] D. Sharp, "Insects, part I," in *The Cambridge Natural History*, S. F. Harmer and A. E. Shipley, Eds., pp. 83–565, MacMillan and Co., London, UK, 1895.

[2] E. O. Wilson, *The Diversity of Life*, W.W. Norton, New York, NY, USA, 1992.

[3] P. J. Mayhew, "Why are there so many insect species? Perspectives from fossils and phylogenies," *Biological Reviews*, vol. 82, no. 3, pp. 425–454, 2007.

[4] D. R. Strong Jr., J. H. Lawton, and R. Southwood, *Insects on Plants: Community Patterns and Mechanisms*, Blackwell Scientific, Oxford, UK, 1984.

[5] C. Mitter, B. Farrell, and B. Wiegmann, "The phylogenetic study of adaptive zones: has phytophagy promoted insect diversification?" *American Naturalist*, vol. 132, no. 1, pp. 107–128, 1988.

[6] B. D. Farrell, "'Inordinate fondness' explained: Why are there so many beetles?" *Science*, vol. 281, no. 5376, pp. 555–559, 1998.

[7] E. A. Bernays and R. F. Chapman, *Host-Plant Selection by Phytophagous Insects*, Chapman and Hall, New York, NY, USA, 1994.

[8] P. D. N. Hebert, E. H. Penton, J. M. Burns, D. H. Janzen, and W. Hallwachs, "Ten species in one: DNA barcoding reveals cryptic species in the neotropical skipper butterfly Astraptes fulgerator," *Proceedings of the National Academy of Sciences of the United States of America*, vol. 101, no. 41, pp. 14812–14817, 2004.

[9] J. B. Joy and B. J. Crespi, "Adaptive radiation of gall-inducing insects within a single host-plant species," *Evolution*, vol. 61, no. 4, pp. 784–795, 2007.

[10] M. A. Condon, S. J. Scheffer, M. L. Lewis, and S. M. Swensen, "Hidden neotropical diversity: Greater than the sum of its parts," *Science*, vol. 320, no. 5878, pp. 928–931, 2008.

[11] C. Darwin, *The Origin of Species*, 1968 reprint, Penguin Books, London, UK, 1859.

[12] B. D. Walsh, "The apple-worm and the apple-maggot," *Tilton's Journal of Horticulture and Florist's Companion*, vol. 2, pp. 338–343, 1867.

[13] J. O. Stireman, J. D. Nason, and S. B. Heard, "Host-associated genetic differentiation in phytophagous insects: General

[14] K. W. Matsubayashi, I. Ohshima, and P. Nosil, "Ecological speciation in phytophagous insects," *Entomologia Experimentalis et Applicata*, vol. 134, no. 1, pp. 1–27, 2010.

[15] S. H. Berlocher and J. L. Feder, "Sympatric speciation in phytophagous insects: Moving beyond controversy?" *Annual Review of Entomology*, vol. 47, pp. 773–815, 2002.

[16] M. Drès and J. Mallet, "Host races in plant-feeding insects and their importance in sympatric speciation," *Philosophical Transactions of the Royal Society B*, vol. 357, no. 1420, pp. 471–492, 2002.

[17] J. O. Stireman III, J. D. Nason, S. B. Heard, and J. M. Seehawer, "Cascading host-associated genetic differentiation in parasitoids of phytophagous insects," *Proceedings of the Royal Society B*, vol. 273, no. 1586, pp. 523–530, 2006.

[18] S. Skúlason, S. S. Snorrason, and B. Jónsson, "Sympatric morphs, populations and speciation in freshwater fish with emphasis on arctic charr," in *Evolution of Biological Diversity*, A. E. Magurran and R. M. May, Eds., Oxford University Press, Oxford, UK, 1999.

[19] A. P. Hendry, S. K. Huber, L. F. De León, A. Herrel, and J. Podos, "Disruptive selection in a bimodal population of Darwin's finches," *Proceedings of the Royal Society B: Biological Sciences*, vol. 276, no. 1657, pp. 753–759, 2009.

[20] J. Antonovics and A. D. Bradshaw, "Evolution in closely adjacent plant populations. 8. Clinal patterns at a mine boundary," *Heredity*, vol. 25, pp. 349–362, 1970.

[21] E. B. Rosenblum and L. J. Harmon, "'same same but different': replicated ecological speciation at white sands," *Evolution*, vol. 65, no. 4, pp. 946–960, 2011.

[22] H. D. Rundle and P. Nosil, "Ecological speciation," *Ecology Letters*, vol. 8, no. 3, pp. 336–352, 2005.

[23] J. Peccoud and J. C. Simon, "The pea aphid complex as a model of ecological speciation," *Ecological Entomology*, vol. 35, no. 1, pp. 119–130, 2010.

[24] A. P. Hendry, D. I. Bolnick, D. Berner, and C. L. Peichel, "Along the speciation continuum in sticklebacks," *Journal of Fish Biology*, vol. 75, no. 8, pp. 2000–2036, 2009.

[25] J. L. Feder, S. H. Berlocher, and S. B. Opp, "Sympatric host-race formation and speciation in *Rhagoletis* (Diptera: Tephritidae): a tale of two species for Charles D," in *Genetic Structure and Local Adaptation in Natural Insect Populations*, S. Mopper, Ed., pp. 408–441, Chapman and Hall, New York, NY, USA, 1998.

[26] J. L. Feder, T. H. Powell, K. Filchak, and B. Leung, "The diapause response of *Rhagoletis pomonella* to varying environmental conditions and its significance for geographic and host plant-related adaptation," *Entomologia Experimentalis et Applicata*, vol. 136, no. 1, pp. 31–44, 2010.

[27] W. G. Abrahamson and A. E. Weis, *Evolutionary Ecology Across Three Trophic Levels: Goldenrods, Gallmakers, and Natural Enemies*, Princeton University Press, Princeton, NJ, USA, 1997.

[28] T. P. Craig and J. K. Itami, "Divergence of *Eurosta solidaginis* in response to host plant variation and natural enemies," *Evolution*, vol. 65, no. 3, pp. 802–817, 2011.

[29] S. H. Berlocher, "Radiation and divergence in the *Rhagoletis pomonella* species group: inferences from allozymes," *Evolution*, vol. 54, no. 2, pp. 543–557, 2000.

[30] J. D. Nason, S. B. Heard, and F. R. Williams, "Host-associated genetic differentiation in the goldenrod elliptical-gall moth,

Gnorimoschema gallaesolidaginis (Lepidoptera: Gelechiidae)," *Evolution*, vol. 56, no. 7, pp. 1475–1488, 2002.

[31] A. M. Dickey and R. F. Medina, "Testing host-associated differentiation in a quasi-endophage and a parthenogen on native trees," *Journal of Evolutionary Biology*, vol. 23, no. 5, pp. 945–956, 2010.

[32] S. P. Carroll and C. Boyd, "Host race radiation in the soapberry bug: natural history with the history," *Evolution*, vol. 46, no. 4, pp. 1052–1069, 1992.

[33] D. J. Funk, "Isolating a role for natural selection in speciation: host adaptation and sexual isolation in *Neochlamisus bebbianae* leaf beetles," *Evolution*, vol. 52, no. 6, pp. 1744–1759, 1998.

[34] S. Via, A. C. Bouck, and S. Skillman, "Reproductive isolation between divergent races of pea aphids on two hosts. II. Selection against migrants and hybrids in the parental environments," *Evolution*, vol. 54, no. 5, pp. 1626–1637, 2000.

[35] I. Emelianov, F. Simpson, P. Narang, and J. Mallet, "Host choice promotes reproductive isolation between host races of the larch budmoth *Zeiraphera diniana*," *Journal of Evolutionary Biology*, vol. 16, no. 2, pp. 208–218, 2003.

[36] I. Ohshima, "Host-associated pre-mating reproductive isolation between host races of *Acrocercops transecta*: mating site preferences and effect of host presence on mating," *Ecological Entomology*, vol. 35, no. 2, pp. 253–257, 2010.

[37] I. Ohshima, "Host race formation in the leaf-mining moth *Acrocercops transecta* (Lepidoptera: Gracillariidae)," *Biological Journal of the Linnean Society*, vol. 93, no. 1, pp. 135–145, 2008.

[38] T. Nyman, "To speciate, or not to speciate? Resource heterogeneity, the subjectivity of similarity, and the macroevolutionary consequences of niche-width shifts in plant-feeding insects," *Biological Reviews*, vol. 85, no. 2, pp. 393–411, 2010.

[39] R. Walton, A. E. Weis, and J. P. Lichter, "Oviposition behavior and response to plant height by *Eurosta solidaginis* Fitch (Diptera: Tephritidae)," *Annals of the Entomological Society of America*, vol. 83, pp. 509–514, 1990.

[40] S. T. How, W. G. Abrahamson, and T. P. Craig, "Role of host plant phenology in host use by *Eurosta solidaginis* (Diptera: Tephritidae) on *Solidago* (Compositae)," *Environmental Entomology*, vol. 22, no. 2, pp. 388–396, 1993.

[41] T. P. Craig, J. K. Itami, W. G. Abrahamson, and J. D. Horner, "Behavioral evidence for host-race formation in *Eurosta solidaginis*," *Evolution*, vol. 47, no. 6, pp. 1696–1710, 1993.

[42] B. Schmid and C. Dolt, "Effects of maternal and paternal environment and genotype on offspring phenotype in *Solidago altissima* L.," *Evolution*, vol. 48, no. 5, pp. 1525–1549, 1994.

[43] J. N. Maloof, "QTL for plant growth and morphology," *Current Opinion in Plant Biology*, vol. 6, no. 1, pp. 85–90, 2003.

[44] M. T. J. Johnson and A. A. Agrawal, "Plant genotype and environment interact to shape a diverse arthropod community on evening primrose (*Oenothera biennis*)," *Ecology*, vol. 86, no. 4, pp. 874–885, 2005.

[45] R. A. Rapp and J. F. Wendel, "Epigenetics and plant evolution," *New Phytologist*, vol. 168, no. 1, pp. 81–91, 2005.

[46] A. A. Winn, "Adaptation to fine-grained environmental variation: an analysis of within-individual leaf variation in an annual plant," *Evolution*, vol. 50, no. 3, pp. 1111–1118, 1996.

[47] M. C. Singer, D. Ng, D. Vasco, and C. D. Thomas, "Rapidly evolving associations among oviposition preferences fail to constrain evolution of insect diet," *American Naturalist*, vol. 139, no. 1, pp. 9–20, 1992.

[48] N. Janz and S. Nylin, "The role of female search behaviour in determining host plant range in plant feeding insects: a test of the information processing hypothesis," *Proceedings of the Royal Society B*, vol. 264, no. 1382, pp. 701–707, 1997.

[49] D. I. Bolnick, R. Svanbäck, M. S. Araújo, and L. Persson, "Comparative support for the niche variation hypothesis that more generalized populations also are more heterogeneous," *Proceedings of the National Academy of Sciences of the United States of America*, vol. 104, no. 24, pp. 10075–10079, 2007.

[50] D. I. Bolnick, R. Svanbäck, J. A. Fordyce et al., "The ecology of individuals: incidence and implications of individual specialization," *American Naturalist*, vol. 161, no. 1, pp. 1–28, 2003.

[51] L. M. Evans, G. J. Allan, S. M. Shuster, S. A. Woolbright, and T. G. Whitham, "Tree hybridization and genotypic variation drive cryptic speciation of a specialist mite herbivore," *Evolution*, vol. 62, no. 12, pp. 3027–3040, 2008.

[52] T. P. Craig, J. K. Itami, T. Ohgushi, Y. Ando, and S. Utsumi, "Bridges and barriers to host shifts resulting from host plant genotypic variation," *Journal of Plant Interactions*, vol. 6, no. 2-3, pp. 141–145, 2011.

[53] D. I. Bolnick, "Sympatric speciation in threespine stickleback: why not?" *International Journal of Ecology*, vol. 2011, Article ID 942847, 15 pages, 2011.

[54] M. R. Servedio and M. A. F. Noor, "The role of reinforcement in speciation: theory and data," *Annual Review of Ecology, Evolution, and Systematics*, vol. 34, pp. 339–364, 2003.

[55] P. Nosil, D. J. Funk, and D. Ortiz-Barrientos, "Divergent selection and heterogeneous genomic divergence," *Molecular Ecology*, vol. 18, no. 3, pp. 375–402, 2009.

[56] P. Nosil, S. P. Egan, and D. J. Funk, "Heterogeneous genomic differentiation between walking-stick ecotypes: "isolation by adaptation" and multiple roles for divergent selection," *Evolution*, vol. 62, no. 2, pp. 316–336, 2008.

[57] X. Thibert-Plante and A. P. Hendry, "When can ecological speciation be detected with neutral loci?" *Molecular Ecology*, vol. 19, no. 11, pp. 2301–2314, 2010.

[58] B. J. Crespi, "Vicious circles: positive feedback in major evolutionary and ecological transitions," *Trends in Ecology and Evolution*, vol. 19, no. 12, pp. 627–633, 2004.

[59] A. P. Hendry, "Selection against migrants contributes to the rapid evolution of ecologically dependent reproductive isolation," *Evolutionary Ecology Research*, vol. 6, no. 8, pp. 1219–1236, 2004.

[60] K. Räsänen and A. P. Hendry, "Disentangling interactions between adaptive divergence and gene flow when ecology drives diversification," *Ecology Letters*, vol. 11, no. 6, pp. 624–636, 2008.

[61] A. F. Agrawal, J. L. Feder, and P. Nosil, "Ecological divergence and the origins of intrinsic postmating isolation with gene flow," *International Journal of Ecology*, vol. 2011, Article ID 435357, 15 pages, 2011.

[62] J. A. Coyne and H. A. Orr, *Speciation*, Sinauer Associates, Sunderland, Mass, USA, 2004.

[63] M. S. Singer, "Evolutionary ecology of polyphagy," in *Specialization, Speciation, and Radiation: The Evolutionary Biology of Herbivorous Insects*, K. J. Tilmon, Ed., pp. 29–42, University of California Press, Berkeley, Calif, USA, 2008.

[64] D. J. Funk and P. Nosil, "Comparative analyses of ecological speciation," in *Specialization, Speciation, and Radiation: The Evolutionary Biology of Herbivorous Insects*, K. J. Tilmon, Ed., pp. 117–135, University of California Press, Berkeley, Calif, USA, 2008.

[65] J. C. Semple and R. E. Cook, "Solidago," in *Flora of North America*, Flora North America Editorial Committee, Ed., pp. 107–166, Oxford University Press, Oxford, UK, 2006.

[66] H. M. Hull-Sanders, R. Clare, R. H. Johnson, and G. A. Meyer, "Evaluation of the evolution of increased competitive ability (EICA) hypothesis: loss of defense against generalist but not specialist herbivores," *Journal of Chemical Ecology*, vol. 33, no. 4, pp. 781–799, 2007.

[67] R. H. Johnson, R. Halitschke, and A. Kessler, "Simultaneous analysis of tissue- and genotype-specific variation in *Solidago altissima* (Asteraceae) rhizome terpenoids, and the polyacetylene dehydromatricaria ester," *Chemoecology*, vol. 20, no. 4, pp. 255–264, 2010.

[68] G. D. Maddox and R. B. Root, "Structure of the encounter between goldenrod (*Solidago altissima*) and its diverse insect fauna," *Ecology*, vol. 71, no. 6, pp. 2115–2124, 1990.

[69] R. B. Root and N. Cappuccino, "Patterns in population change and the organization of the insect community associated with goldenrod," *Ecological Monographs*, vol. 62, no. 3, pp. 393–420, 1992.

[70] E. M. G. Fontes, D. H. Habeck, and F. Slansky Jr., "Phytophagous insects associated with goldenrods (*Solidago* spp.) in Gainesville, Florida," *Florida Entomologist*, vol. 77, no. 2, pp. 209–221, 1994.

[71] C. R. Weaver and D. R. King, "Meadow spittlebug," *Ohio Agricultural Experiment Station Research Bulletin*, vol. 741, pp. 1–99, 1954.

[72] R. B. Root and F. J. Messina, "Defensive adaptations and natural enemies of a case-bearing beetle, *Exema canadensis* (Coleoptera: Chrysomelidae)," *Psyche*, vol. 90, pp. 67–80, 1983.

[73] W. E. Miller, "Biology and taxonomy of three gall-forming species of *Epiblema* (Olethreutidae)," *Journal of the Lepidopterists' Society*, vol. 30, pp. 50–58, 1976.

[74] G. L. Waring, W. G. Abrahamson, and D. J. Howard, "Genetic differentiation among host-associated populations of the gallmaker *Eurosta solidaginis* (Diptera: Tephritidae)," *Evolution*, vol. 44, pp. 1648–1655, 1990.

[75] C. P. Blair, W. G. Abrahamson, J. A. Jackman, and L. Tyrrell, "Cryptic speciation and host-race formation in a purportedly generalist tumbling flower beetle," *Evolution*, vol. 59, no. 2, pp. 304–316, 2005.

[76] K. D. McCrea and W. G. Abrahamson, "Variation in herbivore infestation: historical vs. genetic factors," *Ecology*, vol. 68, no. 4, pp. 822–827, 1987.

[77] G. D. Maddox and R. B. Root, "Resistance to 16 diverse species of herbivorous insects within a population of goldenrod, *Solidago altissima*: genetic variation and heritability," *Oecologia*, vol. 72, no. 1, pp. 8–14, 1987.

[78] S. B. Heard and G. H. Cox, "Plant module size and attack by the goldenrod spindle-gall moth," *Canadian Entomologist*, vol. 141, no. 4, pp. 406–414, 2009.

[79] T. P. Craig, J. K. Itami, C. Shantz, W. G. Abrahamson, J. D. Horner, and J. V. Craig, "The influence of host plant variation and intraspecific competition on oviposition preference and offspring performance in the host races of *Eurosta solidaginis*," *Ecological Entomology*, vol. 25, no. 1, pp. 7–18, 2000.

[80] J. D. Horner and W. G. Abrahamson, "Influence of plant genotype and environment on oviposition preference and offspring survival in a gallmaking herbivore," *Oecologia*, vol. 90, no. 3, pp. 323–332, 1992.

[81] G. D. Maddox and N. Cappuccino, "Genetic determination of plant susceptibility to an herbivorous insect depends on environmental context," *Evolution*, vol. 40, pp. 863–866, 1986.

[82] K. Halverson, S. B. Heard, J. D. Nason, and J. O. Stireman, "Differential attack on diploid, tetraploid, and hexaploid *Solidago altissima* L. by five insect gallmakers," *Oecologia*, vol. 154, no. 4, pp. 755–761, 2008.

[83] K. Halverson, S. B. Heard, J. D. Nason, and J. O. Stireman, "Origins, distribution, and local co-occurrence of polyploid cytotypes in *Solidago altissima* (Asteraceae)," *American Journal of Botany*, vol. 95, no. 1, pp. 50–58, 2008.

[84] N. Moran, "The genus *Uroleucon* (Homoptera, Aphididae) in Michigan—key, host records, biological notes, and descriptions of 3 new species," *Journal of the Kansas Entomological Society*, vol. 57, pp. 596–616, 1984.

[85] S. S. Anderson, K. D. McCrea, W. G. Abrahamson, and L. M. Hartzel, "Host genotype choice by the ball gallmaker *Eurosta solidaginis* (Diptera: Tephritidae)," *Ecology*, vol. 70, no. 4, pp. 1048–1054, 1989.

[86] P. W. Price, *Macroevolutionary Theory on Macroecological Patterns*, Cambridge University Press, Cambridge, UK, 2003.

[87] D. T. Quiring, L. Flaherty, R. Johns, and A. Morrison, "Variable effects of plant module size on abundance and performance of galling insects," in *Galling Arthropods and Their Associates: Ecology and Evolution*, K. Ozaki, J. Yukawa, T. Ohgushi, and P. W. Price, Eds., pp. 189–198, Springer, Sapporo, Japan, 2006.

[88] S. B. Heard and C. K. Buchanan, "Larval performance and association within and between two species of hackberry nipple gall insects, *Pachypsylla* spp. (Homoptera: Psyllidae)," *American Midland Naturalist*, vol. 140, no. 2, pp. 351–357, 1998.

[89] J. T. Cronin and W. G. Abrahamson, "Goldenrod stem galler preference and performance: effects of multiple herbivores and plant genotypes," *Oecologia*, vol. 127, no. 1, pp. 87–96, 2001.

[90] J. T. Cronin and W. G. Abrahamson, "Host-plant genotype and other herbivores influence goldenrod stem galler preference and performance," *Oecologia*, vol. 121, no. 3, pp. 392–404, 1999.

[91] E. A. Lehnertz, *Impacts of herbivory in a goldenrod/insect community: effects of early-season herbivores on host plants and a late-season herbivore*, M.S. thesis, University of Iowa, Iowa, Canada, 2001.

[92] D. C. Hartnett and W. G. Abrahamson, "The effects of stem gall insects on life history patterns in *Solidago canadensis*," *Ecology*, vol. 60, pp. 910–917, 1979.

[93] S. B. Heard and E. K. Kitts, "Impact of attack by *Gnorimoschema* gallmakers on their ancestral and novel *Solidago* hosts," *Evolutionary Ecology*. In press.

[94] P. B. McEvoy, "Niche partitioning in spittlebugs (Homoptera: Cercopidae) sharing shelters on host plants," *Ecology*, vol. 67, no. 2, pp. 465–478, 1986.

[95] S. Gripenberg, P. J. Mayhew, M. Parnell, and T. Roslin, "A meta-analysis of preference-performance relationships in phytophagous insects," *Ecology Letters*, vol. 13, no. 3, pp. 383–393, 2010.

[96] D. Carmona, M. J. Lajeunesse, and M. T. Johnson, "Plant traits that predict resistance to herbivores," *Functional Ecology*, vol. 25, no. 2, pp. 358–367, 2011.

[97] N. Janz and S. Nylin, "The oscillation hypothesis of host-plant range and speciation," in *Specialization, Speciation, and Radiation: The Evolutionary Biology of Herbivorous Insects*, K. J. Tilmon, Ed., pp. 29–42, University of California Press, Berkeley, Calif, USA, 2008.

[98] J. L. Feder, T. A. Hunt, and L. Bush, "The effects of climate, host plant phenology and host fidelity on the genetics of apple and hawthorn infesting races of *Rhagoletis pomonella*," *Entomologia Experimentalis et Applicata*, vol. 69, no. 2, pp. 117–135, 1993.

[99] H. R. Dambroski and J. L. Feder, "Host plant and latitude-related diapause variation in Rhagoletis pomonella: a test for multifaceted life history adaptation on different stages of diapause development," *Journal of Evolutionary Biology*, vol. 20, no. 6, pp. 2101–2112, 2007.

[100] J. L. Feder, "The effects of parasitoids on sympatric host races of *Rhagoletis pomonella* (Diptera: Tephritidae)," *Ecology*, vol. 76, no. 3, pp. 801–813, 1995.

[101] M. S. Singer, B. Wee, S. Hawkins, and M. Butcher, "Rapid natural and anthropogenic diet evolution: three examples from checkerspot butterflies," in *Specialization, Speciation, and Radiation: The Evolutionary Biology of Herbivorous Insects*, K. J. Tilmon, Ed., pp. 311–324, University of California Press, Berkeley, Calif, USA, 2008.

[102] D. J. Funk, "Does strong selection promote host specialisation and ecological speciation in insect herbivores? Evidence from *Neochlamisus* leaf beetles," *Ecological Entomology*, vol. 35, no. 1, pp. 41–53, 2010.

13

Hydrodynamic and Sensory Factors Governing Response of Copepods to Simulated Predation by Balaenid Whales

Alexander J. Werth

Department of Biology, Hampden-Sydney College, Hampden-Sydney, VA 23943-0162, USA

Correspondence should be addressed to Alexander J. Werth, awerth@hsc.edu

Academic Editor: Bruce Leopold

Predator/prey interactions between copepods and balaenid (bowhead and right) whales were studied with controlled lab experiments using moving baleen in still water and motionless baleen in flowing water to simulate zooplankton passage toward, into, and through the balaenid oral cavity. Copepods showed a lesser escape response to baleen and to a model head simulating balaenid oral hydrodynamics than to other objects. Copepod escape response increased as water flow and body size increased and was greatest at distances ≥ 10 cm from baleen and at copepod density = 10,000 m^{-3}. Data from light/dark experiments suggest that escape is based on mechanoreception, not vision. The model head captured 88% of copepods. Results support previous research showing hydrodynamic effects within a whale's oral cavity create slight suction pressures to draw in prey or at least preclude formation of an anterior compressive bow wave that could scatter or alert prey to the presence of the approaching whale.

1. Introduction

Balaenid (bowhead, *Balaena mysticetus*, and right, *Eubalaena* spp.) whales (Cetacea: Mysticeti) feed almost exclusively on large aggregations of tiny (1–8 mm total body length) calanoid copepods using continuous ram hydraulic filtration [1, 2]. Unlike balaenopterid (rorqual, including blue, fin, and humpback) and eschrichtiid (gray) whales that use complex foraging behaviors to accumulate, engulf, and process prey from a single mouthful of seawater, balaenid morphology and ecology is as specialized and constrained as their diet: they merely exploit (by swimming through) existing zooplankton patches [3]. They ingest a steady, unidirectional current of prey-laden water that enters the mouth anteriorly, through a subrostral gap between paired racks of baleen, and then passes through or along the keratinous baleen plates (with approximately 250–350 plates on each side) comprising the filtering apparatus (Figure 1) [1, 4]. Filtered water exits the oral cavity lateral to the pharyngeal orifice, just anterior to the eyes, at the trailing edge of each lip. The enormous, scoop-shaped head, which can measure 1/4 to 1/3 of an adult balaenid whale's 13–18 m body length, continually removes prey from water as a tow net does, although it is not pulled but rather propelled by the whale's forward locomotion (at all levels of the water column, including surface and benthic layers) at feeding speeds of 2–9 km h^{-1} [5].

Whereas intermittently filtering whales commonly consume large invertebrates (e.g., euphausiids) or fish, an estimated 90% or more of the balaenid diet (depending on whale species, stock, and feeding grounds) consists of tiny copepods (Arthropoda: Maxillopoda) with 0.5–5 mm prosome (cephalothorax) and 1–8 mm total body length and <0.01-0.02 mL volume [3] as revealed by stomach content and observation/net sampling studies [6]. Balaenid predators are thus 50 billion times larger than their prey and ingest up to 500 kg of calanoid copepods per day [3, 7], primarily *Calanus finmarchicus* (along with *C. glacialis* and *Pseudocalanus* spp.) in North Atlantic right whales, *Eubalaena glacialis* [7, 8], and *C. hyperboreus* in bowheads [6], with small amounts of other zooplankton taxa, including principally euphausiids, mysids, and hyperiid and gammarid amphipods [6]. Because balaenid whales feed near the bottom of the trophic pyramid in this abbreviated ecological web, they reap benefits of plentiful energy and biomass, allowing them to attain massive size. However, unlike other large aquatic vertebrates that skim zooplankton via ram hydraulic filtration, including whale

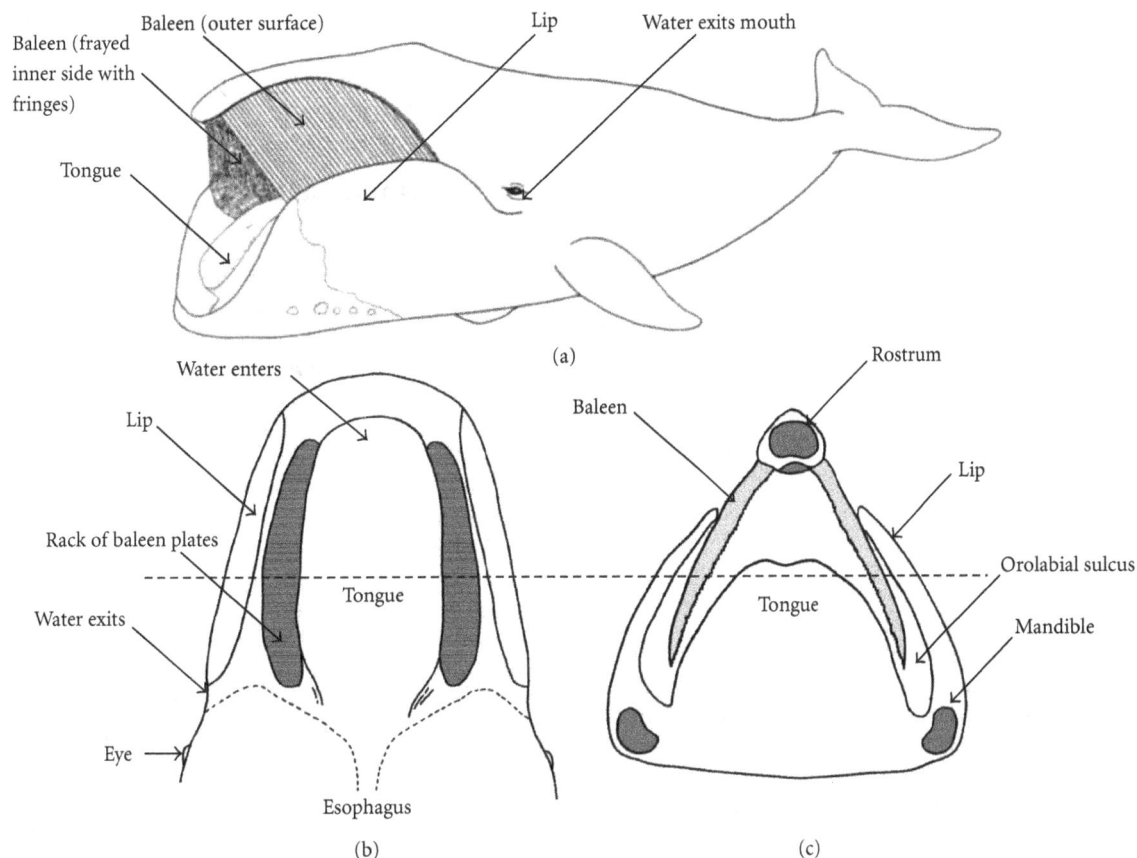

FIGURE 1: Schematic diagrams showing external (a) internal (b), and (c) morphology of the oral region and associated structures in a bowhead whale, *Balaena mysticetus*. At lower left (b) is a frontal section (anterior at top) and at lower right (c) a transverse section (dorsal at top), each at the level of the dashed horizontal line intersecting the other diagram. Prey-laden water enters the mouth anteriorly through a subrostral cleft (between baleen racks) and flows through and around the baleen plates and fringes, exiting, after prey have been removed, behind the lips.

and basking sharks and manta rays, these mammals must sustain a high endothermic metabolic rate, fueled over an annual cycle by alternating periods of feeding and fasting, often of six months each (though the winter fast is typically shorter in *E. glacialis*) [1, 3].

Balaenids are believed to choose copepods as prey for their abundance as well as their oil content and resulting caloric value [9]. There is evidence that copepod stocks affect right whale calving rates and timing [3, 10, 11]. Although bowhead and right whales must engulf huge volumes of copepods, they are constrained more by the density than sheer abundance of copepods, which exhibit an escape response [12–14]. It is therefore essential for optimal foraging that balaenid prey be compacted into dense patches. Engaging and propelling their filter of exceptionally long (up to 4.6 m in *Balaena* and 2.7 m in *Eubalaena*), springy, finely fringed baleen (with porosity similar to a 0.333 mm plankton net [15]) creates substantial drag forces [5]. To optimize caloric intake yet minimize energetic drag costs, balaenids should open their mouths and feed as quickly and efficiently as possible. They have been observed to begin and terminate feeding bouts based on copepod patch density [3], and may alter foraging behavior (notably locomotor speed

and thus hydrodynamics of oral filtration) when forced to prey on larger, more mobile, and evasive zooplankton such as *Euphausia superba* [16]. Balaenids face an enormous challenge of locating an energy-rich yet unreliable, scattered resource—sufficient aggregations of tiny prey that are asymmetrically distributed, temporally and spatially, in highly localized patches throughout a vast ocean—and judging when to open and close their mouths to optimize prey intake, as well as ensuring that prey patches remain concentrated and do not scatter before or during feeding bouts [3, 5]. This is especially crucial when considering that the baleen filter, with its mat of frayed, hair-like filaments, is a passive ram filter that traps items indiscriminately based on the filtration capacity of the baleen plates and fringes, the abundance and density of prey, and the avoidance behavior, if any, of prey species [15].

A previous experimental study of steady-state hydrodynamic forces encountered during balaenid whale foraging [5] focused on pressure effects accompanying the forces and flows of predation on zooplankton. Mathematical and physical modeling (the latter using a 1/15 scale model in a flow tank with pressure transducers) based on morphometric data obtained from bowhead whales harvested by Inupiat

Eskimos revealed that unique features of balaenid oral construction and baleen (subrostral gap, orolabial sulcus, baleen rack curvature, extensive mandibular rotation, and lingual mobility) not only permit unidirectional flow through the mouth, but also establish small-scale pressure effects that improve filtering efficiency. Constriction at the posterior of the pipe-like oral cavity increases fluid flow rate while holding volume of flow constant (as a garden hose nozzle increases flow velocity). This Bernoulli effect in turn generates lower intraoral pressure, producing a Venturi effect, with suction pressures measured at 1300–2000 Pa (10–20 mm Hg) [5]. Significantly, this suction is presumed not only to enhance filtration but also to preclude formation of an anterior compressive bow wave that might alert and scatter nektonic or planktonic prey. These pressure effects likely produce sufficient suction anterior to the oral opening to draw in copepods or other minute prey, or at least to prevent their escape. However, effects of such hydrodynamic influences on actual prey items have not previously been tested.

This biomechanical/ecological study focused on predator/prey interactions between copepods and balaenid whales with controlled lab experiments investigating water flow using a variety of experimental setups (based on morphological data) rather than observation of whales and zooplanktonic prey in their natural habitat. A flow tank (flume) simulated passage of prey-laden water toward, into, and through a portion of the mysticete oral cavity, with motionless baleen and flowing water (at variable flow velocity and volume flow rate); to better recreate a swimming whale, moving baleen was tested in still water. Kinematic sequences were recorded in standard and high-speed video in natural light and in darkness (via laser-induced particle image velocimetry or PIV) to analyze and quantify individual and collective movements of prey items. Studies have investigated movements of individual and aggregated copepods but without the presence of feeding whales [17–20] or have looked at balaenid whales swimming through prey, recording data via digital tags (with hydrophones, depth recorders, and accelerometers) [21, 22] to explore the relationship between whale locomotion and prey engulfment, but without investigating reactions of prey. Crittercams have been used for rorqual lunge feeding to note changes in body position (roll, pitch, and yaw), jaw opening, or other large-scale and sudden events [23, 24], but this is not feasible with right or bowhead whales. Cameras deployed on the back of a feeding balaenid can visualize little due to the tiny size of copepods and great density of their aggregations; cameras used in this way cannot resolve individual prey items or record their movements. Further, none of these approaches offers empirical data on forces, flows, and other events within the whale mouth or addresses ecological questions concerning prey response to whale foraging.

The chief question addressed by this study was (1) how do zooplankters respond to hydrodynamic (flow and pressure) effects simulating an approaching or engulfing whale? In addition, this study asked (2) how does balaenid whale oral flow affect predation on zooplankton? These questions were addressed by flow experiments based on analysis of morphometric and morphological (gross and microscopic)

data from the whale mouth. In summary, this study explored a pair of related ecological issues: how can copepods best avoid becoming prey? How can whales best improve their utilization of this resource?

2. Materials and Methods

2.1. Experimental Subjects. Samples of bowhead baleen from multiple plates were used for flow tank testing; additional gingival and lingual tissues were examined to test hypotheses of sensory abilities impacting whale foraging ecology. Baleen samples were kept submerged in flowing water for at least seven days prior to testing in all trials. All specimens were obtained from adult whales hunted by Inupiat Eskimos of Barrow, Alaska. Tissues were collected under Permit no. 519 issued to T. F. Albert of the North Slope Borough (Alaska) Department of Wildlife Management by the National Marine Fisheries Service. Additional morphometric data, regarding oral dimensions of adult and juvenile whales relating to foraging mechanics and ecology, were taken from previously published studies [5]. No right whale tissues were used; baleen from *Balaena* and *Eubalaena* differs only in plate length, not in fringe number, length, density, or porosity [2].

Initial trials performed to test logistical "proof of concept" (not included in results presented here) used water fleas (*Daphnia magna*) in fresh water and brine shrimp (mixed *Artemia* spp.) in artificial seawater. Experimental trials used live marine pelagic calanoid copepods: *Calanus finmarchicus* (some captured from North Atlantic Ocean, others cultured in lab in pure and artificial seawater at 19°C) and *Acartia tonsa* (some captured from Gulf of Mexico, others cultured in lab). All swam freely, untethered, and unobstructed. Individual copepods used in experiments were in adult and C3–C5 copepodite stages; attempts were made to remove all naupliar larvae from samples placed into testing tanks, and no kinematic/flow data were recorded from the few nauplii that were observed in film sequences. Adult copepods ranged in total body length from 0.9 to 2.7 mm, with mean total lengths of 1.6 mm ($N = 50$, SD = 0.14) for *A. tonsa* and 1.8 mm ($N = 50$, SD = 0.19) for *C. finmarchicus*. Copepods of the two genera were tested separately, not mixed together. Copepods were transferred from holding tanks with fine-mesh nets and put into testing tanks in varying densities ranging from 100 to 50,000 individuals m^{-3}. No attempts were made to separate male, and females, but care was taken to exclude copepod food (phytoplankton) and debris when transferring copepods to testing tanks.

2.2. Flow and Still Tank Testing. Two methods were used to simulate movement of a whale's oral cavity relative to a patch of copepods. In both cases, eight baleen sections (each 20 cm long × 7 cm wide, not including free fringes) were secured by clamping to a metal rod, creating a miniature "rack" of baleen that was submerged just below the surface of the water, with the plates spaced 1 cm apart (as *in vivo*). For the first set of experiments, the baleen filtration apparatus was mounted in a 90-liter circulating flow tank, modeled on a design by Vogel [25]. The tank was made of PVC in a vertical loop with a transparent Plexiglas top in which a completely flat viewing window was installed and through which a ruled grid behind the test chamber could be seen. The working section had a length of 70 cm and cross-sectional area of 900 cm^2 (15% blockage due to the tissue samples, with the rod of baleen attached to the top rim of the testing chamber). Flow through the tank could be adjusted by using impellers of different diameter, by selecting five motor speeds, and with a rheostat to alter input voltage to the motor. Flow velocity for the testing varied from 5 to 140 cm s^{-1}; most trials were performed at flow velocities ranging from 10 to 100 cm s^{-1}, which accords with locomotor velocity of foraging right and bowhead whales [6, 26–28]. Flow velocity was calibrated before and after experimental trials with a digital flow meter (Geopacks model MFP51; Hatherleigh, Devon, UK).

To examine how water flow from the flow tank (flume) was affecting copepod movements, spacing, and interaction with baleen, all trials were duplicated with the same baleen "rack" moving in still water through a patch of copepods (again, with density varying from 100 to 50,000 individuals m^{-3}). Because the working section of the flow tank was too small to accommodate this movement for more than a few seconds, "still water trials" were performed in a 303 L aquarium (125 cm long), with baleen moved manually at speeds equaling water flow in the flume tank (verified via video by time analysis of baleen movement past a ruled grid suspended behind the tank). In some trials, most copepods were lower in this tank's water column than the baleen plates/fringes could reach; data from such trials were not used in the analysis.

In both flowing and still water (with, respectively, motionless and moving baleen "racks"), steady-state, laminar water flow was achieved, as revealed by seeding the water with reflective, neutrally buoyant (1 g cc^{-1}) polymer microspheres with a mean particle size (diameter) of 710 μm. Additional trials tested the influence of other (nonbaleen) submerged objects on copepod orientation and movement. For one set of trials, three items were used (in both 90 L flow tank and 303 L still tank): a sheet of plastic the same size as a section of bowhead baleen (20 × 7 × 0.5 cm), a hollow plastic box (18 × 10 × 2.5 cm), and the same box filled with water. The second set of nonbaleen trials used a detailed 1/15 scale physical model head of a bowhead whale used in an earlier study of balaenid foraging hydrodynamics [5]. The model was made of synthetic plastic clay over an armature of wire, wood, and foam, with paired racks of "baleen" (300 plates each) made of pliable 18 mil high-density polyethylene; it has variable gape, positioned in this study at normal feeding gape (20% of body length). Like the other objects, it was both mounted in the working chamber of the flow tank as well as propelled (suspended on monofilament line) through the large tank of motionless water. Again, care was taken to avoid turbulent flow.

2.3. Kinematic Analysis. In all series of experiments (both flowing and still tank), the "behavior" (spacing, movement, interaction) of copepods relative to baleen fringes was recorded and analyzed. Kinematic sequences were videotaped from the clear viewing window as well as underwater from the testing chamber itself with a digital recording endoscope (VideoFlex SD; Umarex-Laserliner, Arnsberg, Germany) with an illuminated 17 mm camera head (5/25/50 cm focal distances) that recorded AVI video and JPEG still images. Digital sequences (N = 328) were downloaded and analyzed on a Dell Optiplex 745 or Dell Dimension D10 computer using Kinovea 0.8.15 video chronometer and motion analysis software. Sequences were analyzed mainly to detect and measure movement of the zooplankton relative to the stationary or approaching baleen. Principal kinematic variables include copepod locomotor velocity, acceleration, turning radius, and movement of buoyant particles, all tracked relative to observational references (fixed grid background or baleen), with playback at 10–100% of original speed or frame-by-frame, synchronized to time coding. The software allowed for magnification, plane perspective, tracking of path distance, and velocity measurement, which was applied to whole copepods and the baleen fringes/plates/racks or other objects used in flow tank testing.

The camera was mounted in trials using the flow tank, but in trials with moving baleen (or model head or other objects) in the larger tank of motionless water, a fixed camera could not adequately show zooplankton activity. Instead, the camera was moved manually along a track in the same direction and at the same speed as the submerged baleen, such that it could always record, in ambient light or laser illumination, an area of variable distance (10 cm or more) from the closest point of the approaching baleen, at which locations the motion of copepods was measured.

"Capture" of copepods by baleen fringes was noted (with copepods deemed captured not when they made incidental contact with baleen but when they remained in contact for at least 3 s), but trials were performed mainly to assess movements of the copepods relative to the baleen and to other copepods. Four parameters were varied during the trials, for each genus of copepod: flow velocity (10–140 cm s^{-1}), copepod patch density, water temperature (10–29°C), and illumination. Half the trials were performed in natural and artificial light (with and without additional illumination needed from the videocamera system), and half were performed in total darkness to remove visual cues from the copepods' environment. In the dark trials, particle image velocimetry (PIV) was used to analyze copepod movement [18–20], with laser light illumination using a green laser (532 nm, 1W, Nd : YAG/Nd : YVO$_4$) and lens/mirror arrangement to create a single vertical or horizontal plane of green illumination. (It is unknown if *Calanus* or *Acartia* can sense green light, but this is unlikely [29]; in any event, copepods were illuminated by the laser sheet but baleen was not, so copepods could not see it.) For all trials, the water had reflective, neutrally buoyant (1 g cc^{-1}) polymer microspheres. High-speed digital videorecordings of the illuminated and dark (PIV) trials were analyzed to study movement of the copepods and/or of the reflective particles. In non-PIV (i.e., illuminated rather than laser lit) trials, these neutrally buoyant particles were used to provide a scale to measure dimensions of individual copepods; a ruled background with a 1 × 1 cm grid (some squares further ruled into mm) also provided for measurement of copepod movement and copepod size.

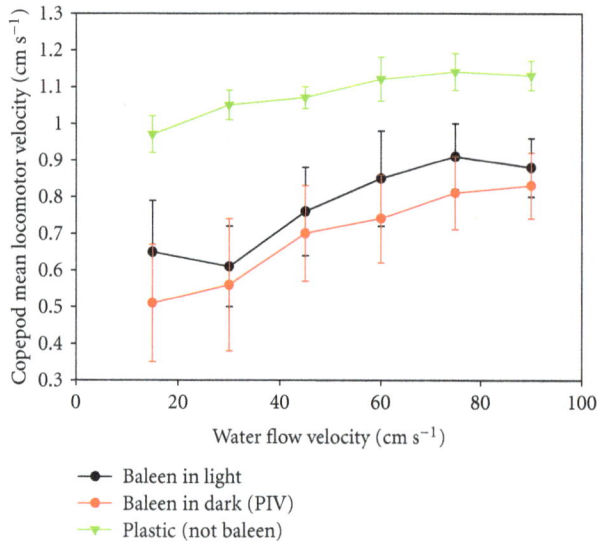

FIGURE 2: Copepod (*Calanus*) locomotor velocity (mean, $N = 50$; error bars = 1 SD) in flow tank, measured at distance = 10 cm from closest baleen, in light and dark (22 degrees C).

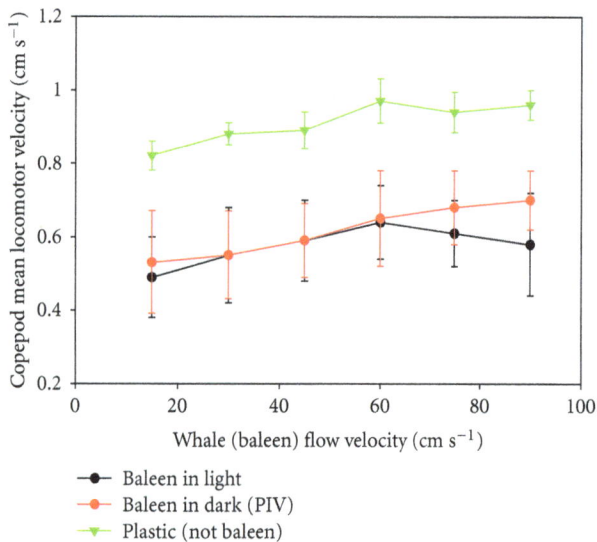

FIGURE 3: Copepod (*Calanus*) locomotor velocity (mean, $N = 35$; error bars = 1 SD) in "still" tank, measured at 10 cm distance from closest point of moving baleen, in light and dark (22 degrees C).

3. Results

3.1. General Response of Copepods to Baleen. Copepods demonstrated an escape response to approaching baleen. This occurred in flow tank testing, with copepods moving relative to fixed baleen (Figure 2), and also in tests involving baleen moving relative to copepods (Figure 3); one-way ANOVAs showed no statistical difference ($P = 0.42$) between tests with fixed and moving baleen. The escape response was occasionally observed in isolated copepods moving/scattering in various directions, but more often involved an apparently coordinated response with most (61%) copepods visible in

videotaped sequences moving in roughly the same direction. This direction was sometimes (<45% of trials) but not exclusively away from (ahead of and below) the approaching baleen, but almost as often (38% of trials) involved movement (beyond that of mere water flow, as indicated by the motion of neutrally buoyant particles in the water column) along with or even toward baleen. Although copepods could often be seen tracking in a general, steady direction over the course of several seconds, individual movements took them in various directions, as if tacking back and forth over an alternating course along a general heading. It was possible to analyze locomotor velocities for copepods moving in water in both light and dark (laser lit) conditions, in the latter case using movement of neutrally buoyant beads suspended in the water for particle image velocimetry, and in light using a ruled background (with beads of known dimension also aiding velocity calculation), with peak velocities sustained for up to 2.5 s.

Trials that involved objects other than the "minirack" of actual baleen tissue (in both the 90 L flow tank and the 303 L stationary tank) produced a greater response than did the baleen. The plastic sheet (same dimensions as baleen, but without its hair-like fringes or bristles), hollow plastic box, and solid (water-filled) plastic box all triggered movements in the copepods that mostly (>75% of instances) involved copepods swimming away from the objects more rapidly (Figures 2 and 3), more directly (i.e., in a straighter line away from the object), and in a more coordinated fashion (with more copepods moving in the same direction rather than going off separately). However, the trials that used the 1/15-scale model of the head in *Balaena* did not produce such an exaggerated response; the model whale head, which was designed to replicate the actual continuous, one-way flow through the balaenid oral cavity, generated copepod response behavior that was much less concerted (both slower and less directed) than with the other nonbaleen objects being placed in the flow tank or moved through the still water (Figure 4). Particles were deemed "captured" by the head when they entered it anteriorly (through the gape at the orolabial sulcus between left and right baleen racks) but did not exit the head through the paired openings behind each lip, posterior to the orolabial sulcus within 10 s (individual copepods were not tracked, but number of copepods entering/exiting the model over time was recorded). A high percentage of copepods were captured (82–95%, depending on the water/head flow rate; mean 88%) by this hydrodynamically correct model, and despite the inherent scaling anomaly (with life-size copepods and a 1/15-scale head), fewer copepods demonstrated an escape response to the model than to the baleen tissue itself, and far fewer copepods reacted to the model head than they did to the approach of another object (the plastic triangle meant to simulate baleen or the plastic box).

All experiments were conducted with either of two pelagic copepod species, *Calanus finmarchicus* or *Acartia tonsa*. No statistically significant quantitative differences (with t tests, $t = 0.36$, and ANOVA, $P = 0.29$) or noteworthzy qualitative differences were observed in trials that tested the influence of prey species on behavior. Thus, for a few analyses, such as influence of copepod body length on locomotor

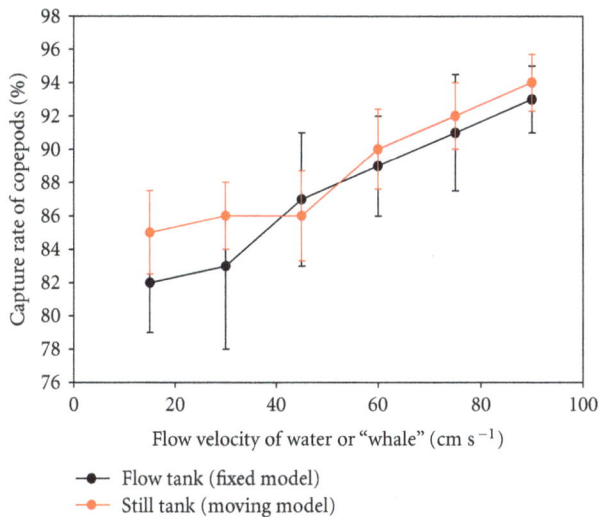

FIGURE 4: Capture rate of copepods (*Acartia*) by 1/15-scale physical model of *Balaena* whale head, including scaled baleen filtering apparatus, in flow and still water tanks (mean, $N = 30$; error bars = 1 SD; water at 20 degrees C), measured by percentage of copepods observed entering "oral cavity" (on videorecordings) that did not emerge from rear of oral cavity after 10 s.

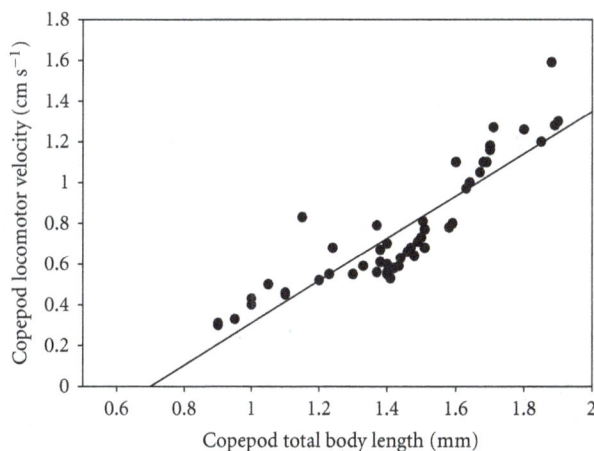

FIGURE 5: Relation of size versus escape speed in copepods (combined data from *Calanus* and *Acartia*) in response to approaching baleen in still tank in light (in water at 20 degrees C).

response (Figure 5), data from *Calanus* and *Acartia* were combined. Both species demonstrated similar behaviors with and without the presence of baleen in the water; they remained throughout the water column and did not gather at the surface or bottom, making them ideal experimental subjects. In general, most individuals of both species were able to move away from baleen that was fixed in the flow tank or moving in the still tank; approximately 23% of copepods overall became entangled in baleen fringes (never >44% in any trial).

3.2. Influence of Flow Velocity on Copepod Response. Copepods of both species (*C. finmarchicus* and *A. tonsa*) showed more movement and greater locomotor velocity of escape

behavior as the flow of water increased (Figure 2), or as the speed of baleen movement in still water similarly increased (Figure 3), with no statistical differences between trials in moving or still water ($P = 0.33$ from one-way ANOVA). Likewise, these data did not reveal generally significant differences ($P = 0.27$) as flow rate varied; however, in some cases (29% of all trials), there was statistical significance ($P = 0.02$) as flow velocity increased from 15 to 90 cm^{-1} s. Even where there were no statistically significant differences, copepod locomotor velocity showed a general direct correspondence with water flow velocity, with greatest escape velocities recorded with water/baleen flow from 60 to 90 cm^{-1} s. As noted above (Section 3.1), copepod locomotor velocities were often sustained for 2-3 s.

3.3. Influence of Light on Copepod Response. The escape response or other movement of the copepods was not found to vary depending on light conditions. The speed, direction, and degree (percentage of individuals in the patch demonstrating movement) of copepod locomotion did not vary whether trials were conducted in natural ambient light or in the dark using laser-illuminated PIV (see Figures 2 and 3). There was slightly more movement of copepods in the light, but statistical ANOVA ($P = 0.24$) and t-testing ($t = 0.18$) reveal that these data did not differ significantly from those conducted in darkness.

3.4. Influence of Temperature on Copepod Response. Trials used seawater at temperatures ranging from 4- to 28°C. No statistically significant differences whatsoever in copepod behavior were observed as temperature varied ($P = 0.48$). All data presented here show tests at a single temperature.

3.5. Influence of Body Size on Copepod Response. Regression analysis of data showing how copepod locomotor velocity varies according to body length (plotted in Figure 5) indicates that size correlates directly with escape velocity ($R^2 = 0.79$). Larger copepods (1.6–1.9 mm) of both *Calanus* and *Acartia* achieved the greatest mean locomotor velocities (1.5–1.7 cm^{-1} s).

3.6. Influence of Distance from Baleen on Copepod Response. Another way to investigate copepod escape behavior, in addition to testing effects of the speed of approaching baleen, was to analyze videotapes to examine how distance of copepods from baleen affected their behavior (Figure 6). Video sequences were set up primarily to record copepod behavior at a distance from baleen of 10 cm, but numerous sequences were recorded to test behavior at different distances. At the least baleen "flow velocity" tested (in still tank, with moving baleen, tested in light) of 30 cm^{-1} s, copepods showed significantly ($P = 0.03$) slower responses than they did at greater flow speeds (60 and 90 cm^{-1} s) when the distance from the approaching baleen increased over 10 cm. At this least approach speed, copepod escape velocity rose as baleen got closer and peaked when baleen was 5 or 10 cm away. At the greater (60 and 90 $cm s^{-1}$) approach speeds, there were no significant differences ($P = 0.41$) as distance

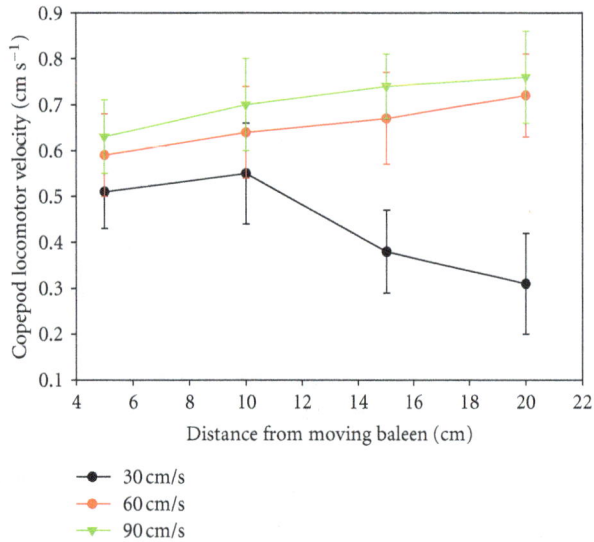

FIGURE 6: Copepod (*Calanus*) locomotor velocity (mean, $N = 35$; error bars = 1 SD) in "still" tank, measured at varying distance from closest point of baleen moving at 30–90 cm s^{-1}, in light (at 22 degrees C).

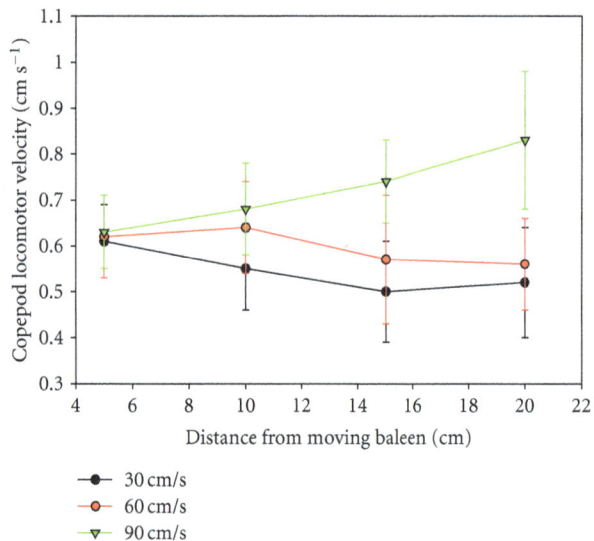

FIGURE 7: Copepod (*Calanus*) locomotor velocity (mean, $N = 30$; error bars = 1 SD) in "still" tank, measured at varying distance from closest point of baleen moving at 30–90 cm s^{-1}, in dark using PIV (at 22 degrees C).

decreased (i.e., as baleen got closer), but the copepod loco-motor velocity was much greater than it was at the 30 cm s^{-1} approach speed. With the same test done in the dark with PIV (Figure 7), copepods demonstrated the same general response (faster escape velocity as baleen approach speed increased), and the influence of the distance from the baleen was the same as in the lighted test conditions, with copepods showing a heightened escape behavior at greater distances at this greater speed (90 cm^{-1} s) than copepods did at the slower speeds (30 and 60 cm s^{-1}). Thus, in both light and dark trials

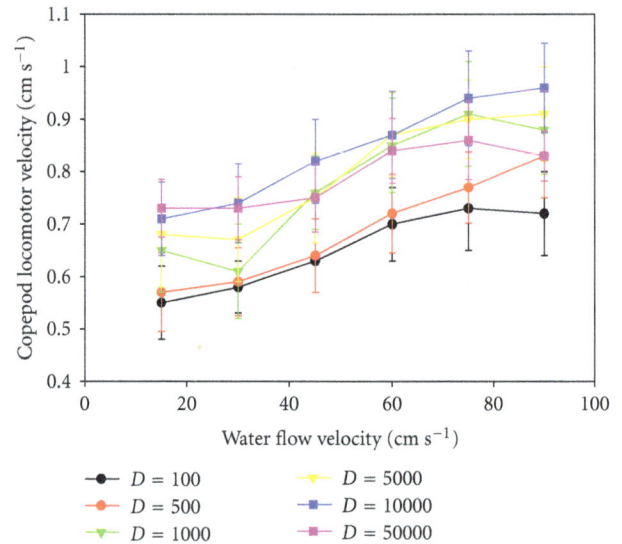

FIGURE 8: Influence of copepod (*Calanus*) density on locomotor velocity (mean, $N = 35$, error bars = 1 SD) in flow tank, measured at distance = 10 cm from closest baleen, in light at 20 degrees C. Density range = 100 to 50,000 copepods m^{-3}.

(Figures 6 and 7), the distance at which copepods reacted to approaching baleen was dependent on flow rate (=speed of approaching baleen). Copepods at extremely short distances from baleen (5 cm) did not demonstrate an escape behavior that differed statistically ($P = 0.37$) from those at greater distances (>5 cm). As noted above, few copepods became captured in the baleen fringe "filter" under the conditions of any experiments.

3.7. Influence of Patch Density on Copepod Response. The final parameter that was studied in these experiments concerned patch density of copepods (*Calanus* and *Acartia*, tested under all experimental conditions). As prey patch density increased from 100 to 50,000 individuals per m^3, it became increasingly difficult to observe and resolve movements of individual copepods relative to each other and relative to calibrating measures (both the ruled grid behind the tank and buoyant particles in the water), but it is clear that loco-motor velocities steadily increase as patch density increases (Figure 8), with significant differences between least (100) and greatest (10,000–50,000 copepods m^{-3}) densities. The escape response was greatest at patch density of 10,000 m^{-3} (Figures 8 and 9); this declined as density increased to 50,000 m^{-3}, and one-way ANOVA testing revealed that the difference was statistically significant ($P = 0.04$; Figure 9).

4. Discussion

4.1. General Response of Copepods to Baleen. Calanoid cope-pods use chemoreception to detect presence and location of their own food (all kinds of phytoplankton as well as microbes and detritus) [13] and have a small eyespot that senses light and darkness, possibly allowing copepods to

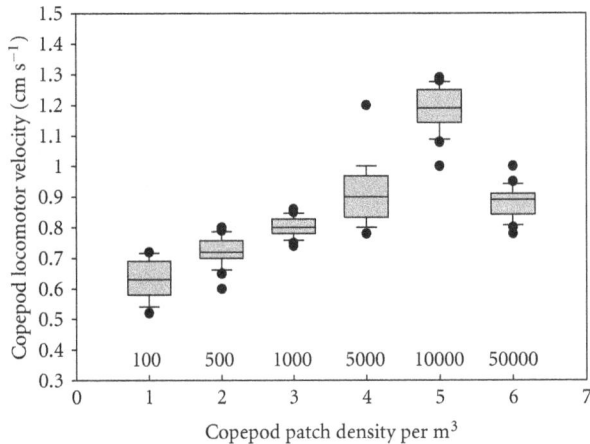

FIGURE 9: Box-and-whisker plot showing range of copepod (*Calanus*) locomotor velocity (N = 35, with median and 25th and 75th percentiles as edges of box, and whiskers as 10th and 90th percentiles), related to patch density (from 100 to 50 K m^{-3}), recorded in light, with flow = 60 cm s^{-1}, at 20 degrees C.

orient their body and track movement (their own and of other objects) [29]. However, copepods mainly sense environmental changes via mechanoreception, by spreading their long, paired antennae to detect vibrations or current changes in adjacent water [12, 13, 30, 31]. Both copepod species tested in this study showed no coordinated movement (motion in the same direction) when on their own in the water, whereas their movements showed coordination when there was baleen or another object nearby and approaching, with the greatest responses observed when the object was 15–20 cm away (Figures 6 and 7). Both *C. finmarchicus* and *A. tonsa* moved more often, more rapidly (Figures 2 and 3), and more efficiently (covering the greatest distance) when they were approached by an object rather than by other copepods. Further, they showed significantly greater escape response to nonbaleen objects than to the actual baleen tissue itself (Figures 2 and 3), perhaps because the fine fringes on this filtering material (which were not present on the plastic objects tested) contribute to greater laminar flow and do not generate a compressive anterior "bow wave" that might alert copepods to the pending approach of a potential predator [30–33]. The capture of high (82–95) percentages of copepods by the model whale head further supports the conclusion that the hydrodynamic effects previously observed [5], namely, the slight suction pressures generated within the balaenid oral cavity due to Bernoulli and Venturi effects, may also be important in capturing planktonic prey and especially in not alerting such prey to presence of a feeding whale [5, 34].

4.2. Influence of Flow Velocity on Copepod Response. The flow velocities tested in this study (both for fixed baleen in the flow tank and moving baleen in the stationary seawater tank) are in accord with those of foraging right and baleen whales [6, 26–28]. Although there were slight differences in copepod locomotor velocity (Figures 2 and 3) as flow velocity increased, these were not significant, and differences with

capture rate of copepods by the scaled model head with changing flow velocity (Figure 4) were barely significant at the $P < 0.05$ level. Thus, the findings presented here are not sufficiently conclusive to address the question of how whale swimming speed will affect prey response. However, given that increased flow velocity (representing a faster swimming whale) did lead to slightly greater escape response by copepods, and especially considering that drag forces from continuous filtration increase substantially as a whale's locomotor velocity increases (as a square of the velocity) [5], it may tentatively be concluded that slower swimming speeds (1 m s^{-1} or less) may be optimal for balaenid whales foraging on copepods, provided no other influences (e.g., currents) are dispersing the plankton.

4.3. Influence of Light on Copepod Response. Calanoid copepods (including both species tested in this study) display a diel vertical migration [35] by altering the density of lipid stores to modulate buoyancy [36, 37] that is presumed to hide them (in deeper, darker water) from predators during daylight hours and allow them to track food sources that may migrate vertically [38, 39]. However, results of this experimental study do not show a difference in copepod behavior in light versus dark environmental conditions, suggesting that the escape response or other reaction to approaching baleen does not depend on vision, either in sighting or otherwise sensing the approaching baleen (or entire whale) or in seeing escape behavior of other, nearby copepods. This supports the conclusion that copepod escape behavior is not visually based but depends on hydrodynamic disturbance. What are the consequences for balaenid foraging ecology? Balaenid whales are most often observed feeding during daylight hours and sleeping (if seen at all) at night, though this likely relates to limitations of human observers. Given the diel vertical migrations of copepods, it is expected that right whales feed at night [35, 37, 40]; furthermore, day/night distinctions may not be meaningful during much of the year for an arctic species such as the bowhead. It is not known how right and bowhead whales locate patches of prey at the surface or at any depth in the water column, but vision is one possibility that has been suggested [37, 41]. If it is true that balaenids use eyesight to locate presence and margins of a patch, and possibly its density, then it appears that bright daylight will not limit ability of whales to forage optimally on this resource, as these data indicate that copepods will not be more prone to scatter in light, decreasing patch density and hence the whale's foraging efficiency, due to their own visual sensation of an approaching whale or baleen filter.

4.4. Influence of Temperature on Copepod Response. Ambient water temperature was not shown to have any effect on copepod behavior in these experiments simulating balaenid filter feeding. Neither normal copepod locomotion nor response to approaching baleen or other objects was affected in any way by temperature (i.e., copepods did not swim more slowly in cold). Most balaenid feeding occurs at high latitudes during the summer feeding season, with bowheads foraging principally on arctic copepods (*Calanus hyperboreus*) [6] and all species of right whale, the Southern (*Eubalaena australis*), North Atlantic (*E. glacialis*), and North Pacific (*E. japonica*), also feeding on cold-water copepods [3, 13, 42, 43]. However, data from satellite-linked transmitters [44] and from stable isotope ratios in baleen [45] suggest that bowheads feed during winter in the Beaufort and Chukchi Seas. Results of this study do not indicate that temperature will make any difference either to copepods in sensing presence of feeding whales or for the foraging whales themselves.

4.5. Influence of Body Size on Copepod Response. Larger copepods (i.e., of longer total body length) showed a significantly greater swimming speed and heightened (faster) escape response to approaching baleen (Figure 5). Given that large patches of copepods and other planktonic prey upon which whales feed include mixed aggregations of different age classes [7, 46], and sometimes species [3, 47, 48], it should make little or no difference for whales seeking to find the densest aggregations of prey. However, from an ecological standpoint, it may be that younger, smaller copepods are more easily captured and hence removed from a population by foraging whales than are older, larger copepods, or that species with smaller body sizes are likewise more easily preyed upon by balaenid whales. It must be emphasized that a determination of whether smaller and more easily captured copepodites are more energetically advantageous prey than larger, more energy-dense [9] yet more elusive copepod age classes and species would require a detailed energetic analysis beyond the scope of this kinematic study.

4.6. Influence of Distance from Baleen on Copepod Response. Because of water's incompressibility, the bow wave generated by a moving whale (if any, given that hydrodynamic effects within the mouth are likely sufficient to cancel such a wave) [5] would be felt by organisms that can sense pressure differences, as can many zooplankton, including calanoid copepods. Findings of these experiments show that at least speed (30 cm s^{-1}), copepods showed the greatest response when baleen was 5–10 cm away (Figure 6), with one-third less escape velocity at greater distances. At greater flow (=whale swimming) speeds, however, copepod swimming velocity rose as distance increased (although this difference was not significant), perhaps because they were alerted to presence of the oncoming baleen by its own compressive force, or by the escape of other copepods. Limited trials showed that there was no coordinated or faster than normal movement at distances from the baleen of greater than 40 cm. No studies in nature have been conducted to investigate at what distance from an actual whale copepods or other zooplankton might initiate an escape response, so this is wholly unknown, but results of these laboratory experiments indicate that at the normal foraging speed of right and bowhead whales (2–9 km hr^{-1}, or 0.5–2.5 m s^{-1}), these distances are almost certainly too small for prey to make any effective escape, with the possible exception of more evasive zooplankton such as larger euphausiids [49, 50], which are occasionally but only rarely taken by balaenids [16]. Balaenids have been observed feeding on elusive prey while swimming at high speeds, however; Hamner et al. [16] reported an account of southern right whales foraging on krill while swimming at 10 knots.

4.7. Influence of Patch Density on Copepod Response. In all conditions tested, copepods displayed a greater response (higher locomotor velocity, presumably of escape response) at greater patch densities, and with significantly greatest response at a density of 10,000 individuals m^{-3} (Figure 9). This may be because as some individuals move in response to the approaching baleen, other copepods sense their movements and are warned to flee. However, it is also possible that increased speeds of copepod movements at greater patch density are due merely to the high number of copepods in a relatively small, enclosed space, with no effect from the simulated predation (i.e., approaching object). Although variable prey densities were tested in this controlled lab setting, copepods were typically in less dense aggregations than have often been measured near feeding right and bowhead whales [7, 46, 47]. Copepods can of course be feeding while they are likewise being fed upon by whales, such that their patch density may depend likewise on abundance and density of phytoplankton [51]. (It was not the case in this study that copepods were moving to collect phytoplankton, which were not placed with copepods in the test chambers.) Still, patch density of copepods and other small zooplankters obviously depends mainly on currents and other oceanographic conditions [52]. However, whereas dense patches afford the most energetic benefit and least energetic (drag) cost for whales feeding via continuous ram filtration, the data from this study indicate that copepods in dense patches are also most likely to display high locomotor velocities that potentially indicate an escape response. This is significant because such patches might disperse to a lesser density and hence become less optimal resources for feeding whales. Yet it is unlikely that the escape response of copepods is fast enough to appreciably scatter patch densities, especially over the scales of times and distances and times needed to affect foraging right or bowhead whales.

4.8. General Conclusions Concerning Balaenid Foraging Ecology and Predation on Copepods. Balaenids are limited by speed and elusiveness of their prey. Copepods are weak swimmers that aggregate only where permitted by currents or other conditions, but this study confirmed that copepods can swim by flicking their antennae for short times (5–10 s) and distances (10–15 cm) at sustained speeds around 1 cm s^{-1}, or over several millimeters in single strokes [20, 31]. To ensure optimal foraging by balaenid whales, movement of a single copepod, however fast or slow, is less important than that of an aggregation with billions to trillions of copepods [7, 53] remain tightly packed.

This study focused on predator/prey interactions between copepod aggregations and balaenid whale filtering tissues using controlled lab experiments (based on morphological analysis) rather than observation of whales and planktonic prey in their natural habitat. Although the simple experimental design involved a different physical scope than a whole whale and full-size patch of zooplankton, it is nevertheless novel and important because it tested aggregations of free-swimming copepods interacting with actual baleen tissue and recorded movements of individual copepods reacting to these conditions of simulated predation. A 2 mm copepod and a 20 m whale occupy vastly different scales, with a difference in Reynolds number of a billion orders of magnitude, from 10^{-1} in the copepod to 10^8 in the whale [25], but their lives are inextricably linked. Previous laboratory studies of copepod locomotor kinematics have used tethered [54] or free single copepods in chambers of stationary water, and it is known that calanoid copepods flee from various stimuli including shadows, currents, and pressure waves that indicate the approach of a predator [12, 30, 31, 55]. Although copepods and other zooplankton are known to avoid nets [56, 57], observers of feeding whales report no bow waves or other hydrodynamic phenomena [47, 58] that would alert or scatter prey.

Likewise, these experiments demonstrated no remarkable hydrodynamic effects from the simulated predation (using baleen tissue and the model head) and little or no consequent response from copepods. Although the plastic sheet and box (hollow and water-filled) showed a distinct compressive wave when moved through water or fixed in the flow tank, as revealed by movement of copepods and especially of buoyant particles, the actual baleen (in miniracks and individually, the latter not used in results here) and model head did not. Instead, particle motion was not disrupted by flow of water toward the baleen/head or by forward motion of the baleen/head through the water. The genuine *in vivo* balaenid filtering apparatus (Figure 1) consists of two racks of 250–350 full-size (up to 4.6 m) plates [59, 60], whereas this study used *ex vivo* smaller collections of eight "plates," each section representing a mere 5% of the length of a real baleen plate. Still, it was instructive to deploy these miniature racks of multiple plates, as copepods often managed to escape the first or second plate of the rack but were swept into more posterior baleen. This is likely how the model head was able to capture so many copepods (nearly 90%), when individual copepods in kinematic sequences were observed avoiding baleen fringes. The ability of the filamentous baleen fringes, and the entire frayed mat they comprise, to sustain laminar flow and to create, based on curvature of the racks plus other oral dimensions, Venturi and other hydrodynamic effects [5, 34] cannot be discounted. This study focused on copepod reactions to baleen, but ongoing experiments are being conducted to determine baleen porosity under varying conditions, and to visualize precise directions of water flow relative to individual plate margins and fringes [61].

It has been estimated that feeding right whales consume 400,000–4,000,000 Calories (or the equivalent of 0.26–26 billion copepods, and between 0.6 and 6.4% of the whale's body weight) per day [53]. How much a whale actually consumes depends on zooplankton concentration: regions with abundant prey may be unsuitable due to their widely dispersed nature, whereas locations with smaller patches of extremely concentrated prey can yield high volumes of ideally distributed food sources. From a sample of dense copepods (331,000 copepods m^{-3}) taken immediately beside a feeding

right whale, Beardsley et al. [7] estimated that the whale was ingesting 1.4 billion copepods hr^{-1}. Balaenids are slow swimmers with a simple, stereotyped foraging mechanism and no repertoire of diverse behaviors (as other whales have), but they have been observed to adjust their precise vertical position and to make sharp turns in response to horizontal clines in prey abundance, so that they optimize foraging by remaining in a path offering the highest potential concentrations of prey items [47], something that rorquals (Balaenopteridae), which are much faster yet feed via lunges and wide turns, are unable to do [58, 62].

How do right and bowhead whales locate copepods and other zooplankton? How do they gauge density of plankton patches? How do they judge where and when best to open and close the mouth? Given their urgent metabolic need to locate and consume dense patches of prey, and indeed the finding that the densest plankton patches ever recorded have been associated with feeding whales [7, 63], Baumgartner et al. [3] ruled out random prospecting. Rather, depending on the scope of the geographic scale, from thousands of kilometers to mere centimeters, these whales likely use a variety of methods, including searching for specific bathymetric contours or topographic landmarks (e.g., seamounts), or perhaps by sensing currents or upwelling or by navigating to known feeding grounds (from instinct, learning, or both) with solar, geomagnetic, or other environmental cues [64, 65]. It is possible that they forage cooperatively, relying on conspecifics to find and relay (actively or passively, intentionally or inadvertently) information on dense sources of prey [3]. At very short distances, whales likely use their own immediate senses to detect prey presence and perhaps also prey density [64]. As with the copepods themselves, foraging whales may use vision or chemosensation (olfaction and gustation are reduced but not absent in balaenids [66, 67], so chemical cues from prey could be detected), but right and bowhead whales probably rely on mechanoreception via the few, scattered hair follicles on the head [68] or with other sensors such as Pacinian and Meissner's corpuscles and free nerve endings. The dermis and subdermis of the balaenid tongue [69] and palatal rete [70] also have sufficient neural networks to relay data from mechanical stimulation, and the vascular systems of these organs, featuring countercurrent blood flow [69], could easily vasodilate to enhance mechanical sensitivity at low water temperatures [71].

Balaenids feed at all levels of the water column, including the surface and benthic layers, with no known differences in prey, in hydrodynamics, or in foraging behavior. The sole difference is that preys at the surface and bottom are restricted by an upper or lower limit on their distribution, whereas midwater prey can be widely distributed vertically [3], although they may also be condensed by oceanographic factors. For example, North Atlantic right whales foraging in the Bay of Fundy have been observed to feed on a dense layer of prey vertically aggregated at the density interface along the upper surface of the bottom mixed layer [8]. Trapping prey at the sea surface would be an effective way for balaenid whales to condense their prey and thus maximize foraging efficiency, though this increases the potential for ship strikes or other collisions [40, 72]. There is evidence of copepods aggregating near the sea floor [3, 34], where active and abandoned fishing gear and lines are sometimes located, thereby creating risks of entanglement [73, 74]. Because balaenid whales are so rigidly adapted morphologically and behaviorally to capturing copepods, these already highly endangered whales are even more vulnerable to extinction when such entanglements occur or should plankton stocks decline due to oceanographic/ecological conditions or human impact.

The past decade has witnessed extraordinary advances in our understanding of mysticete foraging ecology, with data from digital tags deployed on feeding whales providing detailed information on the links between whale locomotion and prey engulfment [22–24], but they tell us nothing about the hydrodynamic forces and flows within the whale mouth. In the continued absence of *in vivo* intraoral data, hopefully to be remedied by placement of tags within the mouth or with swallowed prey, our best approximation of such forces and flows comes from functional morphology and biomechanical investigations of mysticete tissues, especially baleen, tested under controlled conditions presumed to be as realistic as possible. Given that such *ex vivo* tissue studies have not previously been attempted, the study presented here, despite its manifest limitations, advances considerably our understanding of balaenid whale ecology and biomechanics.

Acknowledgments

Samples of baleen are from animals harvested by Inupiat hunters of the Alaska Eskimo Whaling Commission in accordance with their exemption to the U.S. Marine Mammal Protection Act and other national and international restrictions. The author is grateful to these hunters for allowing him to examine and sample their whales. Tissues were collected under Permit no. 519 issued to T. F. Albert of the North Slope Borough (Alaska) Department of Wildlife Management by the National Marine Fisheries Service. He is indebted to I. Robertson for construction of the flow tank, and to S. Vogel for advice on its design. He thanks all his students who assisted with data collection. Numerous constructive comments from anonymous reviewers greatly improved the quality of this paper. Discussions about right and bowhead whale foraging with M. Baumgartner, J. C. George, R. Suydam, R. Payne, C. Mayo, S. Kraus, A. Knowlton, T. Ford, and other scientists also provided valuable information and ideas.

References

[1] A. J. Werth, "Feeding in marine mammals," in *Feeding: Form, Function, and Evolution in Tetrapod Vertebrates*, K. Schwenk, Ed., pp. 487–526, Academic Press, San Diego, Calif, USA, 2000.

[2] A. J. Werth, "How do mysticetes remove prey trapped in baleen?" *Bulletin of the Museum of Comparative Zoology*, vol. 156, pp. 189–203, 2001.

[3] M. F. Baumgartner, C. A. Mayo, and R. D. Kenney, "Enormous carnivores, microscopic food, and a restaurant that's hard to find," in *Urban Whale: North Atlantic Right Whales at the Crossroads*, S. D. Kraus and R. M. Rolland, Eds., pp. 138–171, Harvard University Press, Cambridge, Mass, USA, 2007.

[4] A. Pivorunas, "The feeding mechanisms of baleen whales," *American Scientist*, vol. 67, no. 4, pp. 432–440, 1979.

[5] A. J. Werth, "Models of hydrodynamic flow in the bowhead whale filter feeding apparatus," *Journal of Experimental Biology*, vol. 207, no. 20, pp. 3569–3580, 2004.

[6] L. F. Lowry, "Foods and feeding ecology," in *The Bowhead Whale*, J. J. Burns, J. J. Montague, and C. J. Cowles, Eds., pp. 201–238, Society for Marine Mammalogy, Lawrence, Kan, USA , 1993.

[7] R. C. Beardsley, A. W. Epstein, C. Chen, K. F. Wishner, M. C. Macaulay, and R. D. Kenney, "Spatial variability in zooplankton abundance near feeding right whales in the Great South Channel," *Deep-Sea Research II*, vol. 43, no. 7-8, pp. 1601–1625, 1996.

[8] M. F. Baumgartner and B. R. Mate, "Summertime foraging ecology of North Atlantic right whales," *Marine Ecology Progress Series*, vol. 264, pp. 123–135, 2003.

[9] G. C. Laurence, "Caloric values of some North Atlantic calanoid copepods," *Fishery Bulletin*, vol. 74, pp. 218–220, 1976.

[10] R. D. Kenney, "Right whales and climate change: facing the prospect of a greenhouse future," in *The Urban Whale: North Atlantic Right Whales at the Crossroads*, S. D. Kraus and R. M. Rolland, Eds., pp. 436–459, Harvard University Press, Cambridge, Mass, USA, 2007.

[11] C. H. Greene, A. J. Pershing, T. M. Cronin, and N. Ceci, "Arctic climate change and its impacts on the ecology of the North Atlantic," *Ecology*, vol. 89, supplement 11, pp. S24–S38, 2008.

[12] L. R. Haury, D. E. Kenyon, and J. R. Brooks, "Experimental evaluation of the avoidance reaction of *Calanus finmarchicus*," *Journal of Plankton Research*, vol. 2, no. 3, pp. 187–202, 1980.

[13] L. A. van Duren and J. J. Videler, "The trade-off between feeding, mate seeking and predator avoidance in copepods: behavioural responses to chemical cues," *Journal of Plankton Research*, vol. 18, no. 5, pp. 805–818, 1996.

[14] L. A. van Duren and J. J. Videler, "Escape from viscosity: the kinematics and hydrodynamics of copepod foraging and escape swimming," *Journal of Experimental Biology*, vol. 206, no. 2, pp. 269–279, 2003.

[15] C. A. Mayo, B. H. Letcher, and S. Scott, "Zooplankton filtering efficiency of the baleen of a North Atlantic right whale, *Eubalaena glacialis*," *Journal of Cetacean Research and Management Special Issue*, no. 2, pp. 225–229, 2001.

[16] W. M. Hamner, G. S. Stone, and B. S. Obst, "Behavior of Southern right whales, *Eubalaena australis*, feeding on the Antarctic krill, *Euphausia superba*," *Fishery Bulletin*, vol. 86, no. 1, pp. 143–150, 1988.

[17] M. J. Morris, G. Gust, and J. J. Torres, "Propulsion efficiency and cost of transport for copepods: a hydromechanical model of crustacean swimming," *Marine Biology*, vol. 86, no. 3, pp. 283–295, 1985.

[18] E. J. Stamhuis and J. J. Videler, "Quantitative flow analysis around aquatic animals using laser sheet particle image velocimetry," *Journal of Experimental Biology*, vol. 198, no. 2, pp. 283–294, 1995.

[19] L. A. van Duren, E. J. Stamhuis, and J. J. Videler, "Copepod feeding currents: flow patterns, filtration rates and energetics," *Journal of Experimental Biology*, vol. 206, no. 2, pp. 255–267, 2003.

[20] H. Jiang and T. Kiorboe, "Propulsion efficiency and imposed flow fields of a copepod jump," *Journal of Experimental Biology*, vol. 214, no. 3, pp. 476–486, 2011.

[21] M. F. Baumgartner and F. W. Wenzel, "Springtime foraging ecology of North Atlantic right whales," in *Proceedings of the Ocean Sciences Meeting*, American Geophysical Union and American Society of Limnology and Oceanography, Orlando, Fla, USA, March, 2008.

[22] M. J. Simon, M. Johnson, P. Tyack, and P. T. Madsen, "Behaviour and kinematics of continuous ram filtration in bowhead whales (*Balaena mysticetus*)," *Proceedings of the Royal Society B*, vol. 276, no. 1674, pp. 3819–3828, 2009.

[23] J. A. Goldbogen, N. D. Pyenson, and R. E. Shadwick, "Big gulps require high drag for fin whale lunge feeding," *Marine Ecology Progress Series*, vol. 349, pp. 289–301, 2007.

[24] J. A. Goldbogen, J. Calambokidis, E. Oleson et al., "Mechanics, hydrodynamics and energetics of blue whale lunge feeding: efficiency dependence on krill density," *Journal of Experimental Biology*, vol. 214, no. 1, pp. 131–146, 2011.

[25] S. Vogel, *Life in Moving Fluids: The Physical Biology of Flow 2e*, Princeton University Press, Princeton, NJ, USA, 1996.

[26] R. R. Reeves and S. Leatherwood, "Bowhead whale, *Balaena mysticetus* Linnaeus 1758," in *Handbook of Marine Mammals*, S. H. Ridgway and R. Harrison, Eds., vol. 3 of *The Sirenians and Baleen Whales*, pp. 305–344, Academic Press, San Diego, Calif, USA, 1985.

[27] G. M. Carroll, J. C. George, L. F. Lowry, and K. O. Coyle, "Bowhead whale (*Balaena mysticetus*) feeding near Point Barrow, Alaska, during the 1985 spring migrations," *Arctic*, vol. 40, pp. 105–110, 1987.

[28] D. P. Nowacek, M. Johnson, P. Tyack, K. A. Shorter, W. A. McLellan, and D. A. Pabst, "Buoyant balaenids: the ups and downs of buoyancy in right whales," *Proceedings of the Royal Society B*, vol. 268, no. 1478, pp. 1811–1816, 2001.

[29] J. H. Cohen and R. B. Forward, "Spectral sensitivity of vertically migrating marine copepods," *Biological Bulletin*, vol. 203, no. 3, pp. 307–314, 2002.

[30] D. M. Fields and J. Yen, "The escape behavior of marine copepods in response to a quantifiable fluid mechanical disturbance," *Journal of Plankton Research*, vol. 19, no. 9, pp. 1289–1304, 1997.

[31] E. J. Buskey, P. H. Lenz, and D. K. Hartline, "Escape behavior of planktonic copepods in response to hydrodynamic disturbances: high speed video analysis," *Marine Ecology Progress Series*, vol. 235, pp. 135–146, 2002.

[32] T. Kiorboe and E. Saiz, "Planktivorous feeding in calm and turbulent environments, with emphasis on copepods," *Marine Ecology Progress Series*, vol. 122, no. 1–3, pp. 135–145, 1995.

[33] H. Yamazaki and K. D. Squires, "Comparison of oceanic turbulence and copepod swimming," *Marine Ecology Progress Series*, vol. 144, no. 1–3, pp. 299–301, 1996.

[34] R. H. Lambertsen, K. J. Rasmussen, W. C. Lancaster, and R. J. Hintz, "Functional morphology of the mouth of the bowhead whale and its implications for conservation," *Journal of Mammalogy*, vol. 86, no. 2, pp. 342–352, 2005.

[35] M. F. Baumgartner, N. S. J. Lysiak, C. Schuman, J. Urban-Rich, and F. W. Wenzel, "Diel vertical migration behavior of *Calanus finmarchicus* and its influence on right and sei whale occurrence," *Marine Ecology Progress Series*, vol. 423, pp. 167–184, 2011.

[36] J. T. Enright, "Copepods in a hurry: sustained high-speed upward migration," *Limnology and Oceanography*, vol. 22, no. 1, pp. 118–125, 1977.

[37] M. F. Baumgartner, T. V. N. Cole, R. G. Campbell, G. J. Teegarden, and E. G. Durbin, "Associations between North Atlantic right whales and their prey, *Calanus finmarchicus*, over diel

and tidal time scales," *Marine Ecology Progress Series*, vol. 264, pp. 155–166, 2003.

[38] D. K. Ralston, D. J. Mcgillicuddy, and D. W. Townsend, "Asynchronous vertical migration and bimodal distribution of motile phytoplankton," *Journal of Plankton Research*, vol. 29, no. 9, pp. 803–821, 2007.

[39] K. Haraguchi, T. Yamamoto, S. Chiba, Y. Shimizu, and M. Nagao, "Effects of phytoplankton vertical migration on the formation of oxygen depleted water in a shallow coastal sea," *Estuarine, Coastal and Shelf Science*, vol. 86, no. 3, pp. 441–449, 2010.

[40] Woods Hole Oceanographic Institution, "Prey-tell: why right whales linger in the Gulf of Maine," *Science Daily*, 2012 http://www.sciencedaily.com/releases/2011/04/110426151042.htm.

[41] Q. Zhu, D. J. Hillmann, and W. G. Henk, "Morphology of the eye and surrounding structures of the bowhead whale, *Balaena mysticetus*," *Marine Mammal Science*, vol. 17, no. 4, pp. 729–750, 2001.

[42] V. J. Rowntree, L. O. Valenzuela, P. F. Fraguas, and J. Seger, "Foraging behaviour of southern right whales (Eubalaena australis) inferred from variation of carbon stable isotope rations in their baleen," Report of the International Whaling Commission (IWC) SC/60.BRG23, 2008.

[43] W. C. Cummings, "Right whales—Eubalaena glacialis and Eubalaena australis," in *Handbook of Marine Mammals*, S. H. Ridgway and R. Harrison, Eds., vol. 3 of *The Sirenians and Baleen Whales*, pp. 275–304, Academic Press, San Diego, Calif, USA, 1985.

[44] L. T. Quakenbush, J. J. Citta, J. C. George, R. J. Small, and M. P. Heide-Jorgensen, "Fall and winter movements of bowhead whales (*Balaena mysticetus*) in the Chukchi Sea and within a potential petroleum development area," *Arctic*, vol. 63, no. 3, pp. 289–307, 2010.

[45] S. H. Lee, D. M. Schell, T. L. McDonald, and W. J. Richardson, "Regional and seasonal feeding by bowhead whales *Balaena mysticetus* as indicated by stable isotope ratios," *Marine Ecology Progress Series*, vol. 285, pp. 271–287, 2005.

[46] K. F. Wishner, E. Durbin, A. Durbin, M. Macaulay, H. Winn, and R. Kenney, "Copepod patches and right whales in the Great South Channel off New England," *Bulletin of Marine Science*, vol. 43, no. 3, pp. 825–844, 1988.

[47] C. A. Mayo and M. K. Marx, "Surface foraging behaviour of the North Atlantic right whale, *Eubalaena glacialis*, and associated zooplankton characteristics," *Canadian Journal of Zoology*, vol. 68, no. 10, pp. 2214–2220, 1990.

[48] L. D. Murison, *Zooplankton distributions and feeding ecology of right whales (Eubalaena glacialis glacialis) in the outer Bay of Fundy, Canada*, M.S. thesis, University of Guelph, Guelph, Canada, 1986.

[49] P. F. Brodie, D. D. Sameoto, and R. W. Sheldon, "Population densities of euphausiids off Nova Scotia as indicated by net samples, whale stomach contents, and sonar," *Limnology and Oceanography*, vol. 23, pp. 1264–1267, 1978.

[50] P. H. Wiebe, C. J. Ashjian, S. M. Gallager, C. S. Davis, G. L. Lawson, and N. J. Copley, "Using a high-powered strobe light to increase the catch of Antarctic krill," *Marine Biology*, vol. 144, no. 3, pp. 493–502, 2004.

[51] B. T. Hargrave and G. H. Geen, "Effects of copepod grazing on two natural phytoplankton populations," *Journal of the Fisheries Research Board of Canada*, vol. 27, no. 8, pp. 1395–1403, 1970.

[52] A. W. Epstein and R. C. Beardsley, "Flow-induced aggregation of plankton at a front: a 2-D Eulerian model study," *Deep-Sea Research II*, vol. 48, no. 1–3, pp. 395–418, 2001.

[53] R. D. Kenney, M. A. M. Hyman, R. E. Owen, G. P. Scott, and H. E. Winn, "Estimation of prey densities required by western North Atlantic right whales," *Marine Mammal Science*, vol. 2, no. 1, pp. 1–13, 1986.

[54] K. B. Catton, D. R. Webster, J. Brown, and J. Yen, "Quantitative analysis of tethered and free-swimming copepodid flow fields," *Journal of Experimental Biology*, vol. 210, no. 2, pp. 299–310, 2007.

[55] R. J. Waggett and E. J. Buskey, "Calanoid copepod escape behavior in response to a visual predator," *Marine Biology*, vol. 150, no. 4, pp. 599–607, 2007.

[56] R. A. Barkley, "The theoretical effectiveness of towed-net samplers as related to sampler size and to swimming speed of organisms," *Journal du Conseil Permanent International pour l'Exploration de la Mer*, vol. 29, pp. 146–157, 1964.

[57] P. H. Wiebe, S. H. Boyd, B. M. Davis, and J. L. Cox, "Avoidance of towed nets by the euphausiid *Nematoscelis megalops*," *Fishery Bulletin*, vol. 80, no. 1, pp. 75–91, 1982.

[58] W. A. Watkins and W. E. Schevill, "Right whale feeding and baleen rattle," *Journal of Mammalogy*, vol. 57, pp. 58–66, 1976.

[59] T. Nemoto, "Food of baleen whales with reference to whale movements," *Scientific Reports of the Whales Research Institute*, vol. 14, pp. 149–291, 1959.

[60] A. Pivorunas, "A mathematical consideration on the function of baleen plates and their fringes," *Scientific Reports of the Whales Research Institute*, vol. 28, pp. 37–55, 1976.

[61] A. J. Werth, "Flow-dependent porosity of baleen from the bowhead whale (*Balaena mysticetus*)," in *Proceedings of the Society for Integrative and Comparative Biology*, Salt Lake City, Utah, USA, 2011, abstract.

[62] W. A. Watkins and W. E. Schevill, "Aerial observations of feeding behavior in four baleen whales: *Eubalaena glacialis, Balaenoptera borealis, Megaptera novaeangliae*, and *Balaenoptera physalus*," *Journal of Mammalogy*, vol. 60, pp. 155–163, 1979.

[63] K. F. Wishner, J. R. Schoenherr, R. Beardsley, and C. Chen, "Abundance, distribution and population structure of the copepod *Calanus finmarchicus* in a springtime right whale feeding area in the South Western Gulf of Maine," *Continental Shelf Research*, vol. 15, no. 4-5, pp. 475–507, 1995.

[64] R. D. Kenney, C. A. Mayo, and H. E. Winn, "Migration and foraging strategies at varying spatial scales in Western North Atlantic right whales: a review of hypotheses," *Journal of Cetacean Research and Management Special Issue*, no. 2, pp. 251–260, 2001.

[65] M. J. Simon, *The sounds of whales and their food: baleen whales, their foraging behavior, ecology and habitat use in an arctic habitat*, Ph.D. thesis, Aarhus University, 2010.

[66] J. G. M. Thewissen, J. George, C. Rosa, and T. Kishida, "Olfaction and brain size in the bowhead whale (*Balaena mysticetus*)," *Marine Mammal Science*, vol. 27, no. 2, pp. 282–294, 2011.

[67] T. Kishida and J. G. M. Thewissen, "Evolutionary changes of the importance of olfaction in cetaceans based on the olfactory marker protein gene," *Gene*, vol. 492, no. 2, pp. 349–353, 2012.

[68] J. K. Ling, "Vibrissae of marine mammals," in *Functional Anatomy of Marine Mammals*, R. J. Harrison, Ed., vol. 3, pp. 387–415, Academic Press, London, UK, 1977.

[69] A. J. Werth, "Adaptations of the cetacean hyolingual apparatus for aquatic feeding and thermoregulation," *Anatomical Record*, vol. 290, no. 6, pp. 546–568, 2007.

[70] T. J. Ford and S. D. Kraus, "A rete in the right whale," *Nature*, vol. 359, no. 6397, article 680, 1992.

[71] G. Dehnhardt, B. Mauck, and H. Hyvärinen, "Ambient temperature does not affect the tactile sensitivity of mystacial vibrissae in harbour seals," *Journal of Experimental Biology*, vol. 201, no. 22, pp. 3023–3029, 1998.

[72] S. E. Parks, J. D. Warren, K. Stamieszkin, C. A. Mayo, and D. Wiley, "Dangerous dining: surface foraging of North Atlantic right whales increases risk of vessel collisions," *Biology Letters*, vol. 8, no. 1, pp. 57–60, 2012.

[73] A. J. Johnson, S. D. Kraus, J. F. Kenney, and C. A. Mayo, "The entangled lives of right whales and fishermen: can they co-exist?" in *The Urban Whale: North Atlantic Right Whales at the Crossroads*, S. D. Kraus and R. M. Rolland, Eds., pp. 380–408, Harvard University Press, Cambridge, Mass, USA, 2007.

[74] T. Johnson, *Entanglements: The Intertwined Fates of Whales and Fishermen*, University of Florida Press, Gainesville, Fla, USA, 2005.

14

Factors Influencing Progress toward Ecological Speciation

Marianne Elias,[1] Rui Faria,[2, 3] Zachariah Gompert,[4] and Andrew Hendry[5]

[1] CNRS, UMR 7205, Muséum National d'Histoire Naturelle, 45 Rue Buffon, CP50, 75005 Paris, France
[2] CIBIO/UP—Centro de Investigação em Biodiversidade e Recursos Genéticos, Universidade do Porto, Campus Agrário de Vairão,
R. Monte-Crasto, 4485-661 Vairão, Portugal
[3] IBE—Institut de Biologia Evolutiva (UPF-CSIC), Universitat Pompeu Fabra, PRBB, Avenue Doctor Aiguader N88,
08003 Barcelona, Spain
[4] Department of Botany, 3165, University of Wyoming, 1000 East University Avenue Laramie, WY 82071, USA
[5] Redpath Museum and Department of Biology, McGill University, 859 Sherbrooke Street West Montreal, QC, Canada H3A 2K6

Correspondence should be addressed to Marianne Elias, melias2008@googlemail.com

1. Introduction

Ecological speciation occurs when adaptation to divergent environments, such as different resources or habitats, leads to the evolution of reproductive isolation [1, 2]. More specifically, divergent (or disruptive) selection between environments causes the adaptive divergence of populations, which leads to the evolution of reproductive barriers that decrease, and ultimately cease, gene flow [3, 4]. Supported by a growing number of specific examples, ecological speciation is thought to be a primary driving force in evolutionary diversification, exemplified most obviously in adaptive radiations [5–8].

As acceptance of the importance of ecological speciation has grown, so too has the recognition that it is not all powerful. Specifically, a number of instances of nonecological speciation and nonadaptive radiation seem likely [9], and colonization of different environments does not always lead to speciation [10, 11]. This latter point is obvious when one recognizes that although essentially all species are composed of a number of populations occupying divergent environments [12], only a fraction of these ever spin off to become full-fledged species. Instead, populations occupying divergent environments or using different resources show varying levels of progress toward ecological speciation—and this variation provides the substrate to study factors that promote and constrain progress along the speciation continuum. By studying these factors, we can begin to understand why there are so many species [13] and also why there are so few species [14].

This special issue on ecological speciation puts snapshots of progress toward speciation sharply in focus and then investigates this topic from several angles. First, several papers provide conceptual or theoretical models for how to consider progress toward ecological speciation (Funk; Heard; Lenormand; Liancourt et al.; Agrawal et al.). Second, several papers highlight the noninevitability of ecological speciation through investigations where ecological speciation seems to be strongly constrained (Räsänen et al.; Bolnick) or at least lacking definitive evidence (Ostevik et al.; Scholl et al.). Some of these papers also uncover specific factors that seem particularly important to ecological speciation, such as the combination of geographic isolation and habitat differences (Surget-Groba et al.), the strength of disruptive selection and assortative mating (Bolnick), and host-plant adaptation (Scholl et al.). Third, several particularly important factors emerge as a common theme across multiple papers, particularly parasites/pollinators (Xu et al.; Karvonen and Seehausen), habitat choice (Webster et al.; Feder et al.; Carling and Thomassen; Egan et al.), and phenotypic plasticity (Fitzpatrick; Vallin and Qvarnström).

Here we highlight the most important aspects of these contributions and how they relate to three major topics: (i) models for progress toward ecological speciation; (ii) variable progress toward ecological speciation in nature; and (iii) factors affecting progress toward ecological speciation.

2. Models for Progress toward Ecological Speciation

Terminological issues have long bedevilled communication among researchers working on speciation. D. J. Funk addresses this topic by first clarifying the relationship between sympatric speciation (whereby reproductively isolated populations evolve from an initially panmictic population) and ecological speciation (whereby reproductive isolation evolves as a consequence of divergent/disruptive natural selection). These are orthogonal concepts [15]. First, even if disruptive selection is a common way of achieving sympatric speciation, this can also be caused by other factors, such as changes in chromosome number. Second, ecological speciation can readily occur in allopatry [16, 17]. Funk then introduces four new concepts aiming to reduce confusion in the literature. *Sympatric race* is a generalisation of *host race* (usually used for herbivores or parasites) and refers to any sympatric populations that experience divergent selection and are partly but incompletely reproductively isolated. *Envirotypes* are populations that differ due to phenotypic plasticity. *Host forms* are populations that exhibit host-associated variation, but for which the nature of variation (e.g., envirotype, host race, cryptic species) has not yet been diagnosed. *Ecological forms* are a generalization of *host forms* for nonherbivore or parasitic taxa. The two latter concepts acknowledge the fact that one has an incomplete understanding of speciation. To overcome the problem of overdiagnosing host races, Funk introduces five criteria, based on host association and choice, coexistence pattern, genetic differentiation, mate choice, gene flow, and hybrid unfitness. Funk's maple and willow associated phytophagous populations of *Neochlamisus bebbianae* leaf beetle meet all these criteria and can, therefore, be considered as host races.

Another phytophagy-inspired conceptual model for how an insect species initially using one plant species might diversify into multiple insect species using different host plants is presented by S. Heard. This effort explicitly links variation in host plant use within insect species or races to the formation of different host races and species. In this proposed "gape-and-pinch" model, Heard posits four stages (or "hypotheses") of diversification defined in part by overlap in the plant trait space used by the insect races/species. In the first stage "adjacent errors," some individuals within an insect species using one plant species might "mistakenly" use individuals of another plant species that have similar trait values to their normal host plant species. In the next stage "adjacent oligophagy," populations formed by the insects that shifted plant species then experience divergent selection—and undergo adaptive divergence—leading to a better use of that new host. In the third stage "trait distance-divergence," competition and reproductive interactions cause character displacement between the emerging insect races or species so that they become specialized on particularly divergent subsets of the trait distributions of the two plant species. In the final stage "distance relaxation," the new species become so divergent that they no longer interact, and can then evolve to use trait values more typical of each plant species. Heard provides

a theoretical and statistical framework for testing this model and applies it to insects using goldenrod plants.

Local adaptation is often the first step in ecological speciation, and so factors influencing local adaptation will be critical for ecological speciation. Local adaptation can either increase over time (if more specialized alleles spread), eventually leading to speciation, or it can decrease over time (if more generalist alleles spread). T. Lenormand reviews the conditions that favor these different scenarios and emphasizes the role of three positive feedback loops that favor increased specialization. In the demographic loop, local adaptation results in higher population density, which in turn favors the recruitment of new locally adapted alleles. In the recombination loop, locally adapted alleles are more likely to be recruited in genomic regions already harbouring loci with locally adapted alleles, thereby generating genomic regions of particular importance to local adaptation. In the reinforcement loop, local adaptation selects for traits that promote premating isolation (reinforcement), which in turn increases the recruitment and frequency of locally adapted alleles. Lenormand then details the mechanisms involved in reinforcement, particularly assortative mating, dispersal, and recombination. He highlights that these characteristics represent the three fundamental steps in a sexual life cycle (syngamy, dispersal, and meiosis) and that they promote genetic clustering at several levels (within locus, among individuals, among loci). His new classification is orthogonal to, and complements, the traditional one- versus two-allele distinction [14]. Overall, the rates of increased specialization and reinforcement determine progress toward ecological speciation.

One of the major constraints on ecological speciation is the establishment of self-sustaining populations in new/marginal environments, because the colonizing individuals are presumably poorly adapted to the new conditions. This difficulty might be eased through facilitation, the amelioration of habitat conditions by the presence of neighbouring living organisms (biotic components) [18]. According to this process, the benefactor's "environmental bubble" facilitates the beneficiary's adaptation to marginal conditions, which can result in ecological speciation if gene flow from the core habitat is further reduced. At the same time, however, facilitation might hinder further progress toward ecological speciation by maintaining gene flow between environments and by preventing reinforcement in secondary contact zones. P. Liancourt, P. Choler, N. Gross, X. Thibert-Plante, and K. Tielbörg consider these possibilities from the beneficiary species perspective, through a spatially and genetically explicit modelling framework that builds on earlier models [19, 20]. They find that ecological speciation is more likely with larger patch (facilitated versus harsh) sizes. Liancourt and coauthors further suggest that facilitation can play another important role in evolution by helping to maintain a genetic diversity "storage" in marginal habitats, a process with some parallel to niche conservationism. A deeper understanding of the role of facilitation in diversification is needed (both theoretically and empirically), and the authors suggest that stressful environmental gradients would be useful study systems for this endeavour.

Intrinsic postzygotic isolation, a fundamental contributor to speciation, is often caused by between-locus genetic incompatibilities [21, 22]. The origin of these incompatibilities, particularly in the face of gene flow, remains an outstanding question. A. F. Agrawal, J. L. Feder, and P. Nosil use two-locus two-population mathematical models to explore scenarios where loci subject to divergent selection also affect intrinsic isolation, either directly or via linkage disequilibrium with other loci. They quantified genetic differentiation (allelic frequencies of loci under selection), the extent of intrinsic isolation (hybrid fitness), and the overall barrier to gene flow (based on neutral loci). They find that divergent selection can overcome gene flow and favors the evolution of intrinsic isolation, as suggested previously [23]. Counterintuitively, intrinsic isolation can sometimes weaken the barrier to gene flow, depending on the degree of linkage between the two focal loci. This occurs because intrinsic isolation sometimes prevents differentiation by divergent selection.

3. Variable Progress toward Ecological Speciation in Nature

Threespine stickleback fish, with their diverse populations adapted to different habitats, had provided a number of examples of how adaptive divergence can promote ecological speciation [24, 25]. Indeed, work on this group has fundamentally shaped our modern understanding of ecological speciation [1–3]. At the same time, however, three-spine stickleback also provides evidence of the frequent failure of divergent selection to drive substantial progress toward ecological speciation [25]. This special issue provides two of such examples. In one, K. Räsänen, M. Delcourt, L. J. Chapman and A. P. Hendry report that, despite strong divergent selection, lake and stream stickleback from the Misty watershed do not exhibit positive assortative mate choice in laboratory experiments. These results are in contrast to the strong assortative mating observed in similar studies of other stickleback systems, such as benthic versus limnetic [26] and marine versus fresh water [27]. In addition to providing potential explanations for this discrepancy, Räsänen et al. conclude that the apparent conundrum of limited gene flow but no obvious reproductive barriers could be very informative about the factors that constrain progress toward ecological speciation.

The second paper on three-spine stickleback, by D. I. Bolnick, considers the opposite conundrum: reproductive barriers are seemingly present but gene flow is not limited. Particularly, even though ecologically driven sympatric speciation does not always occur in sticklebacks, its theoretically necessary and sufficient conditions seem often to be present in nature. First, some populations experience strong competition for resources that causes extreme phenotypes to have higher fitness [28]. Second, assortative mating based on diet and morphology is present in some of these same lakes [29]. So how to solve this new conundrum? Using a simulation model, Bolnick demonstrates that the strengths of selection and assortative mating measured in lake populations in nature are too weak to cause sympatric speciation. Instead, lake stickleback appears to respond to disruptive selection through alternative means of reducing competition, such as increased genetic variance, sexual dimorphism, and phenotypic plasticity.

Another classic system for studying ecological speciation, or more generally adaptive radiation, is *Anolis* lizards of the Caribbean. In particular, many of the larger islands contain repeated radiations of similar "ecomorph" species in similar habitats [8]. Contrasting with this predictable and repeatable diversity on large islands, smaller islands contain only a few species. Y. Surget-Groba, H. Johansson, and R. S. Thorpe studied populations of *Anolis roquet* from Martinique. This species contains populations with divergent mitochondrial lineages, a consequence of previous allopatric episodes, and is distributed over a range of habitats. It can, therefore, be used to address the relative importance of past allopatry, present ecological differences, and their combination in determining progress toward ecological speciation. Using microsatellite markers, the authors find that geographic isolation alone does not result in significant population differentiation, habitat differences alone cause some differentiation, and geographic isolation plus habitat differences cause the strongest differentiation. The authors conclude that speciation is likely initiated in allopatry but is then completed following secondary contact only through the action of adaptation to different habitats.

Even though ecological differences are clearly important in the diversification of both plants and animals [1, 30], it remains uncertain as to whether the process is fundamentally the same or different between them. Part of the reason is that typical methods for studying ecological speciation differ between the two groups. In an effort to bridge this methodological divide, K. Ostevik, B. T. Moyers, G. L. Owens, and L. H. Rieseberg apply a common method of inference from animals to published studies on plants. In particular, ecological speciation is often inferred in animals based on evidence that independently derived populations show reproductive isolation if they come from different habitats but not if they come from similar habitats: that is, parallel speciation [31, 32]. Ostevik and coauthors review potential examples of ecological speciation in plants for evidence of parallel speciation. They find that very few plant systems provide such evidence, perhaps simply because not many studies have performed the necessary experiments. Alternatively, plants might differ fundamentally from animals in how ecological differences drive speciation, particularly due to the importance of behaviour in animals.

A current topic of interest in ecological speciation is whether strong selection acting on a single trait (strong selection) or relatively weak selection acting on a greater number of traits (multifarious selection) is more common and more likely to complete the speciation process [11]. Using another well-studied model of ecological speciation, butterflies of the genus *Lycaeides*, C. F. Scholl, C. C. Nice, J. A. Fordyce, Z. Gompert, and M. L. Forister compared host-plant associated larval performance of butterflies from several populations of *L. idas*, *L. melissa*, and a species that originated through hybridization between the two. By conducting a series of reciprocal rearing experiments, they found little to no evidence for local adaptation to the natal hosts. By putting these results into the context of the other previously studied ecological traits (e.g., host and mate preference, phenology, and egg adhesion [33, 34]), the authors constructed a schematic representation of the diversification within this butterfly species complex. They conclude that no single trait acts as a complete reproductive barrier between the three taxa and that most traits reduce gene flow only asymmetrically. The authors suggest the need for further study of multiple traits and reproductive barriers in other taxa.

4. Factors Affecting Progress toward Ecological Speciation

4.1. The Role of Pollinators/Parasites. In many cases of ecological speciation, we think of the populations in question colonizing and adapting to divergent environments/resources, such as different plants or other food types. However, environments can also "colonize" the populations in question that might then speciate as a result. Colonization by different pollinators and subsequent adaptation to them, for example, is expected to be particularly important for angiosperms. A particularly spectacular example involves sexually deceptive Orchids, where flowers mimic the scent and the appearance of female insects and are then pollinated during attempted copulation by males. In a review and meta-analysis of two Orchid genera, S. Xu, P. M. Schlüter, and F. P. Schiestl find floral scent to be a key trait in both divergent selection and reproductive isolation. Other traits, including flower colour, morphology and phenology, also appear to play an important role in ecological speciation within this group. The authors also conclude that although sympatric speciation is likely rare in nature, it is particularly plausible in these Orchids.

Parasites can be thought of as another instance of different environments "colonizing" a focal species and then causing divergent/disruptive selection and (perhaps) ecological speciation. As outlined in the contribution by A. Karvonen and O. Seehausen, differences in parasites could contribute to ecological speciation in three major ways. First, divergent parasite communities could cause selection against locally adapted hosts that move between those communities, as well as any hybrids. Second, adaptation to divergent parasite communities could cause assortative mating to evolve as a pleiotropic by-product, such as through divergence in MHC genotypes that are under selection by parasites and also influence mate choice (see also [35]). Third, sexual selection might lead females in a given population to prefer males that are better adapted to local parasites and can thus achieve better condition. The authors conclude that although suggestive evidence exists for all three possibilities, more work is needed before the importance of parasites in ecological speciation can be confirmed.

4.2. The Role of Habitat Choice. The importance of habitat (or *host*) isolation in ecological speciation is widely recognized. This habitat isolation is determined by habitat choice (preference or avoidance), competition, and habitat performance (fitness differences between habitats) [36]. S. E. Webster, J. Galindo, J. W. Grahame, and R. K. Butlin propose a conceptual framework to study and classify traits involved in habitat choice, based on three largely independent criteria: (1) whether habitat choice allows the establishment of a stable polymorphism maintained by selection without interfering with mating randomness or if it also promotes assortative mating; (2) whether it involves one-allele or two-allele mechanisms of inheritance; (3) whether traits are of single or multiple effect [37], the latter when habitat choice is simultaneously under direct selection and contributes to assortative mating. The combination of these three criteria underlies ten different scenarios, which the authors visit using previously published empirical data. They argue that the speed and likelihood of ecological speciation depends on the mechanism of habitat choice and at which stage of the process it operates, with scenarios of one-allele and/or multiple-effect traits being more favorable. While these scenarios have rarely been distinguished in empirical studies, Webster et al. reason that such distinctions will help in the design of future studies and enable more informative comparisons among systems. In practice, however, the identification of the mechanisms involved and discriminating among different scenarios may sometimes be difficult, as exemplified by the case of the intertidal gastropod *Littorina saxatilis*, a model system for ecological speciation.

Hybrids resulting from the crosses between individuals from populations with different habitat preferences will tend to show interest in both parental habitats. This will increase gene flow between parental species, inhibiting reproductive isolation. Inspired by host-specific phytophagous insects, J. L. Feder, S. P. Egan, and A. A. Forbes ask, what if individuals choose their habitat based on avoidance rather than preference? According to the authors, hybrids for alleles involved in avoidance of alternate parental habitats may experience a kind of behavioral breakdown and accept none of the parental habitats, generating a postzygotic barrier to gene flow. Feder and collaborators determine the reasons why habitat avoidance is underappreciated in the study of ecological speciation (theoretical and empirical), and try to improve this issue. They propose new theoretical models and do not find strong theoretical impediments for habitat avoidance to evolve and generate hybrid behavioral inviability even for nonallopatric scenarios. They also suggest a physiological mechanism to explain how habitat specialists evolve to prefer a new habitat and avoid the original one. Feder et al. also document empirical support for this theory. Accumulated data on *Rhagoletis pomonella* and preliminary results on *Utetes lectoides* strongly suggest that avoidance has evolved in these species, contributing to postzygotic reproductive isolation. A literature survey in phytophagous insects reveals at least ten examples consistent with habitat avoidance, and three cases of behavior inviability in hybrids consistent with this mechanism. The authors also present suggestions and cautionary notes for design and interpretation of results when it comes to experiments on habitat choice.

Hybrid zones are particularly useful systems for determining whether differences in habitat preference or habitat-associated adaptation contribute to reproductive isolation. M. D. Carling and H. A. Thomassen investigate the effect of environmental variation on admixture in a hybrid zone between the Lazuli Bunting (*Passerian amoena*) and the Indigo Bunting (*P. cyanea*). They find that differences in environment explain interpopulation differences in the frequency and genetic composition of hybrids. This is not the first study to document an effect of environmental variation on the production or persistence of hybrids [38, 39] but Carling and Thomassen were also able to associate this pattern with specific environmental variables, particularly rainfall during the warmest months of the year. They discuss possible, complementary mechanistic explanations for these patterns, including habitat avoidance or preference in hybrids and habitat-dependent fitness. Their results indicate that inherent (i.e., non-geographic) barriers to gene flow between *P. amoena* and *P. cyanea* are environment dependent, which means these barriers could be ephemeral and vary in space and time.

S. P. Egan, G. R. Hood and J. R. Ott present one of the first direct tests of the role of habitat (host) isolation driven by host choice. Different populations of the gall wasp *Belonocnema treatae* feed on different oak species. Egan et al. first confirmed that *B. treatae* prefer their native host plant, with a stronger preference for females. They then demonstrated assortative mating among host populations, which was enhanced by the presence of the respective host plants. This enhancement was due to the fact that females usually mate on their host and that males also prefer their natal host plant. Therefore, host preference is directly responsible for reproductive isolation in *B. treatae*, by decreasing the pro-bability of encounter between individuals from different host populations. The mechanism revealed here likely applies to many host/phytophagous or host/parasite systems.

4.3. The Role of Phenotypic Plasticity. Phenotypic plasticity, the ability of a single genotype to express different phenotypes under different environmental conditions, has long been seen as an alternative to genetic divergence, and therefore as potential constraint on adaptive evolution [40, 41]. More recently, however, adaptive phenotypic plasticity has been rehabilitated as a factor potentially favoring divergent evolution by enabling colonizing new niches, where divergent selection can then act on standing genetic variation [42]. B. M. Fitzpatrick reviews the possible effects of phenotypic plasticity on the two components of ecological speciation: local adaptation and reproductive isolation. He finds that both adaptive and maladaptive plasticity can promote or constrain ecological speciation, depending on several factors, and concludes that many aspects of how phenotypic plasticity acts have been underappreciated.

Several other papers in the special issue also provide potential examples of the role of plasticity in ecological speciation. For instance, N. V. Vallin and A. Qvarnström studied habitat choice in two hybridizing species of flycatchers. When the two species occur in sympatry, pied flycatchers are displaced from their preferred habitat due to competition with the dominant collared flycatchers. Cross-fostering experiments showed that rearing environment matters to recruits' habitat choice more than does the environment of the genetic parents: pied flycatcher fledglings whose parents were displaced to pine habitats were more likely to return to nest in pine habitats. Thus, competition-mediated switches between habitats can cause a change of habitat choice through learning, which might then enhance reproductive isolation via ecological segregation. This role of plasticity and learning in habitat choice is also acknowledged in the contribution of Webster and collaborators.

5. Unanswered Questions and Future Directions

Although it is widely recognized that ecological speciation can occur without gene flow between diverging groups of individuals [43], the recognition of its importance has grown because of recent evidence for speciation *with* gene flow [44]. If gene flow commonly occurs during divergence, some mechanism, such as divergent selection must also occur frequently to counteract the homogenizing effect of gene flow. The manuscripts in this special issue, and a plethora of other recent publications [45–50], have made great strides in advancing our understanding of ecological speciation. These allow us to identify several key factors that affect progress toward ecological speciation, such as habitat choice (preference and avoidance), phenotypic plasticity, role of pollinators/parasites, complex biological interactions such as facilitation, as well as geographical context. However, for most cases, our understanding is still incomplete. For instance, the circumstances under which plasticity favors or inhibits adaptation, mate choice, and consequently ecological speciation are still largely unknown. Further insights will certainly arise from a multitude of empirical and theoretical studies, but certain areas of research are particularly likely to yield important results. For example, whereas we can rarely observe the time course of speciation in a single species, we can learn about factors affecting progress toward ecological speciation by studying and contrasting pairs of related populations at different points along the speciation continuum. Similarly, the study of parallel speciation may be highly informative. Such studies exist (e.g., [25, 26, 51, 52]), but we need many more systems where we can examine variation in progress toward ecological speciation. It is important that we also investigate instances where speciation fails, as these cases will advance our understanding of factors that constrain and enhance progress toward speciation. Furthermore, recent advances in DNA sequencing and statistical analysis offer an unprecedented opportunity to study the genetic basis and evolution of reproductive isolation during ecological speciation. The application of these new methods and models to ecologically well-studied systems have been and will be particularly informative [53, 54]. Finally, more studies using experimental manipulations to study the effects of key parameters on ecological speciation are badly needed, especially if they can be combined with an understanding of natural populations.

Acknowledgments

We would like to express our gratitude to all the authors and reviewers that contributed to the successful completion of this special issue. R. Faria research is financed by the Portuguese Science Foundation (FCT) through the program COMPETE (SFRH/BPD/26384/2006 and PTDC/BIA-EVF/113805/2009).

Marianne Elias
Rui Faria
Zachariah Gompert
Andrew Hendry

References

[1] D. Schluter, *The Ecology of Adaptive Radiation*, Oxford University Press, Oxford, UK, 2000.

[2] P. Nosil, *Ecological Speciation*, Oxford University Press, Oxford, UK, 2012.

[3] H. D. Rundle and P. Nosil, "Ecological speciation," *Ecology Letters*, vol. 8, no. 3, pp. 336–352, 2005.

[4] K. Räsänen and A. P. Hendry, "Disentangling interactions between adaptive divergence and gene flow when ecology drives diversification," *Ecology Letters*, vol. 11, no. 6, pp. 624–636, 2008.

[5] D. Lack, *Darwin's Finches*, Cambridge University Press, Cambridge, UK, 1947.

[6] G. G. Simpson, *The Major Features of Evolution*, Columbia University Press, New York, NY, USA, 1953.

[7] P. R. Grant and B. R. Grant, *How and Why Species Multiply: The Radiation of Darwin's Finches*, Princeton University Press, Princeton, NJ, USA, 2008.

[8] J. B. Losos, *Lizards in an Evolutionary Tree: Ecology and Adaptive Radiation of Anoles*, University of California Press, Berkeley, Calif, USA, 2009.

[9] R. J. Rundell and T. D. Price, "Adaptive radiation, nonadaptive radiation, ecological speciation and nonecological speciation," *Trends in Ecology and Evolution*, vol. 24, no. 7, pp. 394–399, 2009.

[10] A. P. Hendry, "Ecological speciation! Or the lack thereof?" *Canadian Journal of Fisheries and Aquatic Sciences*, vol. 66, no. 8, pp. 1383–1398, 2009.

[11] P. Nosil, L. J. Harmon, and O. Seehausen, "Ecological explanations for (incomplete) speciation," *Trends in Ecology and Evolution*, vol. 24, no. 3, pp. 145–156, 2009.

[12] J. B. Hughes, G. C. Daily, and P. R. Ehrlich, "Population diversity: its extent and extinction," *Science*, vol. 278, no. 5338, pp. 689–692, 1997.

[13] G. E. Hutchinson, "Homage to Santa Rosalia or why are there so many kinds of animals?" *The American Naturalist*, vol. 93, pp. 145–159, 1959.

[14] J. Felsenstein, "Skepticism towards Santa Rosalia, or why are there so few kinds of animals?" *Evolution*, vol. 35, pp. 124–138, 1981.

[15] U. Dieckmann, J. A. J. Metz, M. Doebeli, and D. Tautz, "Introduction," in *Adaptive Speciation*, U. Dieckmann, J. A. J. Metz, M. Doebeli, and D. Tautz, Eds., pp. 1–17, Cambridge University Press, Cambridge, UK, 2004.

[16] R. B. Langerhans, M. E. Gifford, and E. O. Joseph, "Ecological speciation in *Gambusia* fishes," *Evolution*, vol. 61, no. 9, pp. 2056–2074, 2007.

[17] T. H. Vines and D. Schluter, "Strong assortative mating between allopatric sticklebacks as a by-product of adaptation to different environments," *Proceedings of the Royal Society B*, vol. 273, no. 1589, pp. 911–916, 2006.

[18] M. D. Bertness and R. Callaway, "Positive interactions in communities," *Trends in Ecology and Evolution*, vol. 9, no. 5, pp. 191–193, 1994.

[19] M. Kirkpatrick and N. H. Barton, "Evolution of a species' range," *American Naturalist*, vol. 150, no. 1, pp. 1–23, 1997.

[20] J. R. Bridle, J. Polechová, M. Kawata, and R. K. Butlin, "Why is adaptation prevented at ecological margins? New insights from individual-based simulations," *Ecology Letters*, vol. 13, no. 4, pp. 485–494, 2010.

[21] T. Dobzhansky, *Genetics and the Origin of Species*, Columbia University Press, New York, NY, USA, 1st edition, 1973.

[22] H. J. Muller, "Isolating mechanisms, evolution, and temperature," *Biology Symposium*, vol. 6, pp. 71–125, 1942.

[23] M. Turelli, N. H. Barton, and J. A. Coyne, "Theory and speciation," *Trends in Ecology and Evolution*, vol. 16, no. 7, pp. 330–343, 2001.

[24] J. S. McKinnon and H. D. Rundle, "Speciation in nature: the threespine stickleback model systems," *Trends in Ecology and Evolution*, vol. 17, no. 10, pp. 480–488, 2002.

[25] A. P. Hendry, D. I. Bolnick, D. Berner, and C. L. Peichel, "Along the speciation continuum in sticklebacks," *Journal of Fish Biology*, vol. 75, no. 8, pp. 2000–2036, 2009.

[26] H. D. Rundle, L. Nagel, J. W. Boughman, and D. Schluter, "Natural selection and parallel speciation in sympatric sticklebacks," *Science*, vol. 287, no. 5451, pp. 306–308, 2000.

[27] J. S. McKinnon, S. Mori, B. K. Blackman et al., "Evidence for ecology's role in speciation," *Nature*, vol. 429, no. 6989, pp. 294–298, 2004.

[28] D. I. Bolnick and L. L. On, "Predictable patterns of disruptive selection in stickleback in postglacial lakes," *American Naturalist*, vol. 172, no. 1, pp. 1–11, 2008.

[29] L. K. Snowberg and D. I. Bolnick, "Assortative mating by diet in a phenotypically unimodal but ecologically variable population of stickleback," *American Naturalist*, vol. 172, no. 5, pp. 733–739, 2008.

[30] J. M. Sobel, G. F. Chen, L. R. Watt, and D. W. Schemske, "The biology of speciation," *Evolution*, vol. 64, no. 2, pp. 295–315, 2010.

[31] D. J. Funk, "Isolating a role for natural selection in speciation: host adaptation and sexual isolation in Neochlamisus bebbianae leaf beetles," *Evolution*, vol. 52, no. 6, pp. 1744–1759, 1998.

[32] L. Nagel and D. Schluter, "Body size, natural selection, and speciation in sticklebacks," *Evolution*, vol. 52, no. 1, pp. 209–218, 1998.

[33] J. A. Fordyce, C. C. Nice, M. L. Forister, and A. M. Shapiro, "The significance of wing pattern diversity in the Lycaenidae: mate discrimination by two recently diverged species," *Journal of Evolutionary Biology*, vol. 15, no. 5, pp. 871–879, 2002.

[34] J. A. Fordyce and C. C. Nice, "Variation in butterfly egg adhesion: adaptation to local host plant senescence characteristics?" *Ecology Letters*, vol. 6, no. 1, pp. 23–27, 2003.

[35] C. Eizaguirre and T. L. Lenz, "Major histocompatability complex polymorphism: dynamics and consequences of parasite-mediated local adaptation in fishes," *Journal of Fish Biology*, vol. 77, no. 9, pp. 2023–2047, 2010.

[36] J. A. Coyne and H. A. Orr, *Speciation*, Sinauer, Sunderland, Mass, USA, 2004.

[37] C. Smadja and R.K. Butlin, "A framework for comparing processes of speciation in the presence of gene flow," *Molecular Ecology*, vol. 20, pp. 5123–5140, 2011.

[38] D. M. Rand and R. G. Harrison, "Ecological genetics of a mosaic hybrid zone: mitochondrial, nuclear, and reproductive differentiation of crickets by soil type," *Evolution*, vol. 43, no. 2, pp. 432–449, 1989.

[39] T. H. Vines, S. C. Köhler, M. Thiel et al., "The maintenance of reproductive isolation in a mosaic hybrid zone between the fire-bellied toads *Bombina bombina* and *B. Variegata*," *Evolution*, vol. 57, no. 8, pp. 1876–1888, 2003.

[40] S. Wright, "Evolution in Mendelian populations," *Genetics*, vol. 16, pp. 97–159, 1931.

[41] D. A. Levin, "Plasticity, canalization and evolutionary stasis in plants," in *Plant Population Ecology*, A. J. Davy, M. J. Hutchings, and A. R. Watkinson, Eds., pp. 35–45, Blackwell Scientific Publications, Oxford, UK, 1988.

[42] D. W. Pfennig, M. A. Wund, E. C. Snell-Rood, T. Cruickshank, C. D. Schlichting, and A. P. Moczek, "Phenotypic plasticity's impacts on diversification and speciation," *Trends in Ecology and Evolution*, vol. 25, no. 8, pp. 459–467, 2010.

[43] D. Schluter, "Evidence for ecological speciation and its alternative," *Science*, vol. 323, no. 5915, pp. 737–741, 2009.

[44] C. Pinho and J. Hey, "Divergence with gene flow: models and data," *Annual Review of Ecology, Evolution, and Systematics*, vol. 41, pp. 215–230, 2010.

[45] P. Nosil, D. J. Funk, and D. Ortiz-Barrientos, "Divergent selection and heterogeneous genomic divergence," *Molecular Ecology*, vol. 18, no. 3, pp. 375–402, 2009.

[46] J. L. Feder and P. Nosil, "The efficacy of divergence hitchhiking in generating genomic islands during ecological speciation," *Evolution*, vol. 64, no. 6, pp. 1729–1747, 2010.

[47] J. L. Feder, G. Gejji, S. Yeaman, and P. Nosil, "Establishment of new mutations under divergence and genome hitchhiking,," *Philosophical Transactions of the Royal Society B*, vol. 367, pp. 461–474, 2012.

[48] J. L. Feder and P. Nosil, "Chromosomal inversions and species differences: when are genes affecting adaptive divergence and reproductive isolation expected to reside within inversions?" *Evolution*, vol. 63, no. 12, pp. 3061–3075, 2009.

[49] A. P. Michel, S. Sim, T. H. Q. Powell, M. S. Taylor, P. Nosil, and J. L. Feder, "Widespread genomic divergence during sympatric speciation," *Proceedings of the National Academy of Sciences of the United States of America*, vol. 107, no. 21, pp. 9724–9729, 2010.

[50] R. Faria, S. Neto, M. A. F. Noor, and A. Navarro, "Role of natural selection in chromosomal speciation," in *Encyclopedia Life Sciences*, Chichester, UK, 2011.

[51] D. Berner, A. C. Grandchamp, and A. P. Hendry, "Variable progress toward ecological speciation in parapatry: stickleback across eight lake-stream transitions," *Evolution*, vol. 63, no. 7, pp. 1740–1753, 2009.

[52] R. M. Merrill, Z. Gompert, L. M. Dembeck, M. R. Kronforst, W. O. Mcmillan, and C. D. Jiggins, "Mate preference across the speciation continuum in a clade of mimetic butterflies," *Evolution*, vol. 65, no. 5, pp. 1489–1500, 2011.

[53] P. A. Hohenlohe, S. Bassham, P. D. Etter, N. Stiffler, E. A. Johnson, and W. A. Cresko, "Population genomics of parallel adaptation in threespine stickleback using sequenced RAD tags," *PLoS Genetics*, vol. 6, no. 2, Article ID e1000862, 2010.

[54] P. Nosil and J. L. Feder, "Genomic divergence during speciation: causes and consequences," *Philosophical Transactions of the Royal Society B*, vol. 367, pp. 332–342, 2012.

15

Flower Density Is More Important Than Habitat Type for Increasing Flower Visiting Insect Diversity

L. A. Scriven, M. J. Sweet, and G. R. Port

School of Biology, Ridley Building, Newcastle University, Newcastle upon Tyne NE1 7RU, UK

Correspondence should be addressed to L. A. Scriven; l.a.scriven@ncl.ac.uk

Academic Editor: J. J. Wiens

Declines in flora and fauna are well documented and highlight the need to manage available habitats to benefit local biodiversity. Between May and September in 2011 the number, composition, and diversity of flower visiting insects were assessed across eight sites, representing a range of habitats within an industrial site in the North East of England, UK. There was no significant difference in insect assemblages between the sites selected, but there was a significant difference between the months surveyed. Flower density was highlighted as the most important factor driving these changes between months and indicates that flower density is more important to a site for insect diversity than the presence of specific habitats. Analysis of the insect communities each month allowed comparison of dominant insects to the flower density data, highlighting sites where management intervention could be initiated to benefit insect diversity, or alternatively specific management plans to encourage target species. Furthermore, this study highlights the importance of correct data interpretation to answer specific management objectives and recommends analysing the insect community interactions to determine the dominant species present prior to undertaking any management of the site in question.

1. Introduction

As a result of human influences, habitats and ecosystems are continually fragmented by factors such as city expansion, and agricultural intensification [1, 2]. Areas such as gardens, parks, brownfield sites, and working industrial sites are becoming important "islands" for wildlife between the ever-increasing urbanised areas. However, these sites require management to conserve biodiversity and to provide an optimum habitat network for species [3]. Sites left unmanaged over long periods of time can become dominated by rank grasses and pernicious weeds, reducing the biodiversity and value of the habitat [4].

Previous studies have highlighted the impacts of environmental management schemes and the effects of habitat fragmentation on biodiversity within agricultural environments [5–7]. Additionally, the importance of gardens and parks for pollinating species within urbanised areas is becoming more apparent [8–10]. Nevertheless, the importance of industrial areas has not yet been considered. Despite industrial sites often being heavily utilised, a large proportion have the potential to create wildlife refuges within an urban and

agriculturally dominant landscape. Industrial sites are found worldwide and frequently cover a large expanse of land, usually incorporating varied unmanaged habitats.

The aluminium smelter site at Lynemouth, UK, incorporates a range of these different habitats including scrub, grassland, and wetland. Although it is widely accepted that increased habitat variety often results in higher species diversity [11–13], factors such as the botanical structure and flower density within the site can be more influential on insect assemblages than distinct habitat types [14, 15]. This highlights further questions as to what drives these trends within such habitats. For example, how variable do habitats need to improve insect diversity? Does habitat connectivity affect habitat quality? And importantly, if a specific management strategy was implemented at sites such as Lynemouth, would it be possible to increase diversity of insect species present?

The success of any programme to enhance biodiversity is dependent on how people manage land and invest in development. Within large companies, where employees have been encouraged to take an interest in biodiversity, site action plans have become successful. However, an important step before management strategies can be employed is to determine the

TABLE 1: Description of the eight sites used within the study, highlighting the name given, location, habitat type, and a description of the dominant plant species present within each site.

Site	Location (British National Grid)	Habitat area (hectares)	Description
Flower Rich (FR)	NZ29350 BNG89685	0.55	Well drained, poor quality soil. Flowering species such as *Lotus corniculatus* (Birds foot Trefoil), *Trifolium pratense* (Red clover), and *Dactylorhiza incarnata* (Marsh orchid) present
Mown Grassland (MG)	NZ29674 BNG89205	0.30	Lawn areas regularly mown and dominated by *Taraxacum officinale* (Dandelions), *Bellis perennis* (daisies). Backing onto long unmanaged grassland surrounded by farmland
New Hedge (NH)	NZ29741 BNG89191	0.20	Running alongside wheat field. Consisting of *Rosa canina* (dogrose), *crataegus* sp. (hawthorn), *Prunus spinosa* (blackthorn), and *Rubus fruticosus* (blackberry)
Old Hedge (OH)	NZ28918 BNG89685	0.20	Single species hedge hawthorn, bordering *Brassica napus* (oil seed rape) field, with weed species such as *Matricaria discoidea* (pineapple weed) and *Leucanthemum vulgare* (oxeye daisy) in the field no flowering plants in the field margin
Plantation Woodland (Pl)	NZ29135 BNG89703	0.62	*Urtica dioica* (Nettles) as understory plants, 2 m separated poplar trees with occasional *Acer pseudoplatanus* (sycamore), *Quercus robur* (oak), and *Sorbus aria* (whitebeam)
Pond (Po)	NZ29384 BNG89278	0.55	Pond surrounded by *Vicia sativa* (vetches), *Anthriscus cerefolium* (chervil), *Oenanthe crocata* (Dropwort water hemlock), *Rubus sp.* (blackberry), and *Centaurea nigra* (common knapweed)
Ridge and Furrow Grassland (RF)	NZ29001 BNG89671	0.60	*Heracleum sphondylium* (Hogweed) dominated grassland with tall dominant grasses such as *Elymus repens* (couch grass) and *Arrhenatherum elatius* (false oat grass) surrounded by farmland
Woodhorn Woodland (WW)	NZ29079 BNG89616	0.55	Older woodland trees including *A. pseudoplatanus* and *Q. robur*, with an understory of *Hyacinthoides nonscripta* (bluebells), *Galanthus sp.* (snowdrops), and *Rubus sp.* (blackberry)

value of the site for biodiversity. By assessing different habitats and understanding which species are utilising these areas, informed decisions on future land management can be made.

Therefore, this study aims to assess flower visiting insects to determine the value of the Lynemouth smelter for pollinators and other flower visiting insects. Specifically we aimed at determining: (i) which flower visiting insects are active at the Lynemouth smelter between May and September, (ii) which site/habitat hosts the highest diversity of flower visiting species, and (iii) what factors influence differences between sites/habitats. Answering these questions allows us to determine which habitats have the highest value for flower visitor biodiversity at present and which sites would benefit from management.

2. Study Design

Eight individual sites representing different habitat types were identified before the start of the study and the diversity of flower visiting insects was assessed within these sites on a monthly basis for one survey season (May–September).

2.1. Study Site. The study was conducted on land surrounding the Rio Tinto Alcan Aluminium smelter in Lynemouth, UK (55.2016°N, 1.5396°W). Covering 82.7 hectares the site is typical of a working industrial site, with intensively managed grassland and shrub borders around offices, access roads, car parks, and production units. However, 20.7 hectares (25%) of the site is predominately scrub, woodland, and wetland forming a buffer zone. To the west of the smelter, hybrid poplar trees have been mixed with native European tree species such as *Sorbus aria* (whitebeam), *Acer pseudoplatanus* (sycamore), and *Quercus robur* (oak) to create a fast growing screen to the smelter. Since this planting, the densely populated woodlands have left a bleak understory, dominated by two species of plant: the blackberry (*Rubus fruticosus*) and the nettle (*Urtica dioica*). Grassland which has been left unmanaged is becoming dominated by thistles (*Cirsium* sp.) and rank grasses which are restricting the growth of other species.

Insect flower visitors were sampled from eight sites around the smelter over the period of May–September 2011. The selected sites represented a range of habitats which include Flower Rich grassland (FR), Mown Grassland (MG), New Hedge (NH), Old Hedge (OH), Plantation Woodland (Pl), Pond (Po), Ridge and Furrow Grassland (RF), and Woodhorn Woodland (WW). All eight sites were contiguous and covered a similar area; therefore patch size is unlikely to be influential (Table 1).

2.2. Insect Sampling

2.2.1. Pan Traps. Three pan traps (17 cm diameter and 6 cm depth) were placed at each site, 1 m apart in a triangle formation. Blue, yellow, and white UV reflective plastic bowls were used to account for colour preference by certain insects [16]. These colours were used for three principal reasons; they represent a range of wavelengths found in the visual spectrum; they are similar to flower colours and have been proven to attract a variety of flower visiting species [17, 18]. Traps were filled to the three-quarter line with water, to which several drops of unscented dishwashing detergent (Ecover Zero) were added to reduce the surface tension. Pan traps were set approximately 0.5 m above the ground at the height of the surrounding vegetation to allow the trap to be visible to flying insects. Wooden posts with brackets and wire were used to secure the pans in place during sampling.

Traps were exposed for a period of 30 hours (traps set at 10.00 and collected at 16.00 the following day) twice a month. On collection, the specimens were transferred into glass vials, labeled, and preserved in 70% ethanol. All flower visiting species shown to be important for pollination were identified to genus or family level using a dichotomous key. From the order Diptera, frequent flower visitors are concentrated in three main families: Syrphidae, Bombyliidae, and Tachinidae [19]. Families such as Empididae and Asilidae, known to frequent flowers for predatory reasons, were also collected. All other Diptera families were not included within this study.

2.2.2. Observation Plots. Data was collected from observation plots at each site to complement the pan trap data as it allowed the monitoring of species less represented within pan trap samples, such as Lepidoptera and Apidae [20]. Initially, net collecting along a transect was proposed; however due to the access restrictions within industrial areas and the topography of the land this method was considered inappropriate. Each observation plot measuring $1 m^2$ was surveyed twice each month during May, June, July, August, and September. Observations were made from a single point for a period of five minutes. Each insect seen to enter the observation plot was recorded; if the insect began to forage, the plant host was also recorded. Bumblebees and butterflies were identified to species, while hoverflies were identified to genus where possible. Due to the similarity between workers of *B. terrestris* and *B. lucorum* these species were treated as an aggregate species, as identification is unreliable in the field [21]. Observations were only initiated between 10:00 and 17:00 h, when weather conformed to Butterfly Monitoring Scheme standards [22].

2.3. Botanical Structure and Flower Density. Estimates of flower density were collected for each site twice monthly. Ten randomly placed $625 cm^2$ quadrats were used within the sample area (area surrounding the trio of pan traps). In each quadrat, the numbers of plants and the number of flower heads per plant species were recorded. The flower density was used as a surrogate measure for nectar availability as direct measurement of nectar parameters in the field is regarded as impractical [14].

2.4. Data Analysis. Data collected from the pan traps and the observation plots was combined to create one dataset. Minitab 16 was used to complete Correlation and ANOVA (Kruskal-Wallis) analysis on the total numbers of insects recorded for each habitat. Shannon Weiner diversity index was used to determine a diversity value for each habitat. This value was derived from species richness and relative abundance of each species and quantifies how well species are represented within a community. Diversity was then compared against month and flower density as individual factors. The statistical program R version 2.15.1 [23] was used to perform the Mantel test function within the "ade4" package [24]. This was conducted to determine whether the differences observed resulted from the factors studied or spatial differences.

As a result of the taxonomic variation between insect samples primary analysis has been conducted with four guilds of flower visiting insect: (1) nectar feeding, (2) parasitic insects, (3) pollen collecting, and (4) predatory insects. Where trends were indicated, further analysis using the detailed dataset was undertaken.

Due to the multispecies nature of the data and the survey design utilised in the study, multivariate analysis was utilised [25, 26]. Insect assemblages within each habitat were compared using PRIMER 6.0, a nonparametric multivariate statistical package. Multidimensional scaling (MDS) plots based on Bray-Curtis similarity measures were used to compare insect assemblages. Similarity percentage (SIMPER) analysis was also run on the data matrix; SIMPER decomposes Bray-Curtis similarities between all pairs of samples to identify those species that contribute most to the differences observed [26].

3. Results

Over the period May–September 2011, a total of 1138 individual insects were sampled across the four guilds, within the eight sites. A nested ANOSIM of sites within month showed there was no significant difference in the numbers of each flower visiting guild between sites ($R = 0.054$, $P = 0.265$), but there was a significant difference between months ($R = 0.387$, $P < 0.001$). The same trend was observed looking at insect assemblage for sites ($R = -0.063$, $P = 0.268$) and month ($R = 0.342$, $P < 0.001$). Despite the lack of significant differences between flower visiting insects between the sites, there were strong patterns noticeable within the dataset. A Mantel test showed that there was no spatial correlation between distance and insect diversity within this study ($r = 0.093$, $P = 0.65$).

3.1. Response of Insect Assemblages to Site. The highest number of individuals was recorded within the Pond site ($n = 234$) and the lowest within the Flower Rich Grassland ($n = 67$). Few insect families recorded were site specific; most species were recorded across all sites; however individuals from Satyridae, Panorpidae, Tipulidae, and Coccinellidae were in isolated populations. Tenthredinidae occurred in all sites except the Plantation Woodland. The most abundant families recorded were the Ichneumonidae ($n = 264$) and the Syrphidae ($n = 375$). Within the family Syrphidae, 47% of records were from a single species *Episyrphus balteatus*, with 176 individuals recorded across the survey period. Individuals from both families were present across all sites.

Removing month as a factor, average count data showed that the Old Hedge, Pond and Ridge and Furrow Grassland, had more individuals compared to the other sites

(a)

(b)

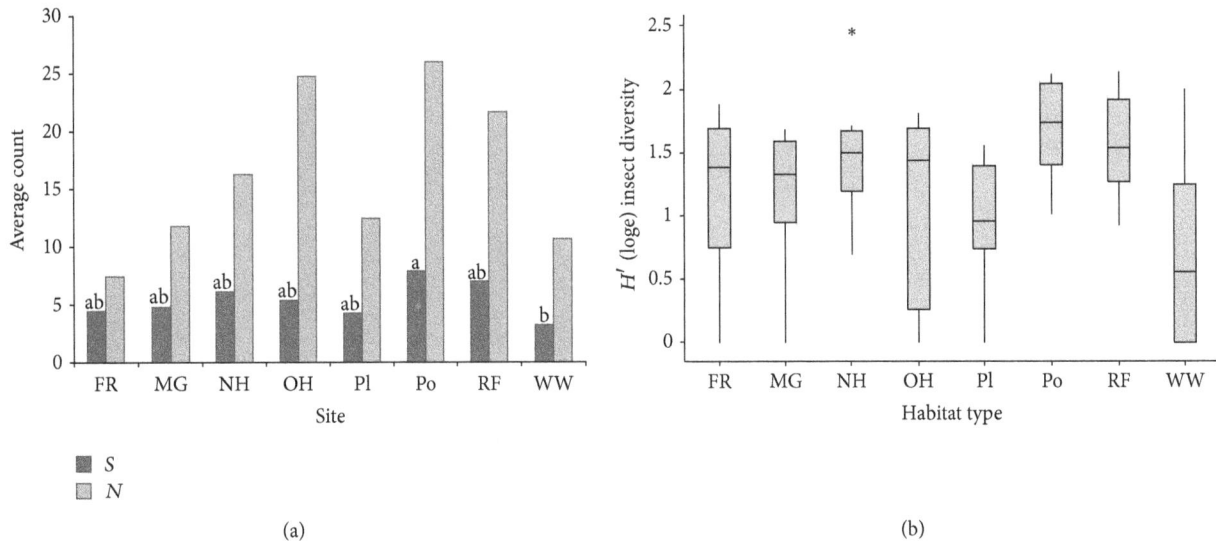

FIGURE 1: Response of insects to site. (a) Mean count for number of species (S) and abundance (N). Letters represent significant differences between sites following Tukey comparisons (b) Whiskered Box Plot showing the range of diversity scores. Box indicates median value and lower and upper quartiles. Whiskers indicate the range. Outliers indicated by an asterisk. Flower Rich Grassland (FR), Mown Grassland (MG), New Hedge (NH), Old Hedge (OH), Plantation Woodland (Pl), Pond (Po), Ridge (RF) and Furrow Grassland, and Woodhorn Woodland (WW).

(Figure 1(a)); however the difference was nonsignificant (Kruskal-Wallis $H = 11.87$, $P = 0.105$). When using the number of species/families recorded for each site there was a significant difference between sites (ANOVA, $F = 2.43$, $P < 0.05$). Tukey comparisons revealed that the differences between Pond and Woodhorn Wood were the cause of this variation ($t = 2.98$, $P < 0.01$) (Figure 1(a)).

When looking at total diversity rather than abundance, a higher diversity of species was present in the sites; New Hedge, Pond and Ridge and Furrow (Figure 1(b)). This changes the result from the abundance data, whereby the Old Hedge was more important than the New Hedge. However, the data is highly variable between sample dates.

A significant difference in the diversity of flower visiting insects based on Shannon diversity scores was recorded across all sites (Kruskal-Wallis, $H = 17.75$, $P < 0.05$) (Figure 1(b)). Repeated Mann Whitney tests revealed that the Pond had a significantly higher species diversity compared to Mown Grassland, Plantation, and Woodhorn Wood. Woodhorn Wood had significantly lower species diversity than Mown Grassland, Pond and Ridge and Furrow (Figure 5).

Bray Curtis Similarity analysis highlighted similarities between sites driven by the guilds (Figure 2(a)). Parasitic insects were found in high numbers within all sites; however, they did not occur in all replicate samples (Figures 2(a), 2(b), 6, and 7). There was a higher dominance of parasitic insects recorded within Plantation and Woodhorn Woodland indicating the importance of a woodland environment for this guild. The families Ichneumonidae and Tenthredinidae behave in a similar manner with records across all sites; however in contrast to Ichneumonidae, Tenthredinidae was underrepresented within the woodland sites (Figure 7).

The pollen collecting, nectar feeding, and predatory insect guilds were also present throughout all habitats. The pollen collecting guild saw a marked reduction in numbers within the Plantation and Old Hedgerow sites, whereas the other guilds were in higher abundance (Figure 6). This difference appeared to be influenced by two *Bombus* species (*B. lapidarius* and *B. lucorum/terrestris*) which were found in similar abundances throughout all sites; however larger numbers were recorded in the Flower Rich site for *B. lapidarius* and the New Hedge for *B. lucorum/terrestris*. There was no record of *B. lucorum/terrestris* within the Old Hedge (Figure 7).

The nectar feeding guild was also influenced heavily by species within the family Pieridae including *Anthocharis cardamines, Pieris napi,* and *P. brassicae* which were observed across six of the eight sites, yet were more dominant within the Pond site (Figure 7).

3.2. Response of Insect Assemblage to Month Surveyed. The numbers within each flower visiting guild were significantly different between months (ANOSIM, $R = 0.387$, $P < 0.001$), with the dominance of each guild affected (Figure 2). Furthermore, insect assemblages were also significantly different between months (ANOSIM, $R = 0.342$, $P = 0.001$), with pairwise comparisons indicating that all months were significantly different to each other with regard to insect assemblage except for the months of May and June (Figure 8).

The parasitic insect guild dominated throughout the entire survey season (Figures 2(c), 9, and 10). Higher numbers of this guild were recorded in May but reduced through the survey season with a marked reduction in September. Ichneumonidae and Tenthredinidae appear to be the most influential families for this trend (Figure 10). The nectar feeding guild also showed significant seasonal changes in abundance; however, the trend was directly opposite to that of the parasitic guild, with records increasing over the survey season

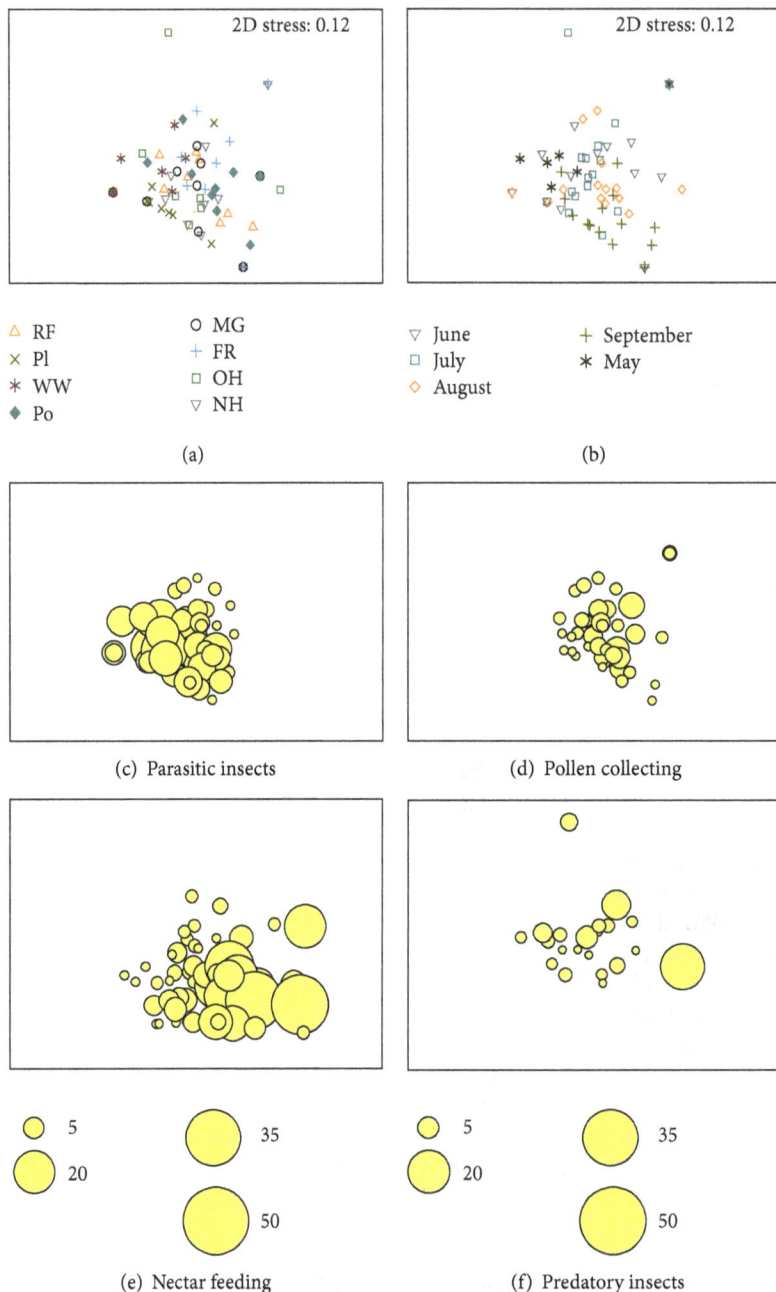

FIGURE 2: (a) Multidimensional Scaling (MDS) plot of flower visiting guilds at each of the eight sites based on Bray-Curtis similarity. (b) MDS plot of flower visiting guilds at each of the five sampling month based on Bray-Curtis similarity. ((c)–(f)) Bubble plot overlaid on the MDS sample points indicating patterns driven by the abundance of each guild. These highlight the role of each guild in shaping the community, bubble size relative to the number of individuals at that point.

with a peak in September (Figure 9). This trend was clearly influenced by late feeding Syrphidae present during this month (Figure 10).

3.3. Response of Insect Assemblages to Flower Density. Mean flower density across the site ranged between 0 and 17.6 per m^2 over the season. Only one site had no flowering plants recorded and this was in the plantation (Figure 3(a)). Flowering plants were available across all other sites during July and August, yet by September half of the eight sites (NH,

Pl, RF, and WW) had no flowering plants available to insects (Figure 3(b)). The Flower Rich site was the only one to have flowering plants available all season (Figure 3(b)), although surprisingly this site had the lowest recorded insect visitation. By contrast, the pond had the largest insect assemblages, but no flowers available within the month of May.

Following an ANOSIM on flower visiting guilds an influence of flower density was observed but not significant ($R = 0.14$, $P = 0.06$). However, when the whole data set was analysed flower density was highlighted as a significant factor

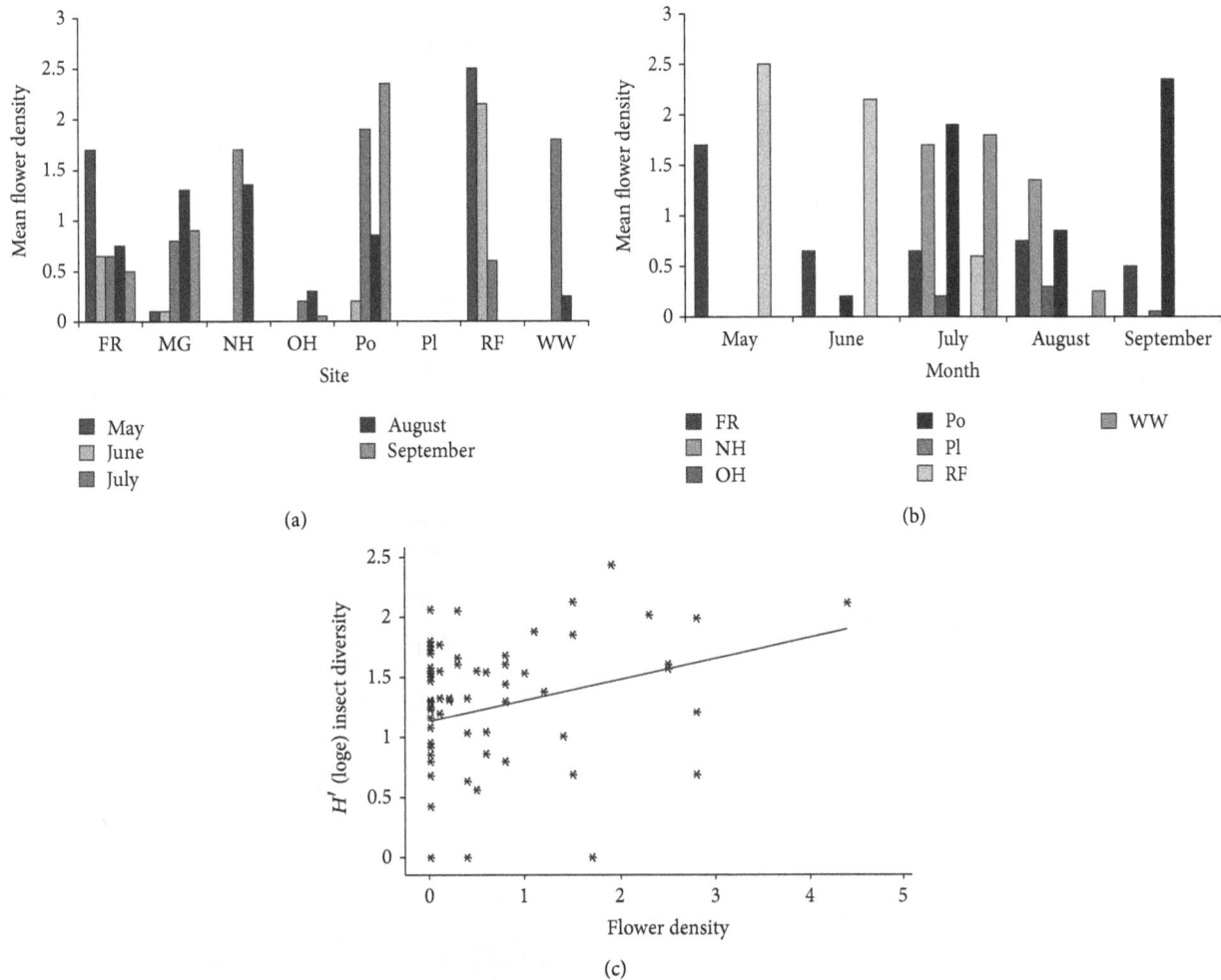

FIGURE 3: Floral resource across the sites: (a) mean flower density for each site subcategorised by month (b). Mean flower density for each month subcategorised by site, (c) Scatterplot with line of best fit showing the correlation between flower density and Shannon Weiner diversity score for insects. Flower rich (FR), Mown grassland (MG), New Hedge (NH), Old hedge (OH), Plantation (Pl), Pond (Po), Ridge and Furrow (RF), and Woodhorn Woodland (WW).

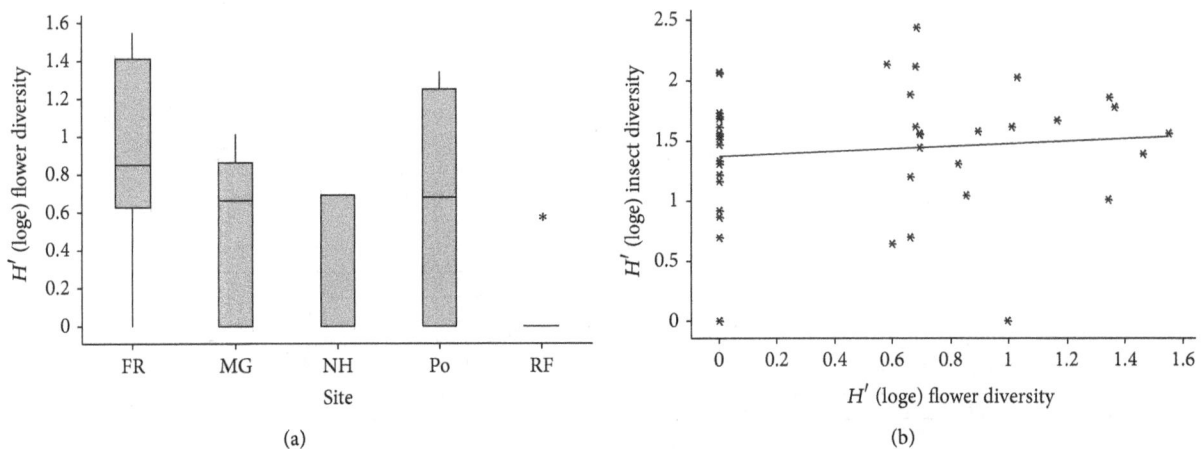

FIGURE 4: (a) Whiskered Box Plot showing the range of diversity scores. Box indicates median value, lower, and upper quartiles. Whiskers indicate the range. Outliers indicated by an asterisk. Flower Rich Grassland (FR), Mown Grassland (MG), New Hedge (NH), Pond (Po), and Ridge and Furrow Grassland (RF). (b) Scatterplot with line of best fit showing the correlation between shannon Weiner diversity score for insects and flowers.

Site	FR	MG	NH	OH	Pl	Po	RF	WW
FR		1a	0.665a	0.961a	0.269a	0.077a	0.289a	0.12a
MG			0.736a	0.923a	0.145a	**0.038b**	0.289a	0.0835a
NH				0.713a	0.075a	0.229a	0.665a	**0.0234b**
OH					0.267a	0.0829a	0.36a	0.2233a
Pl						**0.0047c**	**0.0171b**	0.23a
Po							0.536a	**0.0046c**
RF								**0.0103b**
WW								

FIGURE 5: Comparison of Mann Whitney Statistical test *P* value results for Shannon diversity score between sites. Significant differences are highlighted in bold, and differences in significance threshold are highlighted by a different letter. Flower Rich Grassland (FR), Mown Grassland (MG), New Hedge (NH), Old Hedge (OH), Plantation Woodland (Pl), Pond (Po), Ridge and Furrow Grassland (RF), and Woodhorn Woodland (WW).

Insect guild	FR	MG	NH	OH	Pl	Po	RF	WW
Nectar feeding	3.86	5.27	5.57	5.26	4.53	6.15	5.23	3.2
Parasitic insects	3.85	5.73	5.41	4.86	7.67	4.35	6.03	7.25
Pollen collecting	6.28	2.98	3.31	0.9	0.35	4.16	3.2	1.93
Predatory insects	0.41	0.77	1.06	3.93	1.21	1.2	1.05	1.7

Key %	0.00–1.00	1.01–2.00	2.01–3.00	3.01–4.00	4.01–5.00	5.01–6.00	6.01–7.00	7.01–8.00

FIGURE 6: Heatmap-table summarising the relative abundance (%) of insect guild within each site, based on SIMPER analysis of count data recorded during the survey period. (FR) Flower Rich Grassland, Mown Grassland (MG), New Hedge (NH), Old Hedge (OH), Plantation Woodland (Pl), Pond (Po), Ridge and Furrow Grassland (RF), and Woodhorn Woodland (WW).

	FR	MG	NH	OH	Pl	Po	RF	WW
Icheumonidae	2.27	4.4	3.12	3.98	6.54	2.71	3.87	7.18
Tenthredinidae	0.89	1.25	1.41	0.73	0	2.29	3.42	0.31
Episyrphus sp.	0.3	1.31	1.99	3.08	1.28	2.96	2.38	1.68
Chalcidoidea	1.02	0.9	1.44	1.55	0.84	0.83	1.43	2.36
Syrphus sp.	0.71	0.41	1.31	1.99	1.5	1.38	0.91	0.45
Platycheirus sp.	0.99	0	1.01	1.6	1.1	0.41	0.92	0
Bombus lapidarius	4.77	1.44	0.8	0.46	0.24	1.3	1.61	0.38
Bombus lucorum / terrestris	1.12	0.45	2.47	0	0.24	1.99	1.59	1.25
Asilidae	0.34	0.75	0.98	3.59	1.03	0.65	0.63	1.93
Empididae	0	0	0.25	0.29	0.53	0.78	0.67	0
Eupeodes sp.	0	1.03	0.87	1.69	0	0.51	0.93	0
Melanostoma sp.	0.28	0	0	0.19	0.8	0.17	0.16	0.31
Noctuidae	1.24	0	0.42	0	0.37	0.33	0.6	0
Coccinellidae	0.74	0	0	0	0	0	0.75	0
Hesperiidae	0.32	0	0	0	0	0.53	0.69	0
Megachilidae	0.74	1.49	1.18	0.29	0	0.21	0.51	0
Apis mellifera	0.6	0	0	0	0	0	0.31	0.31
Bombus pascuorum	0.32	0.24	0	0.25	0	1.53	0	0.9
Vespidae	0.3	0.73	0.45	0	0.24	0.72	0	0.45
Pieridae	0.24	0.64	1.03	0.36	0.24	2.19	0	0
Helophilus sp.	0.43	0.71	0.66	0	0	1.27	0.37	0
Epistrophe sp.	0	0	0	0	0	0.57	0	0
Eristalis sp.	1.33	0.79	0.29	0	0	0.44	0.16	0
Sphaerophoria sp.	0	0	0.34	0	0	0	0.33	0
Satyridae	0	0	0.47	0	0	0	0	0
Panorpidae	0	0	0.42	0.34	0	0	0	0

Key %	0	0.01–1.00	1.01–2.00	2.01–3.00	3.01–4.00	4.01–5.00	5.01–6.00	6.01–7.00	7.01–8.00

FIGURE 7: Heatmap-table summarising the relative abundance (%) of dominant taxon within each site, based on SIMPER analysis of count data recorded during the survey period.

Month	May	June	July	Aug	Sept
May		0.086a	**0.014b**	**0.015b**	**0.004c**
June			**0.002c**	**0.003c**	**0.001c**
July				**0.001c**	**0.001c**
Aug					**0.006c**
Sept					

FIGURE 8: Summary of *P*-value results for Pairwise comparisons of insect assemblages between Months. Significant differences are highlighted in bold, and differences in significance threshold are highlighted by a different letter.

Insect guild	May	June	July	August	September
Parasitic insects	7.07	5.24	5.95	6.31	4.4
Pollen collecting	2.72	3.98	3.28	3.23	1.47
Nectar feeding	2.24	3.88	4.41	4.7	7.76
Predatory insects	2.98	1.34	1.98	0.83	0.5

Key %	0.00–1.00	1.01–2.00	2.01–3.00	3.01–4.00	4.01–5.00	5.01–6.00	6.01–7.00	7.01–8.00

FIGURE 9: Heatmap-table summarising the relative abundance (%) of dominant guilds for each month, based on SIMPER analysis of count data recorded during the survey period.

	May	June	July	Aug	Sept
Icheumonidae	6.23	4.05	4.64	4.07	2.87
Bombus lapidarius	2.07	1.45	1.44	1.71	1.2
Asilidae	2.21	1.01	1.89	0.68	0.46
Bombus lucorum / terrestris	0.48	2.26	1.33	0.96	0.34
Tenthredinidae	1.77	1.2	1.08	2.07	0.71
Empididae	1.51	0.36	0.19	0.14	0.13
Chalcidoidea	0.48	0.99	1.19	1.51	1.53
Pieridae	0.93	0.49	1.16	0.35	0.3
Megachilidae	0.4	0.85	0.92	0.62	0
Nymphalidae	0.91	0.32	0.16	0	0
Coccinellidae	0	0.96	0	0	0
Noctuidae	0	0.83	0.4	0.15	0.43
Melanostoma sp.	0.57	0	0.12	0.27	0.44
Helophilus sp.	0	0.53	0	0.16	1.37
Panorpidae	0	0.41	0	0	0
Vespidae	0	0.32	0.46	0.38	0.47
Chrysididae	0	0.32	0.19	0	0
Hesperiidae	0	0.18	0.69	0	0
Bombus pascuorum	0	0.25	0.73	0.74	0
Syrphus sp.	0	0.32	0.88	1.19	2.31
Episyrphus sp.	0	0.32	1.74	2.73	3.08
Apis mellifera	0	0.35	0.34	0	0
Eupeodes sp.	0	0.17	0.11	1.3	1.1
Platycheirus sp.	0	0.24	0.11	0.82	2.17
Eristalis sp.	0	0.17	0	0.48	1.22

Key %	0	0.01–1.00	1.01–2.00	2.01–3.00	3.01–4.00	4.01–5.00	5.01–6.00	6.01–7.00

FIGURE 10: Heatmap-table summarising the relative abundance (%) of dominant taxon for each month, based on SIMPER analysis of count data recorded during the survey period.

Site	FR	MG	NH	Po	RF
FR		0.1971a	**0.0387b**	0.3722a	**0.001d**
MG			0.2212a	0.585a	**0.012c**
NH				0.2736a	0.1741a
Po					**0.0354b**
RF					

FIGURE 11: Comparison of Mann-Whitney Statistical test P value results for Shannon diversity score (flower diversity) between sites. Significant differences are highlighted in bold, and differences in significance threshold are highlighted by a different letter.

with regard to insect assemblage (ANOSIM, $R = 0.233$, $P < 0.01$). Although flower density was not significantly different between months (Kruskal-Wallis, $H = 8.68$, $P = 0.070$), insect assemblages were significantly affected by flower density when nested within month (ANOSIM, $R = 0.178$, $P < 0.01$). Analysis of the entire data set also highlighted a significant correlation ($\rho = 0.245$, $P < 0.01$) between Shannon Wiener Diversity of insects and flower density (Figure 3(c)).

3.4. Response of Insect Assemblages to Flower Diversity. The Shannon Wiener diversity of flowering plants observed within each site was calculated. Three of the eight sites (OH, WW, and Pl) had diversity scores of zero and therefore have been omitted from further analysis. The Flower Rich site had the highest recorded diversity (Figure 4), and a significant difference was observed between sites (Kruskal-Wallis $H = 13.37$, $P < 0.01$). Repeated Mann-Whitney tests highlighted which sites saw the greatest differences (Figure 11). The Flower Rich and Ridge and Furrow Grasslands were the greatest drivers for this result. Flower diversity was not identified as a significant factor for flower visiting guild assemblage (ANOSIM, $R = -0.082$, $P = 0.824$) and there was no correlation of flower diversity with insect diversity ($\rho = 0.162$, $P = 0.298$).

4. Discussion

Kunin [13] highlighted that many habitat types should be incorporated within a reserve area to capture the species variation caused by habitat discontinuities, suggesting that more habitat types are generally better than one to enhance biodiversity. This study, however, shows that flower visiting insects recorded at the Lynemouth Smelter site were influenced predominantly by the flower density, rather than the distinct sites/habitats themselves, a finding which is widely supported [14, 15, 27]. Although, site/habitat type was not a significant factor in insect assemblage within this study, flower density was, suggesting that varied botanical structure and the presence of certain flowering plants such as *Centaurea nigra* (common knapweed) and *Heracleum sphondylium* (common hogweed) are more important than the number of different types of habitats specifically. Conversely, within this study, flower diversity did not impact the insect assemblage. This could be a result of the low numbers of flower species recorded, as flower diversity is indeed an important factor in determining flower visitor presence [14, 28]. The type of

flowerhead available to insects will determine which species feed, hunt, and breed within a particular habitat [29, 30]. Ideally, a variety of host plants for larvae and immature insects are required [31–33]. Furthermore, flower longevity and nectar resource have previously been shown to have an effect on the insect community within habitats [34, 35].

As expected, month had a significant effect on insect assemblages, again complementing results seen in previous studies [36, 37]. Certain taxa were present at different times of the year, likely due to variation in emergence and breeding periods as a result of an insect's dependence on factors such as weather, as well as the availability of host and food plants for different insects groups. Only one of the eight sites had plants flowering for the entire survey period (the Flower Rich site); however due to regular mowing of this area the flower density was low, resulting in this area having the lowest recorded insect visitation. Nevertheless, insects were recorded across the whole season suggesting that this was not the only factor affecting the site's attractiveness. For example, the pan traps could have been less attractive to the foraging insects than the flowers present, resulting in the catch being proportional to the flower density as suggested by [38]. Additionally, despite a relatively high diversity of plant species within the Flower Rich site, it was dominated by one particular flower species, *Lotus corniculatus* (bird's foot trefoil). *L. corniculatus* was predominantly foraged by one particular species of pollinating insect, *Bombus lapidarius*, a species known to have a preference for yellow flowers and the appropriate mouth parts to access the nectar from this flower [30]. This dominance of the site by one flower species could in part explain the low insect visitation rates recorded in this particular study. *L. corniculatus* may not have been an appropriate flower resource for other insect species. Therefore, improving diversity of flowering plants within certain preexisting habitats will undoubtedly have a significant effect on the pollinating species present within the sites. Alternatively, there may have been other factors such as noise pollution and forage distance to consider which were not assessed during this study.

The number and diversity of insects could be influenced by improvements to the assessed sites, not only diversifying the flowering species available for forage, but also increasing diversity with regard to flower head shape. Ensuring a variety of flower heads such as umbels or composite heads will increase the number of flower visiting insects [39]. For example, low abundance of the hoverfly *Episyprhus balteatus* was recorded within the Flower Rich habitat, where hoverflies would normally be expected [40]. The dominant flower, *L. corniculatus*, is not preferred by hoverfly species as their mouthparts do not allow access to the nectar. By contrast larger numbers of hoverflies were observed within the Ridge and Furrow Grassland, Old Hedge, and Pond sites as a result of the flowering species present: umbelled flowers such as *Oenanthe crocata* (dropwort water hemlock) and *Heracleum sphondylium* (hogweed) and simple flowering species such as *Crataegus monogyna* (hawthorn). A reduction in the mowing frequency at the Flower Rich site may diversify the flowering species present, allowing species which are less hardy to germinate from the seedbank [41]. Furthermore, a study by López-Mariño et al. [42] also highlights, that due to the high proportion of perennial grasses often present within semi-improved habitats, only half the species stored within grassland's natural seedbank are present above the soil surface. This suggests that the sward can be diversified by managing the grass species present, proving a beneficial management strategy to enhance insect diversity for this site.

The Pond site had the largest insect assemblages, despite the limited floral resource available within the month of May. Some insect species are locating nest sites around this time, particularly bumblebees which start searching around April and peak in May [43]. Ensuring flowering plants are available throughout April to September is therefore an important management strategy, particularly for insect diversity. The availability of nest and forage sites is essential to pollinating insects and improvement in this area would likely have a positive effect on the overall insect diversity of this site [44, 45]. Conversely, there are few species of plant which flower early, therefore improvement to surrounding hedgerows, or planting of species such as *Salix cinera* and *Malus* sp. complemented with *Laminum album* and *Glechoma hederacea* could be more beneficial to the communities around the Lynemouth smelter [46].

Interestingly, this study highlights the importance of the method in which data sets are analysed, which could result in different interpretation of the data and therefore have a serious impact on the proposed management of a particular site. Firstly, the data was organised into guilds of insects to minimise taxonomic variation between data samples. The trends reported in this study were mirrored in both the guild analysis and complete taxon analysis showing that many of the results reported hold fast when analysing the data in different ways. However, some trends were weaker when looking at guilds, rather than when analysis was completed on the whole data set, which means significant findings could be overlooked if only one type of analysis was utilised. Furthermore, the data in this study showed that two sites, the Plantation and Woodhorn Woodland, had lower mean insect abundance, species richness, and Shannon Weiner diversity score than all the other surveyed sites, suggesting these two sites may not be that important for the flower visiting insect communities (although these differences were not all significant). However, when comparing insect assemblages these habitats were highlighted as important for parasitic insects such as Ichneumonidae. Following the initial analysis comparing mean averages of insect abundance per site, the importance of these habitats could have been overlooked and the site managed inappropriately for the dominant inhabitants. Until the community analysis was performed the dominance of Ichneumonidae and influence on the community may not have been appreciated, highlighting that community analysis is invaluable within biodiversity assessment. Secondly, with regard to Shannon Weiner diversity, a higher diversity of species was found to be present in three particular sites; the New Hedge, the Pond and the Ridge and Furrow Grassland. By contrast when looking at the abundance data, the Old Hedge is a more important site for insect assemblages than the New Hedge highlighted in the diversity data set.

The methods of analysis used in this study highlight the importance of data interpretation before management action

plans are devised. The two variations in interpretation of the results, highlighted above, were both made with the same original data set; however, one utilised abundance data whilst the other used total diversity. The differences are likely caused as the diversity index takes into account both species richness and the relative abundance of each species to quantify how well species are represented within a community. Many management plans and advice provided to site managers contain information collected in a similar manner to this study to achieve specific objectives. Whether these objectives are to maximise species diversity or simply abundance, we recommend that the community interactions are assessed before management plans are drawn up to avoid the potential loss of valuable habitats and species through inappropriate management.

5. Conclusion

In conclusion, this study highlights the importance of data interpretation to determine management objectives and recommends analysing the community structure and identifying the dominant species prior to undertaking any land management. Although no significant difference was found between flower visiting insect diversity at sites when month was taken into account, flower density was highlighted as a factor driving the insect diversity. This result highlights that increasing the number of flowering plants rather than increasing the amount of specific habitats is a more cost-effective management tool for industrial sites. Although, sites such as Flower Rich Grassland, would be expected to attract the highest diversity of flower visiting insects, this was not observed in the case of this study. This is likely a result of the high dominance of one particular species of flower, *L. corniculatus*, which may exclude certain insect assemblages. Importantly, this study highlights that with relatively low cost industrial sites such as the Lynemouth smelter could be improved with regard to insect diversity. This can be achieved by specific seed planting or a refinement of the mowing practices to allow diversification of flora within and between the sites, ultimately improving the overall ecological value of industrial areas.

References

[1] O. E. Sala, F. S. Chapin, J. J. Armesto et al., "Global biodiversity scenarios for the year 2100," *Science*, vol. 287, no. 5459, pp. 1770–1774, 2000.

[2] J. Krauss, R. Bommarco, M. Guardiola et al., "Habitat fragmentation causes immediate and time-delayed biodiversity loss at different trophic levels," *Ecology Letters*, vol. 13, no. 5, pp. 597–605, 2010.

[3] B. S. Law and C. R. Dickman, "The use of habitat mosaics by terrestrial vertebrate fauna: implications for conservation and management," *Biodiversity and Conservation*, vol. 7, no. 3, pp. 323–333, 1998.

[4] S. D. Wratten, M. Gillespie, A. Decourtye, E. Mader, and N. Desneux, "Pollinator habitat enhancement: benefits to other ecosystem services," *Agriculture, Ecosystems & Environment*, vol. 159, pp. 112–122, 2012.

[5] J. Feehan, D. A. Gillmor, and N. Culleton, "Effects of an agri-environment scheme on farmland biodiversity in Ireland," *Agriculture, Ecosystems and Environment*, vol. 107, no. 2-3, pp. 275–286, 2005.

[6] M. J. Whittingham, "Will agri-environment schemes deliver substantial biodiversity gain, and if not why not?" *Journal of Applied Ecology*, vol. 44, no. 1, pp. 1–5, 2007.

[7] R. E. Kenward, M. J. Whittingham, S. Arampatzis et al., "Identifying governance strategies that effectively support ecosystem services, resource sustainability, and biodiversity," *Proceedings of the National Academy of Sciences of the United States of America*, vol. 108, no. 13, pp. 5308–5312, 2011.

[8] K. J. Gaston, R. M. Smith, K. Thompson, and P. H. Warren, "Urban domestic gardens (II): experimental tests of methods for increasing biodiversity," *Biodiversity and Conservation*, vol. 14, no. 2, pp. 395–413, 2005.

[9] M. A. Goddard, A. J. Dougill, and T. G. Benton, "Scaling up from gardens: biodiversity conservation in urban environments," *Trends in Ecology and Evolution*, vol. 25, no. 2, pp. 90–98, 2010.

[10] R. W. F. Cameron, T. Blanuša, J. E. Taylor et al., "The domestic garden—its contribution to urban green infrastructure," *Urban Forestry & Urban Greening*, vol. 11, pp. 129–137, 2012.

[11] T. G. Benton, J. A. Vickery, and J. D. Wilson, "Farmland biodiversity: is habitat heterogeneity the key?" *Trends in Ecology and Evolution*, vol. 18, no. 4, pp. 182–188, 2003.

[12] T. Tscharntke, I. Steffan-Dewenter, A. Kruess, and C. Thies, "Characteristics of insect populations on habitat fragments: a mini review," *Ecological Research*, vol. 17, no. 2, pp. 229–239, 2002.

[13] W. E. Kunin, "Sample shape, spatial scale and species counts: implications for reserve design," *Biological Conservation*, vol. 82, no. 3, pp. 369–377, 1997.

[14] S. G. Potts, B. Vulliamy, A. Dafni, G. Ne'eman, and P. Willmer, "Linking bees and flowers: how do floral communities structure pollinator communities?" *Ecology*, vol. 84, no. 10, pp. 2628–2642, 2003.

[15] J. Ghazoul, "Floral diversity and the facilitation of pollination," *Journal of Ecology*, vol. 94, no. 2, pp. 295–304, 2006.

[16] J. S. Wilson, T. Griswold, and O. J. Messinger, "Sampling bee communities (Hymenoptera: Apiformes) in a desert landscape: are pan traps sufficient?" *Journal of the Kansas Entomological Society*, vol. 81, no. 3, pp. 288–300, 2008.

[17] D. Moroń, H. Szentgyörgyi, M. Wantuch et al., "Diversity of wild bees in wet meadows: implications for conservation," *Wetlands*, vol. 28, no. 4, pp. 975–983, 2008.

[18] C. Westphal, R. Bommarco, G. Carré et al., "Measuring bee diversity in different European habitats and biogeographical regions," *Ecological Monographs*, vol. 78, no. 4, pp. 653–671, 2008.

[19] B. M. H. Larson, P. G. Kevan, and D. W. Inouye, "Flies and flowers: taxonomic diversity of anthophiles and pollinators," *Canadian Entomologist*, vol. 133, no. 4, pp. 439–465, 2001.

[20] S. Vrdoljak and M. Samways, "Optimising coloured pan traps to survey flower visiting insects," *Journal of Insect Conservation*, vol. 16, pp. 345–354, 2012.

[21] O. E. Prys-Jones and S. A. Corbet, *Bumblebees*, Cambridge University Press, Cambridge, UK, 1991.

[22] E. Pollard and T. J. Yates, *Monitoring Butterflies for Ecology and Conservation: The British Butterfly Monitoring Scheme*, Chapman & Hall, London, UK, 1993.

[23] R. C. Team, "R: a language and environment for statistical computing," in *Computing RFfS*, Vienna, Austria, 2012.

[24] S. Dray and A. B. Dufour, "The ade4 package: implementing the duality diagram for ecologists," *Journal of Statistical Software*, vol. 22, no. 4, pp. 1–20, 2007.

[25] K. R. Clarke, "Non-parametric multivariate analyses of changes in community structure," *Australian Journal of Ecology*, vol. 18, no. 1, pp. 117–143, 1993.

[26] K. Clarke and R. Warwick, *Change in Marine Communities: An Approach to Statistical Analysis and Interpretation*, PRIMER-E, Plymouth, UK, 2001.

[27] M. A. Molina-Montenegro, E. I. Badano, and L. A. Cavieres, "Positive interactions among plant species for pollinator service: assessing the "magnet species" concept with invasive species," *Oikos*, vol. 117, no. 12, pp. 1833–1839, 2008.

[28] J. Fründ, K. E. Linsenmair, and N. Blüthgen, "Pollinator diversity and specialization in relation to flower diversity," *Oikos*, vol. 119, no. 10, pp. 1581–1590, 2010.

[29] M. Stang, P. G. L. Klinkhamer, and E. Van Der Meijden, "Size constraints and flower abundance determine the number of interactions in a plant-flower visitor web," *Oikos*, vol. 112, no. 1, pp. 111–121, 2006.

[30] L. Comba, S. A. Corbet, L. Hunt, and B. Warren, "Flowers, nectar and insect visits: evaluating British plant species for pollinator-friendly gardens," *Annals of Botany*, vol. 83, no. 4, pp. 369–383, 1999.

[31] T. Jermy, F. E. Hanson, and V. G. Dethier, "Induction of specific food preference in lepidopterous larvae," *Entomologia Experimentalis et Applicata*, vol. 11, no. 2, pp. 211–230, 1968.

[32] J. H. Lawton, "Plant architecture and the diversity of phytophagous insects," *Annual Review of Entomology*, vol. 28, pp. 23–39, 1983.

[33] N. M. Haddad, D. Tilman, J. Haarstad, M. Ritchie, and J. M. H. Knops, "Contrasting effects of plant richness and composition on insect communities: a field experiment," *American Naturalist*, vol. 158, no. 1, pp. 17–35, 2001.

[34] S. G. Potts, B. A. Woodcock, S. P. M. Roberts et al., "Enhancing pollinator biodiversity in intensive grasslands," *Journal of Applied Ecology*, vol. 46, no. 2, pp. 369–379, 2009.

[35] M. Albrecht, B. Schmid, Y. Hautier, and C. B. Muller, "Diverse pollinator communities enhance plant reproductive success," *Proceedings of the Royal Society B*, vol. 279, pp. 4845–4852, 2012.

[36] C. McCall and R. B. Primack, "Influence of flower characteristics, weather, time of day, and season on insect visitation rates in three plant communities," *American Journal of Botany*, vol. 79, no. 4, pp. 434–442, 1992.

[37] M. W. Brown and J. J. Schmitt, "Seasonal and diurnal dynamics of beneficial insect populations in apple orchards under different management intensity," *Environmental Entomology*, vol. 30, no. 2, pp. 415–424, 2001.

[38] J. H. Cane, R. L. Minckley, and L. J. Kervin, "Sampling bees (Hymenoptera: Apiformes) for pollinator community studies: pitfalls of pan-trapping," *Journal of the Kansas Entomological Society*, vol. 73, no. 4, pp. 225–231, 2000.

[39] F. S. Gilbert, "Foraging ecology of hoverflies: morphology of the mouthparts in relation to feeding on nectar and pollen in some common urban species," *Ecological Entomology*, vol. 6, pp. 245–262, 1981.

[40] A. Stubbs and S. Faulks, *British Hoverflies: An Illustrated Identification Guide: British Entomological and Natural History Society*, 2002.

[41] E. Gaujour, B. Amiaud, C. Mignolet, and S. Plantureux, "Factors and processes affecting plant biodiversity in permanent grasslands. A review," *Agronomy for Sustainable Development*, vol. 32, pp. 133–160, 2011.

[42] A. López-Mariño, E. Luis-Calabuig, F. Fillat, and F. F. Bermúdez, "Floristic composition of established vegetation and the soil seed bank in pasture communities under different traditional management regimes," *Agriculture, Ecosystems and Environment*, vol. 78, no. 3, pp. 273–282, 2000.

[43] A. R. Kells and D. Goulson, "Preferred nesting sites of bumblebee queens (Hymenoptera: Apidae) in agroecosystems in the UK," *Biological Conservation*, vol. 109, no. 2, pp. 165–174, 2003.

[44] P. Westrich, "Habitat requirements of central European bees and the problems of partial habitats," in *The Conservation of Bees*, A. Matheson, S. L. Buchmann, C. O'Toole, P. Westrich, and I. H. Williams, Eds., pp. 1–16, Academic Press for the Linnean Society of London and IBRA, London, UK, 1996.

[45] R. Winfree, "The conservation and restoration of wild bees," *Annals of the New York Academy of Sciences*, vol. 1195, pp. 169–197, 2010.

[46] R. F. Pywell, W. R. Meek, L. Hulmes et al., "Management to enhance pollen and nectar resources for bumblebees and butterflies within intensively farmed landscapes," *Journal of Insect Conservation*, vol. 15, no. 6, pp. 853–864, 2011.

16

Larval Performance in the Context of Ecological Diversification and Speciation in *Lycaeides* Butterflies

Cynthia F. Scholl,[1] **Chris C. Nice,**[2] **James A. Fordyce,**[3]
Zachariah Gompert,[4] **and Matthew L. Forister**[1]

[1] *Department of Biology, University of Nevada, Reno, NV 89557, USA*
[2] *Department of Biology, Population and Conservation Biology Program, Texas State University, San Marcos, TX 78666, USA*
[3] *Department of Ecology and Evolutionary Biology, University of Tennessee, Knoxville, TN 37996, USA*
[4] *Department of Botany, Program in Ecology, University of Wyoming, Laramie, WY 82071, USA*

Correspondence should be addressed to Cynthia F. Scholl, cynthia.scholl@gmail.com

Academic Editor: Rui Faria

The role of ecology in diversification has been widely investigated, though few groups have been studied in enough detail to allow comparisons of different ecological traits that potentially contribute to reproductive isolation. We investigated larval performance within a species complex of *Lycaeides* butterflies. Caterpillars from seven populations were reared on five host plants, asking if host-specific, adaptive larval traits exist. We found large differences in performance across plants and fewer differences among populations. The patterns of performance are complex and suggest both conserved traits (i.e., plant effects across populations) and more recent dynamics of local adaptation, in particular for *L. melissa* that has colonized an exotic host. We did not find a relationship between oviposition preference and larval performance, suggesting that preference did not evolve to match performance. Finally, we put larval performance within the context of several other traits that might contribute to ecologically based reproductive isolation in the *Lycaeides* complex. This larger context, involving multiple ecological and behavioral traits, highlights the complexity of ecological diversification and emphasizes the need for detailed studies on the strength of putative barriers to gene flow in order to fully understand the process of ecological speciation.

1. Introduction

Understanding the processes underlying diversification is a central question in evolutionary biology. Lineages diversify along multiple axes of variation, including morphological, physiological, and ecological traits. With respect to diversification in ecological traits, many recent studies have found that ecological niches can be highly conserved from a macroevolutionary perspective [1–3]. In other words, closely related species tend to utilize similar resources or occupy similar environments. In contrast, the field of ecological speciation suggests that ecological traits can evolve due to disruptive selection and drive the process of diversification [4–8]. In herbivorous insects, evolution in response to habitat or host shifts is often thought to be a first step in the evolution of reproductive isolation [9, 10]. In most

well-studied systems, although exceptions exist [6, 11], niche conservatism and niche evolution are often characterized by a small number of ecological traits, such as habitat preference or physiological performance [12–14]. To understand the causes and consequences of evolution in ecological traits, more studies are needed of groups in which diversification is recent or ongoing and multiple ecological traits are studied. The study of multiple traits is particularly important for our understanding of ecological speciation. For example, it has been suggested that weak selection acting on a multifarious suite of traits could be as important for speciation as strong selection acting on a single ecological trait [15].

The butterfly genus *Lycaeides* (Lycaenidae) includes a complex of taxa in North America that has been the focus of studies investigating the evolution and ecology of host use, mate choice, and genitalic morphology, among other subjects [16–19]. In the context of diversification, this group is interesting because hybridization has been documented among multiple entities, with a variety of consequences [20, 21], including the formation of at least one hybrid species in the alpine of the Sierra Nevada mountains [22]. The *Lycaeides* taxa in western North America (specifically *L. idas*, *L. melissa*, and the hybrid species) differ in many traits, some of which have been implicated in the evolution of ecological reproductive isolation in this system. For example, there is variation in the strength of host preference, which is often linked to reproductive isolation in herbivorous insects that mate on or near their host plants, as *Lycaeides* do [17]. There are also potentially important differences in mate preference, phenology, and egg adhesion [16, 23]. The latter trait is interesting with respect to the evolution of the hybrid species, which lacks egg adhesion [22, 23]. The eggs of the hybrid species fall from the host plants. This is presumed to be an adaptation to the characteristics of the alpine plants, for which the above-ground portions senesce and are blown by the wind away from the site of next year's fresh growth (thus eggs that fall off are well positioned for feeding in the spring). Since the eggs of lower-elevation *Lycaeides* taxa do adhere to hosts, this trait could serve as a barrier to gene flow with respect to individuals immigrating from lower elevations.

Our state of knowledge for *Lycaeides* is unusual for well-studied groups of herbivorous insects in that we know a great deal about a diversity of traits, as discussed above, but we have not heretofore investigated larval performance across taxa in the context of ecological speciation, which is often one of the first traits studied in other insect groups [24]. This study has two goals, first to investigate larval performance and then to put this information in the context of other already-studied traits to investigate which traits might be important for reducing gene flow between populations and species in this system. We have focused on performance of caterpillars from both *L. idas* and *L. melissa* populations as well as from populations of the hybrid species. Beyond the inclusion of the hybrid taxon, of added interest is the fact that *L. melissa* has undergone a recent expansion of diet, encompassing exotic alfalfa (*Medicago sativa*) as a larval host plant across much of its range. Thus we are able to investigate variation in the key ecological trait of larval performance across multiple levels of diversification, including the differentiation of *L. idas* and *L. melissa*, the formation of a hybrid species, and a host expansion that has occurred within the last two hundred years [19, 22, 25].

Using individuals from two *L. idas*, three *L. melissa*, and two hybrid species populations, we conducted reciprocal rearing experiments using all five of the host species found at these focal populations. We assessed larval performance by examining survival, time to emergence (eclosion), and adult weight, and by comparing survival curves from different populations on the different plants. For each population, we contrasted larval performance on a natal host to performance on the plants of other populations. Higher larval performance on natal host plants would support the hypothesis of local adaptation to host plant species. In the second part of the paper, these results are discussed both within the light of local adaptation in a diversifying group and also within the context of possible reproductive isolation related to variation in ecological traits.

2. Methods

Two of our focal taxa, *L. idas* and *L. melissa*, are widely distributed across western North America. Our study focused on populations of these two species and the hybrid species in northern California and Nevada (Figure 1). In this area, *L. idas* is found on the west slope of the Sierra Nevada, *L. melissa* is found on the eastern side, and the hybrid species is only found in the alpine zone. *Lycaeides* species use a variety of plants in the pea family, Fabaceae, as hosts, although (with few exceptions) specific *Lycaeides* populations generally utilize a single host plant species. The two *L. idas* populations studied were Yuba Gap (YG), which uses *Lotus nevadensis* as a host, and Leek Springs (LS) which uses the host *Lupinus polyphyllus* (Table 1). Both populations of the hybrid species, Mt. Rose (MR) and Carson Pass (CP), use *Astragalus whitneyi*. At Washoe Lake (WL), *L. melissa* uses the native host *Astragalus canadensis*; at Beckwourth Pass (BP), butterflies use both *A. canadensis* and alfalfa, *Medicago sativa*; at Goose Lake (GLA), the only available host is alfalfa (Table 1).

Lycaeides idas and the hybrid species are univoltine, while *L. melissa* populations have at least three generations per year. Eggs from the univoltine populations have to be maintained under winter conditions (i.e., cold temperatures and darkness) in the lab for experiments in the following spring. Females and eggs from univoltine populations were collected in the summer of 2008 to be reared in the summer of 2009, while *L. melissa* females and eggs were collected during the 2009 summer. Females were collected from the *L. idas* and hybrid species populations (32 from Yuba Gap, 50 from Leek Springs, 45 from Carson Pass, and 40 from Mt. Rose) and caged individually or in small groups with host plants for a period of three days after which eggs were collected. Eggs were washed with a dilute (2%) bleach solution and held over the winter at 4–6°C. Eggs were removed from cold storage on May 27th, 2009, and the majority hatched within several hours. The number of caterpillars hatching synchronously required that the larvae be moved in groups of twenty to standard-sized petri dishes (100 mm diameter) with fresh plant material on the 27th and 28th. On the 29th and 30th of May, the groups of twenty were split into three dishes each containing three to seven individuals. An average of 6 caterpillars was added to 9 dishes per treatment (plant/population combination); in some cases fewer (but not less than three) caterpillars were added per dish to try to maximize the number of dishes, which is the unit of replication (see below). Once all the larvae in a petri dish reached the 3rd or 4th instar they were moved to larger petri dishes (170 mm diameter). These groups of individuals were considered a "rearing dish," and dish was used as a random factor in statistical analyses (see below).

Females and eggs from the three *L. melissa* populations were collected following similar protocols (though without the necessity of overwintering). Seventeen females were

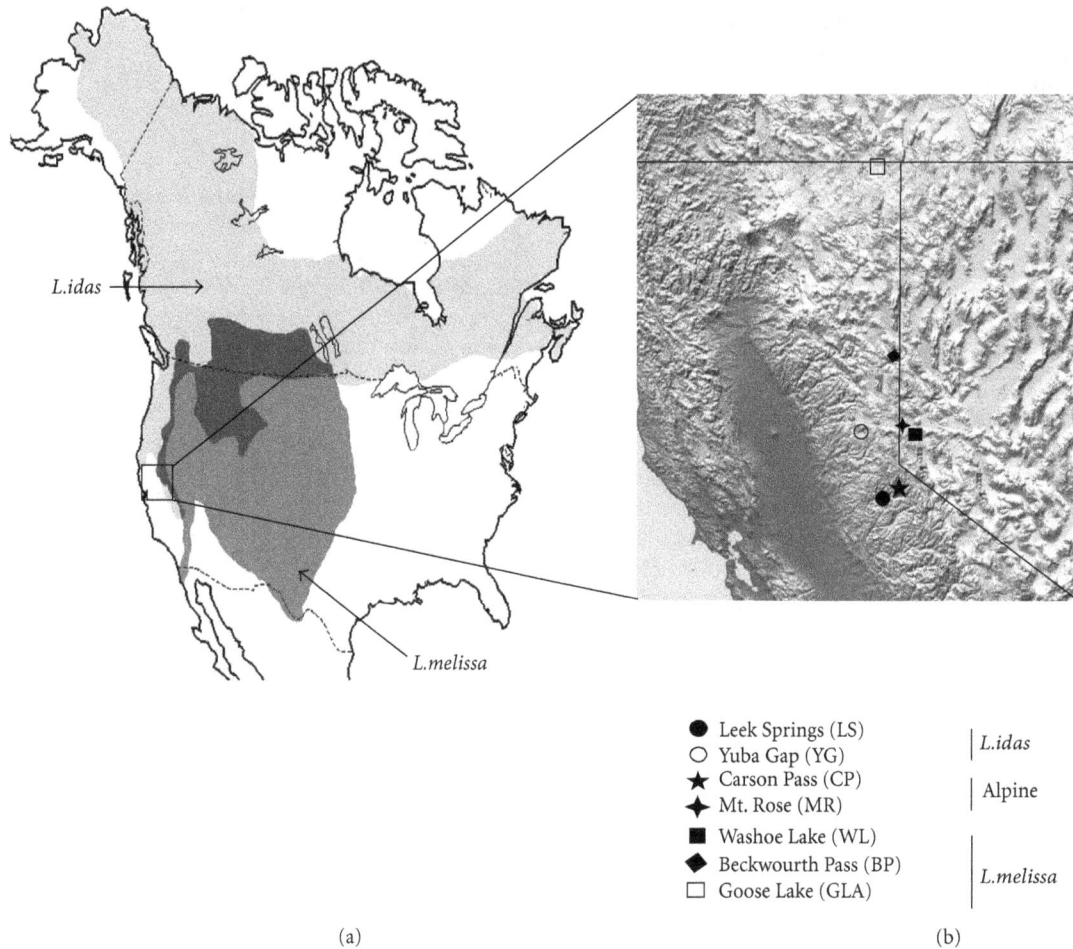

FIGURE 1: (a) Ranges of *L. idas* and *L. melissa* across North America, darker shaded regions correspond to ranges of overlap, which includes alpine populations of the hybrid species considered here (CP and MR). (b) Map of sampled *Lycaeides* populations in Northern California and Nevada, USA. Symbols correspond to populations and taxa.

TABLE 1: Locations of populations (see also Figure 1) and hosts associated with the seven populations studied.

Taxon	Location	Latitude/longitude	Host
L. idas	Leek Springs (LS)	38°38′8″ N/120°14′25″ W	*Lupinus polyphyllus*
	Yuba Gap (YG)	39°19′24″ N/120°35′60″ W	*Lotus nevadensis*
Hybrid species	Carson Pass (CP)	38°42′28″ N/120°0′28″ W	*Astragalus whitneyi*
	Mt. Rose (MR)	39°19′21″ N/119°55′47″ W	*Astragalus whitneyi*
	Washoe Lake (WL)	39°13′59″ N/119°46′46″ W	*Astragalus canadensis*
L. melissa	Beckwourth Pass (BP)	39°46′55″ N/120°4′23″ W	*Astragalus canadensis* and *Medicago sativa*
	Goose Lake (GLA)	41°59′9″ N/120°17′32″ W	*Medicago sativa*

collected from Beckwourth Pass during the last week of June, 27 were collected from Goose Lake the third week of July, and 14 were collected from Washoe Lake the last week of July. For these populations, larvae were added to standard-sized petri dishes in groups of four to six individuals as soon as the eggs hatched. Again, caterpillars were transferred to a large petri dish once all the individuals in a dish reached the 3rd or 4th instar.

Larvae from each population were reared on all five plants, *Astragalus canadensis*, *Astragalus whitneyi*, *Lotus nevadensis*, *Lupinus polyphyllus*, and *Medicago sativa*, with individual rearing dishes being assigned exclusively to a single plant throughout development. Caterpillars in the wild consume both vegetative and reproductive tissues, but only leaves were used in this study, as flowers would be difficult to standardize across plants (not being available synchronously for most species). For a study of this kind, ideally all plant material to be used in rearing would be collected from focal locations (where butterflies are flying) or grown in a common environment. However, many of these species are not easily propagated, and moreover our focal locations are widely dispersed geographically; these factors necessitated some compromise in collecting some of the plants. *A. canadensis* cuttings were obtained at the site of the butterfly populations at Beckwourth Pass and Washoe Lake and from the greenhouse (plants were grown from seeds collected at Washoe Lake). *Astragalus whitneyi* was collected from the site of the Mt. Rose hybrid species population and on a hillside adjacent to Carson Pass (38°42′23″/120°00′23″). All *Lotus nevadensis* were collected from Yuba Gap (YG). The only case in which plant material was collected from a site where the butterfly is not found is *Lupinus polyphyllus*. These plants were collected off I-80 at the Soda Springs exit (39°19′29″/120°23′25″) and seven miles north of Truckee CA, off State Route 89 (39°25′59″/120°12′13″). *Medicago sativa* was obtained from Beckwourth Pass (BP) and from plants grown in the greenhouse with seeds from BP. *M. sativa* was also collected from south of Minden, NV on State Route 88 (38°48′60″/119°46′46″) and off of California State Route 49 in Sierra Valley, CA (39°38′35″/120°23′10″). Plant material was kept in a refrigerator and larvae were fed fresh cuttings whenever the plant material in petri dishes was significantly reduced or wilted, which was approximately every two to seven days. Each time caterpillars were given fresh plant material, the number of surviving caterpillars was recorded along with the date. All dishes were kept at room temperature, 20° to 23° Celsius, on lab benches. Newly emerged adults were individually weighed to the nearest 0.01 mg on a Mettler Toledo XP26 microbalance and sex was recorded.

2.1. Analyses. The strengths of our experiment were that we reared a large number of individuals from multiple taxa across five plants, but a weakness of our design was that not all rearing could be done simultaneously. As discussed further below, flowers were not included in the rearing, and plant material was collected from most but not all focal populations. Experiments were conducted in two phases, first involving the populations of the hybrid species and *L. idas*, being reared together and earlier in the spring, and second involving the three low-elevation *L. melissa* populations being reared later in the summer. This division into two rearing groups was largely a consequence of being constrained by the total number of caterpillars that could be handled and reared in the lab at any one time. Considering the possibility that phenological variation in plants could have implications for larval performance, we conducted analyses separately for the three butterfly species. Postemergence adult weight, time to emergence as adult, and survival to adult were recorded. Mortality (reflected in the survival data) included death associated with caterpillars that died while developing, individuals that pupated but failed to emerge, and disease; we did not distinguish between these sources of mortality. Data were standardized (Z transformed) within populations to facilitate comparisons among

populations and taxa that may have inherent differences, such as in size or in development time. Z scores were used in analyses described below unless otherwise noted.

Dish was considered the unit of replication, thus percent survival was calculated per dish. For analyses of adult weight and time to emergence, dish was used as a random factor. Percent survival was analyzed using analysis of variance (ANOVA) with plant, population and the interaction between the two as predictor variables. Time to emergence and adult weight were both analyzed with ANOVA, using population, plant, the interaction between the two and sex as predictor variables, along with dish as a random factor nested within plant and population. For all of these analyses, ANOVA was performed a second time without the plant/population interaction if it was not significant at $\alpha < 0.05$. These analyses were performed using JMP software version 8.0.2 (SAS Institute).

Differences in survival were also investigated by generating and comparing survival curves. To create survival curves, individual caterpillars were assumed to be alive until the date they were found dead. Rather than analyzing survival curves on an individual-dish basis (where sample sizes were small), the number of individuals surviving on a given day was calculated for each plant/population combination, giving one curve per combination, as is often done in survival analysis [26]. Survival curves were generated in R (2.12.2) using the packages *splines* and *survival*. The shapes of the curves were investigated within population using the packages *MASS* and *fitdistrplus*. Weibull distributions are commonly used to model survival using two parameters, shape and scale. The shape parameter measures where the inflection point occurs or practically whether individuals are lost more at the beginning or end of a given time period, and the scale parameter characterizes the depth of the curve. We estimated the two Weibull parameters, shape and scale, that characterized the fitted curves using maximum likelihood. One-thousand bootstrap replicates were then used to generate 95% confidence intervals for the shape and scale parameters, so that they could be compared across plants within a given population.

3. Results

We began the larval performance experiments with 2040 caterpillars in 357 dishes. Average survival to eclosion across all experiments was 23.4%. In general, differences in larval performance among plants were greater than differences between populations, which can be seen both in Figure 2 and also by comparing variation partitioned by plants and population in Table 2. For example, survival was highest on *Lotus nevadensis* across all populations for all three taxa, with an average survival of 56.5% (survival on *Lupinus polyphyllus* was comparable for two of the three *L. melissa* populations). Survival on alfalfa was consistently the lowest of any plant across populations: only two caterpillars survived to eclosion (Figures 2 and 3). Because survival was so low on alfalfa, it was excluded from most analyses and figures. The inferior nature of alfalfa as a host plant is consistent with previous studies, particularly when caterpillars do not have access to

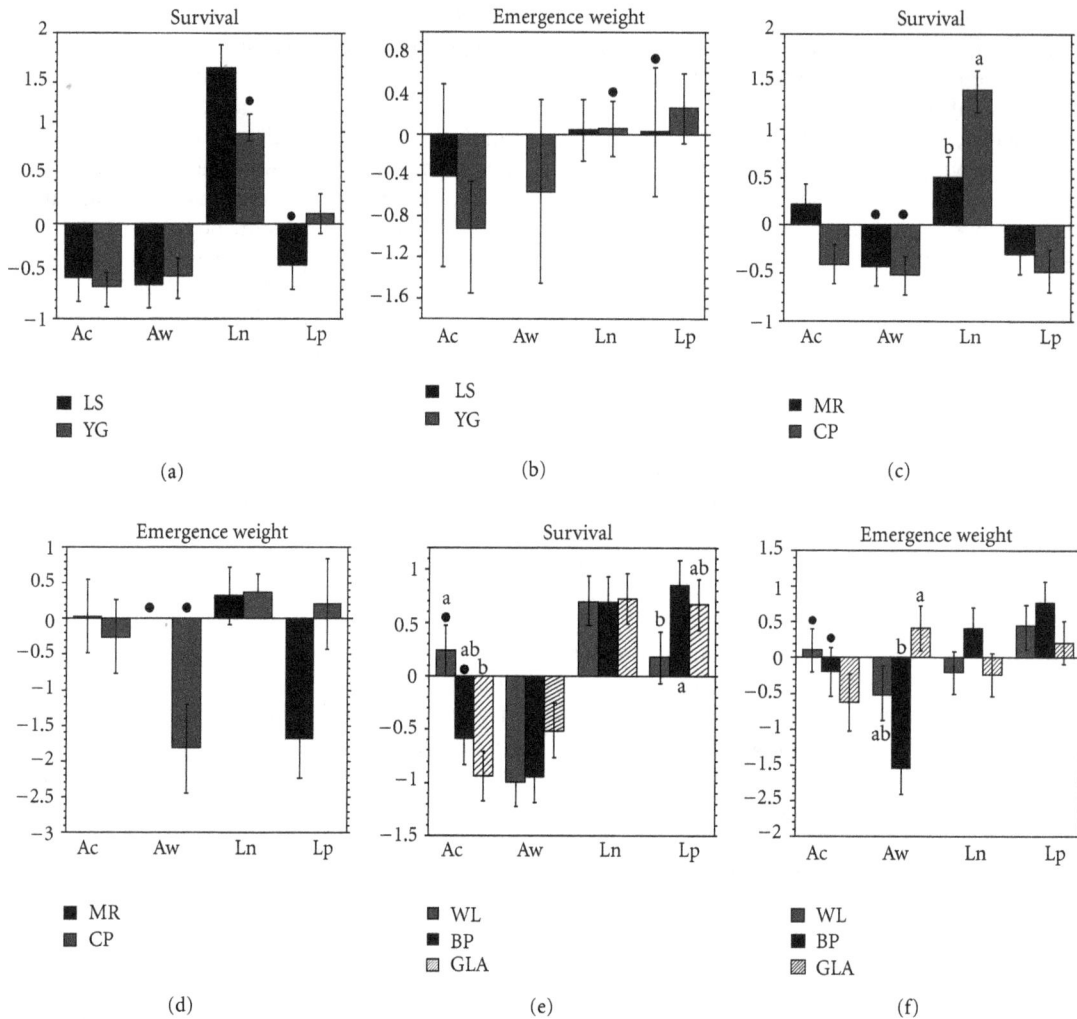

FIGURE 2: Survival and emergence weights from rearing experiments: (a and b) *L. idas*, (c and d) hybrid species, (e and f) *L. melissa*. Values are standardized (*Z* transformed). Shading of bars indicates plant species; see legends and Figure 1 for population locations. See Table 2 for associated model details, including plant and population effects. Significant differences ($P < 0.05$) for population effects within plants are indicated here with small letters near bars. Black dots identify natal host associations for each population. Host plant abbreviations as follows. Ac; *Astragalus canadensis*; Aw; *Astragalus whitneyi*; Ln; *Lotus nevadensis*; Lp; *Lupinus polyphyllus*. Results from alfalfa, *M. sativa*, are not shown here because survival was very low; see main text for details.

flowers and flower buds. When flowers have been included in performance experiments, survival of *L. melissa* on alfalfa and *A. canadensis* was equal, although those individuals reared on alfalfa were significantly smaller adults [19].

Plant and the interaction between plant and population were significant predictors of survival for *L. idas*, *L. melissa*, and the hybrid species (Table 2). Within taxa, there were differences among populations on certain plants. For example, survival for *L. idas* on *Lotus nevadensis* was greater for LS compared to YG, but the pattern was reversed for the host *Lupinus polyphyllus* (Figure 2); in other words, each population had higher survival on the natal host of the other population. A different pattern can be seen across populations of *L. melissa* on *Astragalus canadensis*, where survival was highest for individuals from WL, a population whose natal host is *A. canadensis*. *L. melissa* survival on *A. canadensis* was lowest for GLA, which is a population associated with the

exotic host alfalfa, and survival on *A. canadensis* is intermediate for BP, where both *A. canadensis* and alfalfa are utilized. Thus host use by *L. melissa* populations predicts variation in larval performance. Effects of plant and population were generally not as pronounced for either adult weight or time to emergence (for *L. idas*, the only significant predictors of adult weight were dish and sex); exceptions to this include the significant population by plant interaction for adult weight of *L. melissa*. As with survival, *L. melissa* performance (adult weight) was greater on *A. canadensis* for the population that is associated with that plant, WL (Figure 2(f)).

Consistent with results for survival to emergence as an adult, survival curves through time also showed pronounced differences among plants (Figure 3; Table 4). For example, the Weibull scale parameter for alfalfa was generally different compared to the other plants, reflecting early and pervasive mortality for individuals reared on that plant. Most but not

Table 2: Results from analyses of variance for the three measures of performance: percent survival, adult weight, and time to emergence. In all cases dish was used as the unit of replication. Most population/plant combinations had 9 dishes, except for the following: YG/Ac 12 dishes, YG/Ln 12 dishes, YG/Lp 12 dishes, YG/Ms 12 dishes, and all plant combinations for CP and MR had 12 dishes. The total number of dishes was 357.

	SS	F Ratio$_{df}$	P
Survival *L. idas*			
Plant	48.62	$42.02_{3,73}$	<0.0001
Population	0.02	$0.06_{1,73}$	0.81
Plant × population	4.00	$3.46_{3,73}$	0.02
Survival hybrid species			
Plant	30.93	$16.33_{3,88}$	<0.0001
Population	0.00	$0.00_{1,88}$	1.00
Plant × population	7.50	$3.96_{3,88}$	0.01
Survival *L. melissa*			
Plant	44.95	$29.32_{3,96}$	<0.0001
Population	0.03	$0.03_{2,96}$	0.97
Plant × population	9.99	$3.26_{6,96}$	0.006
Adult weight *L. idas*			
Plant	2.11	$0.52_{3,33.47}$	0.67
Population	0.49	$0.14_{1,19.38}$	0.71
Dish (plant, population) random	47.78	$4.04_{26,74.00}$	<0.0001
Sex	10.81	$11.87_{2,74.00}$	<0.0001
Adult weight hybrid species			
Plant	7.25	$2.81_{3,48.09}$	0.049
Population	0.53	$0.61_{31,48.45}$	0.44
Dish (plant, population) random	14.90	$0.74_{20,30.00}$	0.76
Sex	1.29	$1.28_{1,30.00}$	0.27
Adult weight *L. melissa*			
Plant	21.22	$7.25_{3,132.13}$	0.0002
Population	8.62	$4.54_{2,154.10}$	0.01
Dish (plant, population) random	89.56	$1.44_{84,217.00}$	0.02
Plant × population	11.24	$15.15_{1,217.00}$	0.0001
Sex	25.93	$4.39_{6,126.13}$	0.0005
Time to emergence *L. idas*			
Plant	20.32	$8.91_{3,59.17}$	<0.0001
Population	0.69	$0.99_{1,4.11}$	0.37
Dish (plant, population) random	19.34	$0.94_{26,75.00}$	0.55
Sex	3.87	$2.45_{2,75.00}$	0.09
Time to emergence hybrid species			
Plant	5.30	$3.28_{2,25.82}$	0.05
Population	0	$0_{0,28.00}$	1.00
Dish (plant, population) random	16.06	$0.95_{20,28.00}$	0.54
Sex	1.20	$1.41_{1,28.00}$	0.25
Plant × population	6.04	$3.61_{2,42.94}$	0.04
Time to emergence *L. melissa*			
Plant	9.36	$2.41_{3,106.23}$	0.07
Population	1.66	$0.61_{2,99.01}$	0.54
Dish (plant, population) random	125.50	$2.34_{84,223.00}$	<0.0001
Sex	24.18	$37.87_{1,223.00}$	<0.0001

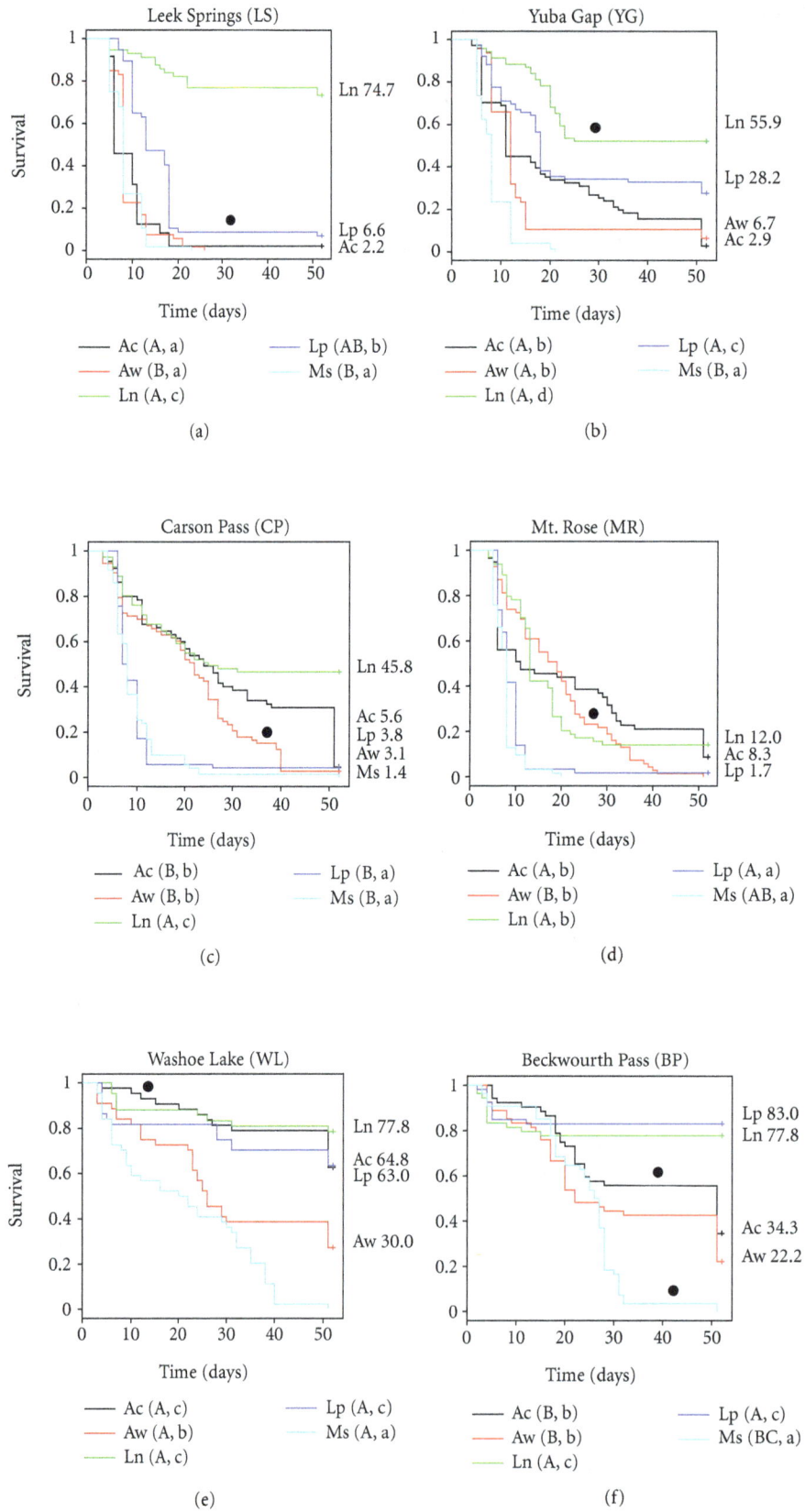

Figure 3: Continued.

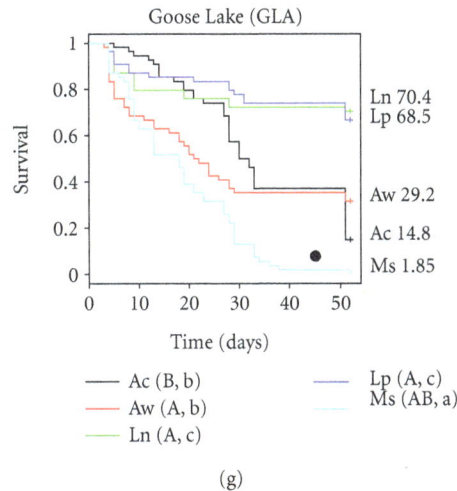

FIGURE 3: Survival curves for the seven populations studied. Colors indicate survival associated with a given plant; letters next to each plant in legends correspond to differences in the shape (upper case) and scale (lower case) of each curve indicated by nonoverlapping 95% confidence intervals from bootstrapped parameter values (see Table 4 for more details). Black dots indicate native host association for each population. Average, final survival is shown to the right of each graph for nonzero results. Plant abbreviations as follows. Ac; *Astragalus canadensis*; Aw; *Astragalus whitneyi*; Ln; *Lotus nevadensis*; Lp; *Lupinus polyphyllus*; Ms; *Medicago sativa*.

all mortality was manifested quite early in development, particularly for the alpine and *L. idas* populations across plants. Mortality was more evenly distributed through time for *L. melissa* on all plants. In some cases, patterns of survival vary among plants within populations, even when overall survival was low. For example, survival curves for CP drop much less rapidly for three plants, one of which is the natal host *A. whitneyi* and another is the congeneric *A. canadensis* (the third is *L. nevadensis*). For *A. canadensis*, it is interesting to note that across all populations there was a drop-off in survival near the end of development: many individuals made it to the pupal stage, but failed to emerge, perhaps suggesting a subtle nutritional challenge for successful completion of development presented by that plant.

4. Discussion

The reciprocal rearing experiment detected strong host plant effects and limited evidence of local adaptation to natal host plant species in the *Lycaeides* species complex. For example, development on *L. nevadensis* (the host of the *L. idas* population at YG) resulted in relatively high survival throughout the experiment, while development on *Medicago sativa* (the exotic host of *L. melissa* at GLA and BP) led to extremely low survival in all cases. These plant effects that transcend populations could be indicative of larval traits (such as high survival on *L. nevadensis*) that are conserved in the group and are not particularly labile. Our results could be influenced by the use of leaves but not flowers in larval rearing. Previous work has shown that survival is improved on *Medicago sativa* for larvae that have access to flowers, but this is not true on *Astragalus canadensis* [19]. We do not know if flowers are or are not important for larvae developing on the other plants.

In general, the survival that we report (23.4% throughout the experiment) could reflect the absence of flowers or other unfavorable lab conditions, and we do not at this time have life history data from the field for *Lycaeides* with which to compare our results. However, in interpreting results here and elsewhere (e.g., [19]), we make the assumption that lab experiments are informative with respect to relative performance across hosts. In other words, the consumption of *M. sativa* by *Lycaeides* caterpillars is associated with development into adults that are small relative to adults that develop on other plants. Without artifacts of lab rearing, it is possible that performance would generally be higher in the wild, but we would predict that performance on *M. sativa* would still be lower relative to performance on native hosts. An alternate possibility, which we cannot test at this time, is that lab rearing has plant-specific effects (i.e., *M. sativa* is a poor host only when used under artificial conditions).

For all the performance results, it is also important to note that phenological effects of changes in plant quality or suitability could be pronounced, but are not addressed by our experimental design. In particular, as noted above, our rearings were conducted in two phases due to logistical constraints: first including *L. idas* and populations of the hybrid species and second including all three *L. melissa* populations. This is not a completely unnatural situation, as *L. idas* and the hybrid species are univoltine, while *L. melissa* populations are multivoltine. Thus *L. idas* and hybrid species caterpillars are more likely to be exposed to only the early spring vegetation, as in our experiment. The consideration of phenological effects in plants is most relevant when comparing performance among butterfly taxa (e.g., the performance of *L. melissa* versus *L. idas* on a particular plant) but is less important when making comparisons within a taxon (e.g., the performance of *L. melissa* on different plant species).

In contrast to the general result of strong plant effects across taxa and populations, one result suggestive of local adaptation is the performance (survival and adult weight) of *L. melissa* on the native host *A. canadensis* [27]. Performance was highest on the native for the population that utilizes that host in the wild and lowest for a population associated with the exotic host alfalfa. Performance on *A. canadensis* is intermediate for the population where both hosts are used. These results raise a number of possibilities, including a scenario in which genetic variants associated with higher performance on an ancestral host were lost in the transition to the exotic host, which could be a consequence of relaxed selection or a population bottleneck in the new environment. Another explanation could involve a change in gene regulation associated with performance, rather than a loss of alleles. In any event, the transition to the novel host has apparently not been accompanied by an increase in performance on alfalfa. One caveat to this conclusion is that the *M. sativa* used in experiments was collected at one of the focal locations (BP), but could not, for logistical reasons, be collected from GLA. The latter population (GLA) is the population associated only with *M. sativa*, thus the conclusion that performance has not increased following the colonization of the novel host could have been different if local plant material from that location had been used in experiments; however, we have found consistently low performance on *M. sativa* in other experiments [19], suggesting generality to the result of low performance on that plant.

Variation in host preference has previously been documented among populations of *Lycaeides* butterflies, with populations of the hybrid species in particular exhibiting strong preferences for their natal host, *A. whitneyi* [17, 22], relative to the hosts of other *Lycaeides* populations. However, we found low survival and low adult weights for individuals of the hybrid species reared on *A. whitneyi* (Figure 2). It is possible that laboratory conditions were a poor reflection of appropriate abiotic conditions for the hybrid species individuals adapted to an alpine environment. It is also possible that other factors, such as the absence of flowers in experiments or induced defenses in leaves, could be important in *A. whitneyi*, which supported poor growth for larvae from all populations. In any event, the patterns of performance that we report are not consistent with an expected preference-performance paradigm for host shifts leading to the evolution of reproductive isolation [28]. Variation in both adult preference and larval performance is discussed further in the following section considering ecological traits and hypotheses relating traits to reproductive isolation.

4.1. Ecology and Diversification. Although many studies of herbivorous insects have focused on larval performance with respect to local adaptation and ecological speciation, populations of herbivorous insects (or of any organism) can of course differ in numerous ways, some related to resource use but also to other aspects of the environment. Nosil et al. [15] have suggested a number of scenarios in which multiple traits could be important in the evolution of reproductive isolation. In particular, natural selection acting on a single

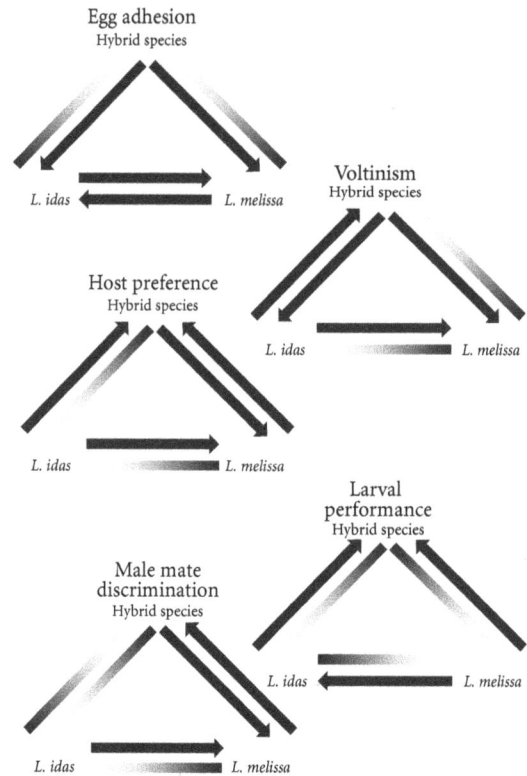

FIGURE 4: Summary of hypotheses relating ecological traits to reproductive isolation between taxa based on the current and past studies [16, 17, 19, 22, 23]. Arrows joining two taxa correspond to greater gene flow and those that are faded represent gene flow that could be prevented or reduced by a given trait. For details see text.

ecological trait or a single niche dimension could be important for initiating speciation, while the evolution of differences along multiple ecological axes might often be needed for complete reproductive isolation [15]. Multiple ecological axes could be different aspects of, for example, resource use [29, 30], or they could be more disparate traits, such as mate finding or predator avoidance. In either case, the idea is that selection along one axis might be insufficient for reproductive isolation, but selection acting along multiple axes might confer a high, overall level of reproductive isolation.

Considering the potential importance of multiple traits in ecological speciation, Figure 4 and Table 3 summarize information from this and other studies in *Lycaeides* and present hypotheses regarding multiple ecological and behavioral traits and how these might interact with ecological, reproductive isolating processes in this system. Specifically, Figure 4 explores hypotheses about reduced gene flow, represented by faded arrows, between the taxa due to the ecological differences of a given trait. For example, the model shown for egg adhesion posits that variation in adhesion could be a barrier to gene flow going from *L. idas* and *L. melissa* populations into populations of the hybrid species. *L. idas* and *L. melissa* females lay eggs that adhere to plants. As discussed above, the alpine host plants senesce and are blown from the area, thus removing any attached eggs from the site

TABLE 3: Details of behavioral and ecological variation among taxa, specifically as variation relates to potential barriers to gene flow; see Figure 4 for a graphical interpretation of these traits in relation to gene flow. The descriptions of male mate discrimination refer to preferences of males for females of the other taxa. Details for larval performance refer to performance on the hosts of the other taxa relative to performance of the other taxa on those same plants.

	L. idas	Hybrid species	L. melissa
Egg adhesion	Yes	No	Yes
Preference for natal host(s)	Moderate	High	Low to high
Male mate discrimination	Against L. melissa and hybrid species	Against L. idas	None
Voltinism	Univoltine	Univoltine	Multivoltine
Larval performance	Poor on hosts of L. melissa; superior on host of hybrid species	Poor on hosts of both L. melissa and L. idas	Equivalent on hosts of L. idas and superior on host of hybrid species

Further information on specific behavioral and ecological variables (other than larval performance, reported here) can be found as follows: egg adhesion [23], preference for natal hosts [17, 19, 22], male mate discrimination [16], and voltinism [23].

of next-spring's fresh plant growth [23]. Differences in host preference might also affect patterns of gene flow between the species. The host of the hybrid species populations is readily accepted by ovipositing females from all Lycaeides examined thus far [17], thus it would likely be accepted by females from L. idas and L. melissa populations arriving at a population of the hybrid species. In contrast, the hosts of the L. idas populations are not preferred by females of either the hybrid species or L. melissa [17, 19]. The arrows pointing towards L. melissa assume the presence of only the native host A. canadensis, not the exotic M. sativa (excluding the exotic is a simplifying assumption for Figure 4, but also appropriate given that ecological diversification occurred before the recent introduction of alfalfa). Astragalus canadensis and A. whitneyi are equally acceptable for oviposition by hybrid species individuals (Forister, unpublished data), and we assume the same equivalence for L. idas (i.e., we assume L. idas females would readily accept A. canadensis, just as they do with A. whitneyi, thus an arrow without a barrier pointing from L. idas to L. melissa in the host preference diagram).

Similar to host preference, variation in male mate discrimination potentially presents barriers only between hybrid species and L. idas populations and between L. idas and L. melissa populations but not between hybrid species and L. melissa populations [17]. L. melissa males will readily approach either L. idas or females of the hybrid species, while L. idas males discriminate against females from the other two taxa, and hybrid species males discriminate against L. idas females. This behavioral variation among taxa, reported in Fordyce et al. [16], comes from choice tests involving dead and paper-model females presented in experimental arrays in the field. It is important to note that being a less-preferred mate is of course not the same as not being mated. In other words, a virgin L. melissa female that immigrated into an L. idas population might be a low-ranked mate for male L. idas relative to local females, but it is possible that she would eventually find a mate. However, the patterns of gene flow shown in Figure 4 are meant to be hypotheses for potential barriers to gene flow within a given trait. An L. melissa female immigrating into a population of the hybrid species would be mated more readily (and thus be more likely to contribute

to the gene pool) relative to the dynamic just described (an L. melissa female arriving at an L. idas population). It is possible to imagine all of the traits depicted in Figure 4 being involved in either pre- or postzygotic isolation. For example, mate preference could act as just described on immigrant, virgin females, as in the immigrant inviability concept of Nosil et al. (2005) [31]. An alternative but similar scenario could involve the offspring of an immigrant; in this case, wing-pattern alleles (related to mate choice) would interact with mate choice in the next generation.

Variation in voltinism could affect gene flow from L. melissa into the other univoltine taxa. Because L. melissa populations are multivoltine, it is possible that an L. melissa female moving into populations of the other taxa would lay eggs that failed to diapause in habitats where the univoltine strategy is superior (such as in the alpine habitat where there is a short window for larval development [32]). Alternatively, diapause could be plastic, in which case the patterns of connectivity (hypothesized patterns of gene flow) pictured would be different.

We can now add larval performance to the suite of hypotheses linking ecology and gene flow in Lycaeides. In generating hypotheses relating larval performance to gene flow, we have used this criterion (focusing on survival, rather than adult weight, as the most straightforward metric of performance): if foreign larvae (i.e., the offspring of a recently arrived female) have lower survival on the local host relative to local individuals, we hypothesize a relative reduction in gene flow associated with performance. For gene flow between L. idas and L. melissa, the survival of L. idas larvae on the host of L. melissa is lower than the survival of L. melissa caterpillars on the same plant (see mean survival values in Figure 3). Interestingly in the context of hybrid speciation, our results suggest that gene flow from both L. idas and L. melissa would be unimpeded into populations of the hybrid species relative to the reverse, meaning that the two parental species had higher survival (relative to hybrid individuals) on the alpine host and that hybrid individuals had relatively inferior performance on the two parental species' hosts. Of course, this could be different if another trait, for example, egg adhesion, had a stronger effect or

TABLE 4: Survival curves for each host-population combination. A weibull distribution was fitted to each combination with 1000 bootstrap replicates. We report shape and scale parameters along with bootstrapped confidence intervals. Upper case and lower letters following shape and scale values correspond to 95% confidence intervals that do not overlap (upper case letters for shape and lower case letters for scale) within populations based on the thousand bootstrapped replicates. See Figure 3 for graphical representation of survival curves.

Population	Host	Shape		Scale	
CP	Ac	1.47 (1.25–1.75)	B	31.16 (26.22–36.37)	b
CP	Aw	1.61 (1.35–1.97)	B	23.53 (20.27–27.09)	b
CP	Ln	0.82 (0.71–0.95)	A	63.75 (42.86–110.98)	c
CP	Lp	1.38 (1.12–2.62)	B	11.73 (9.61–14.66)	a
CP	Ms	1.67 (1.34–2.42)	B	10.86 (9.41–12.76)	a
BP	Ac	1.50 (1.17–2.01)	B	51.38 (41.18–64.28)	b
BP	Aw	1.31 (1.07–1.61)	B	39.56 (31.44–50.49)	b
BP	Ln	0.44 (0.38–0.52)	A	1128.76 (307.38–7477.83)	c
BP	Lp	0.44 (0.39–0.51)	A	2152.31 (501.95–21856.68)	c
BP	Ms	2.65 (2.00–4.07)	BC	26.36 (23.90–28.98)	a
GLA	Ac	2.154(1.76–2.69)	B	40.31 (34.73–46.00)	b
GLA	Aw	0.88 (0.74–1.07)	A	39.16 (25.84–61.45)	b
GLA	Ln	0.66 (0.53–0.86)	A	236.85 (115.08–750.35)	c
GLA	Lp	0.94 (0.68–1.50)	A	134.35 (82.65–321.98)	c
GLA	Ms	1.70 (1.44–2.10)	AB	20.27 (16.98–23.59)	a
LS	Ac	1.57 (1.21–2.86)	A	10.70 (8.96–13.28)	a
LS	Aw	2.31 (1.97–3.09)	B	10.43 (9.22–11.71)	a
LS	Ln	0.88 (0.67–1.24)	A	189.07 (104.65–511.29)	c
LS	Lp	1.58 (1.33–2.17)	AB	19.51 (16.28–23.62)	b
LS	Ms	2.70 (2.20–3.70)	B	9.52 (8.62–10.63)	a
MR	Ac	1.11 (0.97–1.23)	A	24.07 (18.38–31.14)	b
MR	Aw	1.84 (1.59–2.16)	B	21.37 (18.59–24.33)	b
MR	Ln	1.29 (1.08–1.59)	A	22.56 (17.80–29.12)	b
MR	Lp	1.788 (1.38–5.18)	A	10.23 (8.83–11.96)	a
MR	Ms	2.73 (2.26–4.42)	AB	8.54 (7.71–9.45)	a
WL	Ac	1.65 (1.03–3.29)	A	82.61 (62.56–145.41)	c
WL	Aw	1.20 (0.92–1.57)	A	41.69 (31.24–56.54)	b
WL	Ln	0.77 (0.57–1.38)	A	302.19 (125.45–1912.44)	c
WL	Lp	0.77 (0.56–1.24)	A	144.13 (79.93–404.52)	c
WL	Ms	1.43 (1.22–1.89)	A	22.96 (18.23–28.27)	a
YG	Ac	1.29 (1.17–1.45)	A	22.20 (18.23–27.03)	b
YG	Aw	1.34 (1.12–1.88)	A	17.23 (13.41–22.86)	b
YG	Ln	1.05 (0.89–1.21)	A	71.74 (50.50–111.48)	d
YG	Lp	1.07 (0.94–1.23)	A	36.16 (27.35–49.30)	c
YG	Ms	2.45 (2.12–3.32)	B	9.25 (8.30–10.29)	a

acted before larval performance in restricting gene flow (see [6, 11] for examples of the complexities of estimating components of reproductive isolation associated with a suite of traits). We stress that these are hypotheses that bear further investigation, as we know that larval performance is complex, being affected not only by variation in host quality (i.e., the availability of flowers [19]) but also by the presence of mutualistic ants and natural enemies [33].

5. Conclusion

We conducted a performance experiment for seven populations from three species within the *Lycaeides* species complex, *L. idas, L. melissa,* and the hybrid species, on five different plants. Our primary results include large plant effects, with *L. idas* hosts being generally superior for larval development and the exotic host of *L. melissa* being extremely poor, both for *L. melissa* and the other taxa. In general, there is little evidence of local adaptation in these performance data. This conclusion is perhaps consistent with the fact that these butterfly taxa are associated with multiple hosts throughout their geographic ranges. Thus gene flow could limit local adaptation to any particular plant species. As a consequence, variation in larval performance across multiple hosts is unlikely to be the dominant mechanism of reproductive isolation between populations and taxa.

Our results (including some evidence for local adaptation among *L. melissa* populations for their native host, *A. canadensis*) together with previously published data [16, 17, 19, 22, 23] were integrated to build a hypothetical model relating ecology to reproductive isolation and diversification in *Lycaeides*. The model presented in Figure 4 describes a system that is well poised for a test of the "multifarious selection" hypothesis [15]. One hypothesis that can be generated from Figure 4 is that there might not be one single trait that could act as a barrier to gene flow between all three taxa, and most traits might only act to reduce gene flow asymmetrically. For example, egg adhesion could affect gene flow from both *L. idas* and *L. melissa* into the hybrid species, but would not necessarily be effective in the opposite directions (from the hybrid species into *L. idas* and *L. melissa*). More generally in the context of ecological speciation, a greater number of traits might increase the possibility that a hybrid "falls between" the niches represented by the two adaptive peaks occupied by the species or incipient species [34, 35]. In a relatively simple example involving two traits, hybrids between populations of *Mitoura* butterflies associated with different host plant species inherit a maladaptive mismatch of traits: hybrid individuals have higher performance on one of the parental hosts, but express an oviposition preference for the other host [36].

However, the importance of multiple traits for ecological speciation in *Lycaeides* must wait on estimates of historical and contemporary gene flow between pairs of populations and analyses of those estimates in light of variation in ecological and behavioral traits [37]. The inclusion of such comparative data, particularly for a larger suite of populations, would perhaps reveal the influence of a single trait for explaining a majority of the variation in reproductive isolation. It is also possible that a key trait for reproductive isolation remains unstudied in this system. For future studies in this group, it will also be important to sample populations widely throughout the geographic ranges of the focal butterfly species (Figure 1), as dynamics of local adaptation and diversification can be affected by geographic context, particularly proximity to the edge of a range and potentially marginal habitats. Beyond the details of ecological diversification in *Lycaeides*, our results should generally stress the importance of delving deeper than the traditional "preference-performance relationship" when investigating ecological speciation in herbivorous insects.

Acknowledgments

The authors would like to thank Bonnie Young and Temba Barber for assistance in rearing caterpillars. This work was supported by the National Science Foundation (IOS-1021873 and DEB-1050355 to C. C. Nice; DEB-0614223 and DEB-1050947 to J. A. Fordyce; DEB 1020509 and DEB 1050726 to M. L. Forister; DEB-1011173 to Z. Gompert). C. F. Scholl was supported by the Biology Department at the University of Nevada, Reno.

References

[1] A. T. Peterson, J. Soberón, and V. Sánchez-Cordero, "Conservatism of ecological niches in evolutionary time," *Science*, vol. 285, no. 5431, pp. 1265–1267, 1999.

[2] A. Prinzing, W. Durka, S. Klotz, and R. Brandl, "The niche of higher plants: evidence for phylogenetic conservatism," *Proceedings of the Royal Society B: Biological Sciences*, vol. 268, no. 1483, pp. 2383–2389, 2001.

[3] K. H. Kozak and J. J. Wiens, "Does niche conservatism promote speciation? A case study in North American salamanders," *Evolution*, vol. 60, no. 12, pp. 2604–2621, 2006.

[4] U. Dieckmann and M. Doebeli, "On the origin of species by sympatric speciation," *Nature*, vol. 400, no. 6742, pp. 354–357, 1999.

[5] D. Schluter, "Ecology and the origin of species," *Trends in Ecology and Evolution*, vol. 16, no. 7, pp. 372–380, 2001.

[6] J. Ramsey, H. D. Bradshaw, and D. W. Schemske, "Components of reproductive isolation between the monkeyflowers *Mimulus lewisii* and *M. cardinalis* (Phrymaceae)," *Evolution*, vol. 57, no. 7, pp. 1520–1534, 2003.

[7] D. J. Funk, P. Nosil, and W. J. Etges, "Ecological divergence exhibits consistently positive associations with reproductive isolation across disparate taxa," *Proceedings of the National Academy of Sciences of the United States of America*, vol. 103, no. 9, pp. 3209–3213, 2006.

[8] K. W. Matsubayashi, I. Ohshima, and P. Nosil, "Ecological speciation in phytophagous insects," *Entomologia Experimentalis et Applicata*, vol. 134, no. 1, pp. 1–27, 2010.

[9] M. Drès and J. Mallet, "Host races in plant-feeding insects and their importance in sympatric speciation," *Philosophical Transactions of the Royal Society B: Biological Sciences*, vol. 357, no. 1420, pp. 471–492, 2002.

[10] P. Nosil, "Divergent host plant adaptation and reproductive isolation between ecotypes of *Timema cristinae* walking sticks," *American Naturalist*, vol. 169, no. 2, pp. 151–162, 2007.

[11] N. H. Martin and J. H. Willis, "Ecological divergence associated with mating system causes nearly complete reproductive isolation between sympatric *Mimulus* species," *Evolution*, vol. 61, no. 1, pp. 68–82, 2007.

[12] C. Wiklund, "The evolutionary relationship between adult oviposition preferences and larval host plant range in *Papilio machaon* L.," *Oecologia*, vol. 18, no. 3, pp. 185–197, 1975.

[13] S. S. Wasserman and D. J. Futuyma, "Evolution of host plant utilization in laboratory populations of the Southern cowpea weevil, *Callosobruchus maculatus fabricius* (Coleoptera: Bruchidae)," *Evolution*, vol. 35, no. 4, pp. 605–617, 1981.

[14] T. P. Craig, J. D. Horner, and J. K. Itami, "Hybridization studies on the host races of *Eurosta Solidaginis*: implications for sympatric speciation," *Evolution*, vol. 51, no. 5, pp. 1552–1560, 1997.

[15] P. Nosil, L. J. Harmon, and O. Seehausen, "Ecological explanations for (incomplete) speciation," *Trends in Ecology and Evolution*, vol. 24, no. 3, pp. 145–156, 2009.

[16] J. A. Fordyce, C. C. Nice, M. L. Forister, and A. M. Shapiro, "The significance of wing pattern diversity in the Lycaenidae: mate discrimination by two recently diverged species," *Journal of Evolutionary Biology*, vol. 15, no. 5, pp. 871–879, 2002.

[17] C. C. Nice, J. A. Fordyce, A. M. Shapiro, and R. Ffrench-Constant, "Lack of evidence for reproductive isolation among ecologically specialised lycaenid butterflies," *Ecological Entomology*, vol. 27, no. 6, pp. 702–712, 2002.

[18] L. K. Lucas, J. A. Fordyce, and C. C. Nice, "Patterns of genitalic morphology around suture zones in North American *Lycaeides* (Lepidoptera: Lycaenidae): implications for taxonomy and historical biogeography," *Annals of the Entomological Society of America*, vol. 101, no. 1, pp. 172–180, 2008.

[19] M. L. Forister, C. C. Nice, J. A. Fordyce, and Z. Gompert, "Host range evolution is not driven by the optimization of larval performance: the case of *Lycaeides melissa* (Lepidoptera: Lycaenidae) and the colonization of alfalfa," *Oecologia*, vol. 160, no. 3, pp. 551–561, 2009.

[20] Z. Gompert, C. C. Nice, J. A. Fordyce, M. L. Forister, and A. M. Shapiro, "Identifying units for conservation using molecular systematics: the cautionary tale of the Karner blue butterfly," *Molecular Ecology*, vol. 15, no. 7, pp. 1759–1768, 2006.

[21] Z. Gompert, L. K. Lucas, J. A. Fordyce, M. L. Forister, and C. C. Nice, "Secondary contact between *Lycaeides* idas and *L. melissa* in the Rocky Mountains: extensive admixture and a patchy hybrid zone," *Molecular Ecology*, vol. 19, no. 15, pp. 3171–3192, 2010.

[22] Z. Gompert, J. A. Fordyce, M. L. Forister, A. M. Shapiro, and C. C. Nice, "Homoploid hybrid speciation in an extreme habitat," *Science*, vol. 314, no. 5807, pp. 1923–1925, 2006.

[23] J. A. Fordyce and C. C. Nice, "Variation in butterfly egg adhesion: adaptation to local host plant senescence characteristics?" *Ecology Letters*, vol. 6, no. 1, pp. 23–27, 2003.

[24] J. Jaenike, "Host specialization in phytophagous insects," *Annual Review of Ecology and Systematics*, vol. 21, no. 1, pp. 243–273, 1990.

[25] Z. Gompert, J. A. Fordyce, M. L. Forister, and C. C. Nice, "Recent colonization and radiation of North American *Lycaeides* (*Plebejus*) inferred from mtDNA," *Molecular Phylogenetics and Evolution*, vol. 48, no. 2, pp. 481–490, 2008.

[26] J. P. Klein and M. L. Moeschberger, *Survival Analysis: Techniques for Censored and Truncated Data*, Springer, New York, NY, USA, 2nd edition, 2003.

[27] T. J. Kawecki and D. Ebert, "Conceptual issues in local adaptation," *Ecology Letters*, vol. 7, no. 12, pp. 1225–1241, 2004.

[28] S. Gripenberg, P. J. Mayhew, M. Parnell, and T. Roslin, "A meta-analysis of preference-performance relationships in phytophagous insects," *Ecology Letters*, vol. 13, no. 3, pp. 383–393, 2010.

[29] M. L. Forister, A. G. Ehmer, and D. J. Futuyma, "The genetic architecture of a niche: variation and covariation in host use traits in the Colorado potato beetle," *Journal of Evolutionary Biology*, vol. 20, no. 3, pp. 985–996, 2007.

[30] P. Nosil and C. P. Sandoval, "Ecological niche dimensionality and the evolutionary diversification of stick insects," *PLoS ONE*, vol. 3, no. 4, Article ID e1907, 2008.

[31] P. Nosil, T. H. Vines, and D. J. Funk, "Perspective: reproductive isolation caused by natural selection against immigrants from divergent habitats," *Evolution*, vol. 59, no. 4, pp. 705–719, 2005.

[32] L. Somme, "Adaptations of terrestrial arthropods to the alpine environment," *Biological Reviews: Cambridge Philosophical Society*, vol. 64, no. 4, pp. 367–407, 1989.

[33] M. L. Forister, Z. Gompert, C. C. Nice, G. W. Forister, and J. A. Fordyce, "Ant association facilitates the evolution of diet breadth in a lycaenid butterfly," *Proceedings of the Royal Society B: Biological Sciences*, vol. 278, no. 1711, pp. 1539–1547, 2011.

[34] H. A. Orr, "Adaptation and the cost of complexity," *Evolution*, vol. 54, no. 1, pp. 13–20, 2000.

[35] S. Gavrilets, *Fitness landscapes and the origin of species*, Princeton University Press, Princeton, NJ, USA, 2004.

[36] M. L. Forister, "Independent inheritance of preference and performance in hybrids between host races of *Mitoura* butterflies (Lepidoptera: Lycaenidae)," *Evolution*, vol. 59, no. 5, pp. 1149–1155, 2005.

[37] G. Lu and L. Bernatchez, "Correlated trophic specialization and genetic divergence in sympatric lake whitefish ecotypes (*Coregonus clupeaformis*): support for the ecological speciation hypothesis," *Evolution*, vol. 53, no. 5, pp. 1491–1505, 1999.

17

Interactions between a Top Order Predator and Exotic Mesopredators in the Australian Rangelands

Katherine E. Moseby,[1,2] **Heather Neilly,**[2] **John L. Read,**[1,2] **and Helen A. Crisp**[2]

[1] *School of Earth and Environmental Sciences, The University of Adelaide, SA 5005. Arid Recovery, P.O. Box 147, Roxby Downs, South Australia 5725, Australia*
[2] *Arid Recovery, P.O. Box 147, Roxby Downs, SA 5725, Australia*

Correspondence should be addressed to Katherine E. Moseby, katherine.moseby@adelaide.edu.au

Academic Editor: Cajo J. F. ter Braak

An increase in mesopredators caused by the removal of top-order predators can have significant implications for threatened wildlife. Recent evidence suggests that Australia's top-order predator, the dingo, may suppress the introduced cat and red fox. We tested this relationship by reintroducing 7 foxes and 6 feral cats into a 37 km^2 fenced paddock in arid South Australia inhabited by a male and female dingo. GPS datalogger collars recorded locations of all experimental animals every 2 hours. Interactions between species, mortality rates, and postmortems were used to determine the mechanisms of any suppression. Dingoes killed all 7 foxes within 17 days of their introduction and no pre-death interactions were recorded. All 6 feral cats died between 20 and 103 days after release and dingoes were implicated in the deaths of at least 3 cats. Dingoes typically stayed with fox and cat carcasses for several hours after death and/or returned several times in ensuing days. There was no evidence of intraguild predation, interference competition was the dominant mechanism of suppression. Our results support anecdotal evidence that dingoes may suppress exotic mesopredators, particularly foxes. We outline further research required to determine if this suppression translates into a net benefit for threatened prey species.

1. Introduction

Introduced feral cats (*Felis catus*) and red foxes (*Vulpes vulpes*) have been implicated in the historical extinction and decline of many Australian mammal species [1–4] as well as the failure of several recent attempts to reintroduce threatened species to the wild [5–9]. Effective control of the red fox and feral cat is a core objective of many Australian mammal and terrestrial bird recovery programs. Although the red fox has been successfully controlled in some areas of Australia using poison meat baits [10], the efficacy of long-term baiting can attenuate due to high selection pressure for toler-ance to 1080 [11] and bait shyness attributable to receiving a sublethal dose of poison. Control of foxes is also thought to lead to an increase in cat density [10, 12, 13] which could negate any positive biodiversity benefits. Poisoning feral cats is often ineffective owing to poor bait uptake [14–20] and a cost

effective, large scale control mechanism for feral cats is currently not available [21].

Interspecific killing between carnivores is common [22], and recent studies have highlighted the possible role of top-order predators in controlling second-tier carnivores (mesopredators) [23–25]. The mesopredator release hypothesis predicts that reduced abundance of top-order predators results in increased abundance or activity of smaller subordinate predators [23]. This hypothesis has most support in North America, where studies have found that when coyote (*Canis latrans*) abundance declines, red fox numbers increase [23, 26]. The removal of the grey wolf, *Canis lupus*, has also been linked to an increase in coyote populations [27], and the removal of coyotes has resulted in changes in bobcat (*Lynx rufus*) and gray fox (*Urocyon cinereoargenteu*) populations [28]. In Scandinavia, the pine marten (*Martes martes*) was found to increase after a decline in red fox populations [29].

Where predation efficiency or prey specificity of smaller predators is superior or different to that of the top-order predator then changes in prey abundance can result [23]. Glen and Dickman [24] outlined complex interactions between carnivores in Australia and suggested that mesopredator release is an important mechanism shaping current prey populations in Australia. The post-European extinction of some Australian mammal species is thought to at least be partly attributable to mesopredator release through the removal or control of the dingo (*Canis lupus dingo*) [4]. Stable dingo populations are still found in many arid areas of Australia and may provide a net benefit to some threatened wildlife species through a decrease in predation rates by the red fox and/or feral cat [2, 29–32]. Smith and Quin [2] found lower rates of conilurine rodent extinction in areas where dingoes were abundant, and Johnson et al. [4] has suggested that mammal extinctions and decline are less severe in areas where dingoes are still present. Letnic et al. [33, 34] also favour the mesopredator release hypothesis as well as the trophic cascade theory, which suggests that top predators such as dingoes have either positive or negative effects on lower trophic levels and may indirectly enhance plant biomass [35]. The removal of dingoes may thus allow herbivores [36] and smaller introduced predators to increase, depleting plant biomass and increasing predation pressure.

Unfortunately little empirical data exist to support the perceived role of the dingo in suppressing fox and cat abundance at landscape scales [37] with evidence relying on correlations using historical or observational data (see [34, 38]). However, dingoes have been recorded occasionally killing or eating foxes [39] and cats [40, 41], and remains of both have been recorded in dingo scats, although usually at a very low occurrence [39, 42–45]. Dingoes are thought to exclude foxes from resource points such as carcasses during drought [44], and fox abundance has also been found to be higher in areas where dingoes are absent or controlled [30, 34, 43, 46]. Dingoes could potentially suppress fox, and cat populations through intraguild predation, interference, and/or exploitative competition. Interference competition may include direct attack, exclusion from resource points, causing a change in habitat use or activity times, or by increasing stress levels through frequent avoidance behaviour.

Dingoes are currently excluded or controlled over most of the Australian pastoral zone for the protection of commercial stock. Understanding any role that dingoes play in controlling introduced predators could assist in seeking a balance between the control of dingoes for pastoral production and the protection of dingoes for broader biodiversity benefits.

This study aimed to test the hypothesis that dingoes can suppress feral cats and foxes by examining their interactions within a landscape scale enclosure. A pair of dingoes was reintroduced to a 37 square km fenced paddock in northern South Australia. Feral cats and foxes were reintroduced 4 months later, and all animals were monitored for up to 12 months using GPS datalogger collars. Interactions between species, mortality rates, and postmortems were used to determine if suppression was due to interference or exploitative competition and/or intraguild predation. Cats, and foxes were also introduced to an adjacent unfenced control area where dingoes were removed. Indices of cat, fox, and rabbit spoor were compared between the two areas. Two factors were critical to the study: firstly, that densities of dingoes, cats, foxes, and prey species were typical of those found in the wider environment, and secondly that all study animals were local inhabitants and familiar with the habitats present in the study area.

2. Study Area

A 37 km^2 "Dingo Paddock" was fenced between July and November 2008 (30.27°S, 136.93°E) on Stuart Creek Pastoral Station. The paddock is situated approximately 35 km north of Roxby Downs in northern South Australia and is enclosed on three sides by a 1.6 m high netting fence (50 mm holes) with a 50 cm floppy top curving inwards to keep dingoes, cats, and foxes within the paddock but allowing cats and foxes to climb in. The netting fence was based on the Arid Recovery fence design [47] but was built from 50 mm netting to allow small rabbits to pass through the fence. The southern boundary of the paddock is shared with the Arid Recovery Reserve's Red Lake exclosure and is a 1.15 m high netting fence made from 30 mm netting with a floppy top overhang facing the dingo paddock. This study was conducted between December 2008 and December 2009 and formed part of a larger predator behaviour study which began in January 2008.

The southern section of the Dingo Paddock comprised a clay interdunal swale more than 2 km wide and vegetated with chenopod shrubs, bladder saltbush (*Atriplex vesicaria*), Oodnadatta saltbush (*A. omissa*), and low bluebush (*Maireana astrotricha*). Longitudinal orange sand dunes supporting sandhill wattle (*Acacia ligulata*) and sticky Hopbush (*Dodonaea viscosa*) shrublands were present in the northern sections, separated by 100 to 400 m wide swales. Other habitats include mulga (*Acacia aneura*) sandplains, patches of dune canegrass (*Zygochloa paradoxa*), and a breakaway range comprising silcrete capped hills with colourful eroding shale slopes in the western section of the paddock. Three ephemeral creeklines dissected the paddock from south to north and were characterised by denser vegetation cover and shallow sandy beds usually 1-2 m in width. Creeks flowed after rain into a near-permanent dam, a bulldozed depression in the soil located in the northern section of the paddock. The dam contained water throughout the study, and water was also present at three minor pipeline leaks along the southern boundary.

We chose an unfenced control area south of the dingo fence, a man-made wire netting fence erected to exclude dingoes from southern sheep grazing areas. The control area was on adjoining Mulgaria Pastoral Station and situated 5 km east of the Dingo Paddock, a distance considered sufficient to ensure independence but close enough to contain similar habitat types and reflect similar climatic events. Habitats within the control area were similar to the Dingo Paddock with a large clay swale, an area of closely spaced sand dunes, a pastoral dam, and an area of breakaways. The dam within the control area was stocked with domestic cattle (*Bos taurus*).

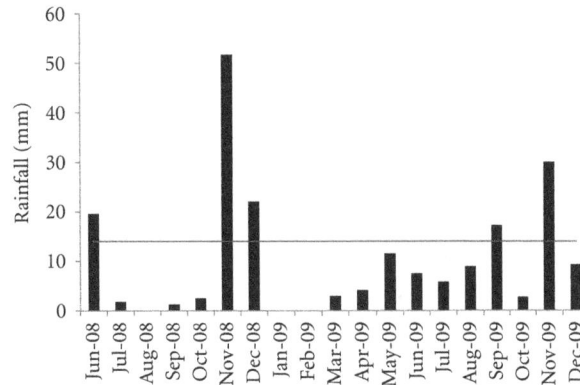

FIGURE 1: Rainfall recorded 6 months prior and during the study period. The line indicates the monthly average.

The Roxby Downs' climate is arid, failing to reach its long-term average rainfall of 166 mm in 60% of years [48]. Rainfall is aseasonal and with equal likelihood of rain during any month. Productivity within arid zone ecosystems is driven by unpredictable rainfall events, and only 100 mm of rainfall was recorded in both 2008 and 2009, leading to prolonged dry conditions. A significant rainfall event occurred just prior to the study in November 2008 (Figure 1), which filled the dam within the paddock and led to a flush of grass growth. However, conditions then remained relatively dry until the end of the study period.

3. Methods

3.1. Dingoes. In December 2008, a male and female dingo were captured from Stuart Creek Pastoral Station and released into the paddock. The wild dingoes were captured using soft catch Jake foot hold traps set around a cattle carcass located approximately 50 km north of the dingo paddock. Traps were fitted with springs to reduce injury and were checked in the late evening and again at dawn. No teeth damage was recorded after capture. We lightly anaesthetised the captured adult dingoes using a mixture of 1 mL of Medetomidine Hydrochloride and 0.5 mL of Ketamine, administered intramuscularly.

The anaesthetic was reversed using 0.5 mL of Atipamezole Hydrochloride. Anaesthetic and reversal doses for all animals were prepared in advance by a qualified veterinarian who also trained all animal handlers in correct administration of the preprepared doses. An anaesthesia procedure was developed and approved by the Wildlife Ethics Committee, including monitoring of rectal temperature during anaesthesia. Dingoes were weighed, checked for reproductive condition, and fitted with Global Positioning System (GPS) datalogger ARGOS satellite collars with VHF (SIRTRACK, Havelock, New Zealand) that nominally recorded fixes every 2 hours. Collars weighed 640 g and were no more than 4% of dingo body weight, less than the manufacturer's, and South Australian Wildlife Ethics Committee's maximum approved proportional collar weight of 5%. Dingoes were transported in an air-conditioned car and released at the dam within the dingo paddock on the same morning as

capture. Dingoes were checked after two hours and were then radiotracked daily for the first week. Radiotracking fixes indicated that both dingoes began moving throughout the paddock within a few hours of release. Although the number of dingoes placed in the paddock mirrored regional density, we provided a food subsidy to determine whether the availability of prey was limited in the paddock and could have influenced study outcomes. Between December 2008 and October 2009, kangaroo or rabbit carcasses and occasionally meat offcuts were placed at least fortnightly at a carcass dump established near the dam within the dingo paddock. Two remote motion sensor cameras (DVR Eye, Pix Controller, PA, USA) were placed at the carcass dump to record activity.

Weekly ARGOS satellite downloads were used to check whether the dingoes were in the paddock, and we conducted daily fence checks during the first month to repair any attempts to dig out under the fence. We recaptured the male and female dingoes in January and March 2010, respectively, to replace the GPS collars before the VHF batteries expired. No collar-related injuries such as rubbing or ulcerations were recorded. The male was captured using a single soft catch Jake trap set under an *Acacia ligulata* bush using a cat's head as bait, and the female was captured along the fenceline using a single Victor Soft-Catch (No. 1.5) trap. During the study, the pair of dingoes successfully raised a single male dingo pup born in June 2009. The female started using the breeding den in the northern sand dunes of the paddock on June 1 and continued to use it until July 16. After this time, the female and pup moved around the paddock and frequently changed shelter sites.

3.2. Cats and Foxes. Feral cats that remained in the paddock after construction were trapped in August 2008 and fitted with GPS data logger radiocollars with VHF (SIRTRACK, Havelock North, New Zealand) for a separate study comparing cat behaviour before and after dingo reintroduction. Cats were fitted with a small hind foot ring made from a cable tie with a 10 cm length of light chain attached. The chain dragged behind the cat when it moved and left a small indentation in soft substrate where tracks could be detected indicating that the cat had been fitted with a radiocollar. All cat tracks recorded during quad bike traverses of the paddock immediately after trapping were from collared cats suggesting that most, if not all, cats within the paddock had been captured and radiocollared. Only two of these cats still remained alive when additional cats and foxes were placed in the paddock in April 2009 and by then foot rings had been removed.

Between April and October 2009, 4 to 10 months after the dingoes were released into the paddock, we captured six feral cats and seven foxes, fitted them with radiocollars, and released them inside the dingo paddock. The majority of these cats and foxes were captured outside the paddock within 10 km of the dingo paddock in similar habitat (Table 1). However, one feral cat and one fox were captured inside the paddock, after the remote cameras detected that new animals had breached the fence and were visiting the carcass dump. The two radiocollared cats from a previous experiment that were resident in the paddock when the study began in April 2009 were also monitored during the study. Four of the seven foxes and all cats were captured in areas where dingo tracks are regularly observed suggesting they were not naive to dingo presence. Animals were captured using Victor Soft-Catch (no. 1.5) rubber jawed leg-hold traps fitted with springs to prevent injury (Coast to Coast Vermin Traps). Two lures were used in association with the traps; "pongo" (cat urine) and occasionally a Felid Attracting Phonic "FAP" (Westcare Electronics). Traps were checked in the evening or early each morning and captured feral cats and foxes were restrained using gloves and towels and anaesthetised with a mixture of medetomidine hydrochloride and ketamine administered intramuscularly. It was not possible to weigh animals before sedation so doses were preprepared for small (less than 5 kg—0.32 mL of medetomidine and 0.2 mL of ketamine) and large (more than 5 kg— 0.4 mL medetomidine and 0.25 mL of ketamine) cats and adult foxes (1 mL of medetomidine and 0.5 mL of ketamine).

We weighed and sexed the cats and foxes and noted the condition of their teeth, body, and reproductive organs. Only animals weighing at least 2.7 kg were used in the study to ensure radiocollars remained less than 5% of body weight (Table 1). The 135 g GPS data logger collars with VHF transmitter (SIRTRACK, Havelock North, New Zealand) were constructed from synthetic belting, and recorded GPS fixes every 2 hours. The units were housed in epoxy resin and contained 2 antennas, micromouse GPS and 220 mm, 2NC gauge whip antenna. The VHF transmitter (40/80 ppm) was equipped with a mortality sensor, triggered after longer than 24 hours without movement. The rectal temperature was taken every 3 minutes whilst under anaesthetic and cold packs were placed between the hind legs if the body temperature rose above 39 degrees Celsius. Animals were then administered the reversal drug atipamezole hydrochloride, placed in a cage trap covered by a towel in a vehicle and only released when they had fully recovered. If the ambient temperature was over 30 degrees Celsius, the towel was moistened and the vehicle air conditioned. We released animals at the dam or within sand dunes in the dingo paddock and watched to ensure they ran off after the cage trap was opened. They were radiotracked either later that day or early the following day to ensure they had moved.

Between September and December 2009, we captured and radiocollared an additional three foxes and three feral cats and released them into the unfenced Mulgaria control area to act as controls. One control cat was trapped in the Mulgaria control area, and the other two control cats were captured within 15 km of the control area. All control foxes were captured on Roxby Downs Station, 50 km south of the dingo paddock, two in October 2009, and one in December 2009.

Between April and December 2009, we radiotracked all collared animals within the dingo paddock and control area weekly or fortnightly on foot, quadbike, or from a Cessna 172 aeroplane with a wing-mounted aerial. If an animal was found dead, its location was recorded and a thorough search of the death location ensued. Habitat, tracks, scats, bones, fur, warrens, or any other signs of interest were recorded. Any fresh carcasses were sent to Zoos South Australia where postmortems were performed by qualified veterinarians.

3.3. Data Analysis. We converted collar downloads from Greenwich Mean Time to Australian Central Standard Time (nondaylight saving) and plotted them using Arc GIS software. Collar accuracy varied according to the number of satellites available at the time of the GPS fix, but precision was usually less than 10 m. For deceased animals, GPS fix locations were used to confirm the point of death by identifying clusters of points in the same location indicating no movement for an extended period. The time of death was estimated as the time interval between the first GPS fix at the death location and the time of the last GPS fix recorded in an area prior to the death location, which typically permitted time of death to be estimated to be within 2 hrs. In cases where multiple clusters of fixes were evident at a number of localities within a 1.5 km radius, ground searches revealed that carcasses had been dragged after death, and the first cluster was identified as the kill site. Time and location of death of all cats and foxes within the paddock were compared to dingo GPS fix locations for the same period to determine whether the dingoes were present at the death location within the correct time interval. Other factors were also considered when determining the cause of death, including the results of any autopsy and presence of dingo tracks.

The distances between all fox and dingo GPS fix locations at each 2 hr interval was used to determine if any possible encounters had occurred between the two species prior to death. Given that the approximate dimensions of the paddock were 7 km by 5 km, distances of less than 500 m between animals within a 4 hr time interval were conservatively considered possible encounters. Additionally, all GPS fix locations within 24 hrs of death were closely compared to dingo locations to determine if the dingoes had followed the foxes prior to death. GPS fix locations of cats and dingoes were also compared but only for the 24 hr period prior to death as cats remained alive longer than foxes and produced significantly more GPS fix locations for analysis.

To investigate the influence of fox presence on dingo activity, each dingo's minimum daily distance moved was compared on days when foxes were present and absent in

TABLE 1: Details, location, and fate of cats and foxes captured and radiocollared during the experiment. Distance refers to how far away from the release area the animal was initially captured. Time of death is the time interval between the first GPS fix recorded at the death location and the last fix recorded at a different location prior to death.

Animal	Distance from paddock	Release location	Sex	Release Wt	Release date	Death date	Days alive	Time of death	Habitat of death location	Cause of death
Fox 31	5 km	Paddock	M	3500	1/6/09	4/6/09	3	04:56–20:56	Swale	Dingo
Fox 32	5 km	Paddock	M	4500	9/6/09	13/6/09–14/6/09	5	08:30–08:30	Dune	Dingo
Fox 33	In paddock	Paddock	F	3500	24/6/09	24/6/09–25/6/09	1	23.03–01.02	Dune	Dingo
Fox 34	5 km	Paddock	M	3500	27/6/09	12/7/09–13/7/09	17	22:31–00:31	Swale	Dingo
Fox 35	50 km	Paddock	F	5005	15/8/09	20/8/09	6	23:04–03:05	Swale	Dingo
Fox 36	50 km	Paddock	F	4400	18/10/09	28/10/09	10	22:54–02:54	Dune	Dingo
Fox 37	50 km	Paddock	F	5200	16/10/09	18/10/09–9/10/09	3	08:30–08:30	Dune	Dingo
Cat 22b	In paddock	Paddock	M	4050	26/4/09	14/6/09–13/7/09	48–78	unknown	Swale	Unknown
Cat 23b	7 km	Paddock	F	2950	22/7/09	2/11/09	103	20:54–22:54	Swale	Unknown
Cat 24b	10 km	Paddock	F	3750	28/8/09	30/11/09–7/12/09	94–112	unknown	Swale	Unknown
Cat 25b	10 km	Paddock	F	4050	16/9/09	6/10/09	20	8.59–10.58	Creekline	Dingo
Cat 28	10 km	Paddock	F	2950	3/4/09	30/4/09	27	16:56–18:56	Dune	Dingo
Cat 29	10 km	Paddock	F	2750	26/4/09	30/5/09	34	19:19–07:18	Dune	Unknown
Cat 21	In paddock	Paddock	F	2700	20/8/08	21/6/2009	300	17:04–19:04	Dune	Dingo
Cat 23	In paddock	Paddock	M	4200	28/8/08	8/4/09	210	17:00–19:00	Dune	Dingo
Fox 38	50 km	Control	F	4400	18/10/09					Fate unknown
Fox 39	50 km	Control	M	4800	18/10/09					Fate unknown
Fox 30	10 km	Control	M	4000	14/12/09					Fate unknown
Cat 27b	15 km	Control	M	4650	3/10/09					Fate unknown
Cat 28b	15 km	Control	F	3950	31/10/09					Fate unknown
Cat 26	In control	Control	M	3600	29/9/08	2/7/09	276			Euthanased

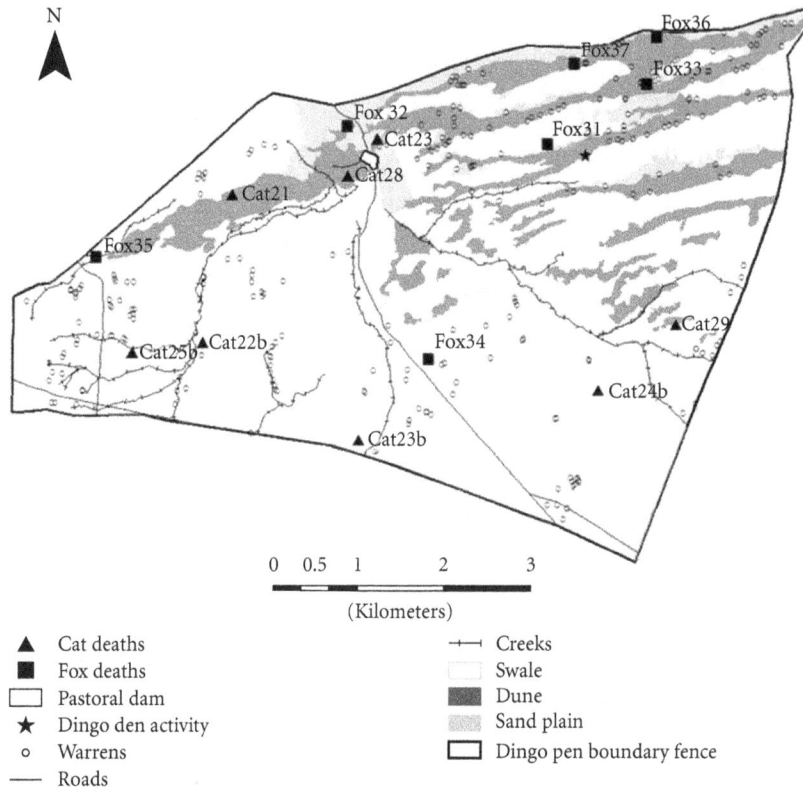

FIGURE 2: Location of animal deaths attributed to dingoes within the Dingo Pen. Habitat types, rabbit warrens, and the dingo den site are also marked.

the paddock. Minimum daily distance was calculated as the total distance between successive GPS fix locations over a 24 hr period. At least one fox was in the paddock over three different periods between June and October for 34, 6, and 13 consecutive days, respectively. Minimum daily distances during these times were compared with the remaining 161 days when foxes were absent during the study period. Male and female dingoes were analysed separately using one-way ANOVAs.

3.4. Prey Abundance.

Red kangaroos (*Macropus rufus*) remained present in both the dingo paddock and control area throughout the study. European rabbit (*Oryctolagus cuniculus*) warrens were common throughout the sandy dunes, and sandplains and clusters of larger, more permanent rabbit warrens were located in calcrete outcrops throughout the clay swales. Warren systems of the spinifex hopping mouse (*Notomys alexis*) were present throughout the sand dunes and, along with the Bolam's mouse (*Pseudomys bolami*), have consistently been the most common small mammal present on regional sand dunes over the preceding decade [49]. Other small mammals including the introduced house mouse (*Mus musculus*) and dunnarts, *Sminthopsis* spp. also occur in the region but at low densities and are usually restricted to the clay interdunal swales [49].

Indices of dingo, fox, cat, and rabbit activity were derived from the presence of spoor along 200 m track transects established in both the control and dingo paddock in the three

main habitat types: sanddune, swale, and creekline. Swale transects were all placed on roads where suitable substrate for tracking existed. Transects were swept clean using a metal bar dragged behind a quadbike the night before the first of two consecutive mornings of track counts. Data from the two mornings were combined to give a presence/absence score for each transect for each monitoring period. A total of 39 transects (20 sand dune, 10 creeklines, and 9 swale) were established in the dingo paddock and 38 (20 dune, 8 creekline, and 10 swale) in the control area. All transects were sampled every 4 months from February 2008 until February 2010. Sampling began 11 months prior to dingo reintroduction and continued for 3 months after the completion of the experiment.

4. Results

4.1. Foxes.

All seven foxes released into the dingo paddock died within 17 days of release (Table 1). GPS fix locations, kill site inspections, and autopsies suggested that all seven animals were killed by dingoes. One fox appeared to have been killed by the female on her own when the male was at the den site. All other deaths occurred when the dingoes were travelling together. Where the time of death was known, foxes died between 10.30 pm and 3 am (Table 1). Four of the animals died on sand dunes and three on swales (Figure 2). None of the deaths occurred in areas of dense vegetation. Deaths were recorded at various locations around the paddock with no apparent association with the breeding den or resource points (Figure 2). Additionally, the deaths were recorded both during and after the female whelped. There was no indication that any of the foxes had been eaten and most exhibited little external sign of injury. Some carcasses were mauled and parts dragged up to 1500 m after death. In the four cases where the fox carcass was not retrieved for more than 12 hrs after death, the dingoes either remained with or returned to the carcass for up to 6 days after death (Table 2).

Three of the seven foxes (Fox 32, 36, and 37) were found within a few hours of death and could be necropsied (Table 2). Injuries sustained included ruptured leg muscles and/or trauma to the lumbar region and ribs with herniation of the abdominal muscles resulting in extensive and terminal haemorrhaging. Veterinarians from Zoos South Australia indicated that the injuries were consistent with an attack by dingo or dingoes. In one instance, the fox had been chased several times at high speed around a bush. In another, scrape marks and diggings suggested that the fox had been flushed out of a warren on a sand dune.

Tracks and GPS fix locations from the dingoes and foxes suggested that they also killed three other foxes (Table 2), with one or both dingoes recorded less than 10 m from the death points during the time of death. The remaining fox, Fox 31, was within 110 m of the male dingo when it died. After death, the fox and male dingo GPS fix locations were within 10 m of each other at two different cluster locations up to 1.5 km from the kill site suggesting that the carcass was dragged after death. Unfortunately the collar failed to record most of the female dingo fixes taken during the 17 h death period, but she was travelling with the male just prior to the death period.

There were no recorded interactions between the foxes and dingoes prior to fox deaths. The only instance when dingo and fox fixes were recorded within 500 m of each other within a 4 h time interval was at the time of fox deaths. Furthermore, outside this 4 hr window, more than 450 m and 12 h were recorded between any fox and dingo locations suggesting that the first physical encounter between dingo and fox was also the last. There was also no indication that dingoes were following foxes prior to death as both species were moving in different directions, and the distance between fox and dingo GPS fix locations recorded just prior to death was between 1703 and 3000 m (Table 2). No fox deaths were recorded along fencelines or roads despite both foxes and dingoes regularly using these features during the study. One fox collar did not store any fixes during its time in the dingo paddock so predeath interactions with the dingoes could not be determined.

There was a strong trend towards longer daily movements in male and female dingoes when foxes were present in the paddock compared with when foxes were absent (female $F = 3.847$, $df = 1,213$, $P = 0.051$; male $F = 3.434$, $df = 1,213$, $P = 0.065$) but results were not significant. The average minimum daily distance moved by the female dingo increased from 2782 m to 3617 m when foxes were present in the paddock and the male average increased from 3375 m to 4267 m.

4.2. Cats. All six feral cats released into the paddock died between 20 and 123 days after being translocated into the paddock, and we recorded evidence that at least three cats were killed by dingoes. An additional two cats already present and radiocollared in the paddock when the experiment began also appeared to have been killed by dingoes. Where dingoes were implicated in deaths, three occurred in the early evening and one in the mid morning. When the female dingo killed two cats on her own, the male dingo was at the den site, more than 1 km from the death points. Deaths occurred before, during, and after denning and were in different habitat types and locations around the paddock (Figure 2). Dingoes displayed similar postdeath behaviour to that shown with killed foxes, staying with and/or returning to carcasses after death.

A postmortem confirmed death by dingo attack in one cat (cat 28, Table 2), but the 4 remaining cats were too decomposed for autopsy, so tracks, dingo behaviour, and GPS fix locations were used to determine if the dingoes may have been involved in the cat deaths. Although in two instances (cat 25b and cat 23) the dingo fixes were several hundred metres from the cats during the death period, other factors such as direction of predeath movement, postdeath dingo behaviour, and tracks and saliva marks suggested that the cause of death was dingo attack.

The cause of death could not be determined for three of the cats (Table 1). Cat 23b was several kilometres from the dingoes when it died out on a swale. Its remains were found under a wedge-tailed eagle (*Aquila audax*) nest suggesting that it may have been killed or scavenged by an eagle. The other two cats were within 350 m and 400 m of the dingoes during the death period, and it is possible that the dingoes were involved in these deaths. Of the five cats that remained in the paddock long enough to be recaptured and recollared during the study, two had lost weight, one had maintained weight and one had gained weight.

Collars were removed from dead cats and foxes, and no rubbing or collar-induced injuries were detected. The dingoes were recaptured 12–18 months after initial capture, and no collar injuries were detected.

4.3. Control Animals. Only one cat and no foxes could be relocated after release into the control area. The cat that was captured within the Mulgaria control area remained in the control area for 276 days before it was recaptured and euthanased at the end of the experiment. This cat sheltered extensively in rabbit warrens on rocky swales, and, although usually staying within a 12 km linear area, it was known to travel more than 35 km to the south and back again within a two week-period. This cat was recaptured three times over the study and, its weight remained between 3350 and 3600 g.

All other control animals were transferred to the control area from surrounding areas, and, despite more than five attempts to locate them using a light aircraft, they could not be found. Searches from the air included a 20 km radius around the control site, all of the original capture locations and 1 km traverses across the control area. The fate of these

TABLE 2: Details of animal deaths attributed to dingoes during the experiment. Evidence for dingo attack in bold. M: Male, F: Female.

Animal	Dingo distance (m) from fox/cat during death period		At carcass hours after death	Distance from closest dingo at fix preceding kill	Both dingoes together prior to kill	Dingo tracks at death site	Carcass dragged	Saliva on carcass	Autopsy confirmed dingo attack	Fox/cat movement in 2 hours prior to death (m)
	Male	Female								
Fox 31	**114**	1200[2]	2–14, 22–24, 64, 96–98, 122–130 (M)	2014	Yes		Yes		n/a	100
Fox 32	**<10**	470,680	48 (M)	3000	Yes	**Yes**	**Yes**	**Yes**	**Yes**	
Fox 33	900 den	**<10**	4.5–6.5 (F)	2124	No				n/a	218,230
Fox 34	**<10**	**<10**	2–16[1] (MF)	1703	Yes	**Yes**		**Yes**	n/a	843
Fox 35	**<10**	**<10**	20–22 (MF)	1668	Yes	**Yes**			n/a	2833
Fox 36	**<10**	**<10**	2[1] (F)		Yes	**Yes**		**Yes**	**Yes**	1662
Fox 37	**110**	**110**	[1]		Yes	**Yes**		**Yes**	**Yes**	
Cat 21	3446 den	**<10**	2,12,14,16 (F)	1885	No		**Yes**		n/a	574
Cat 22b	4878 den	**<10**	2–6 (F)	2300	No				n/a	
Cat 23	**650**	**700**	26 (M)	800	Yes				n/a	270,670
Cat 25b	200	130	3–5 (M) 14–18 (F)	1726	Yes	**Yes**		**Yes**	n/a	1609
Cat 28	**300**	**160**	2–4,12,24 [1](M) 2–4 (F)	1886	Yes	**Yes**	**Yes**		**Yes**	949

[1] Body removed within 12–24 hours of death.
[2] Female collar failed to record 5 fixes during death period.

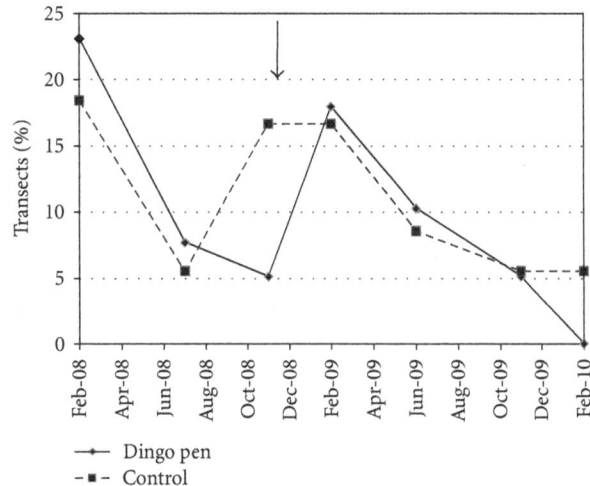

FIGURE 3: The percentage of transects (Dingo Pen $n = 39$, control $n = 38$) with cat tracks recorded at sites within the Dingo Pen and control area. Cats were added to the Dingo Pen between April and October 2009 and to the control area in October 2009. The pen was completed in November 2008, and the arrow indicates when dingoes were released.

animals remains unknown, but it is likely that they moved away from the control area.

4.4. Track Transects. Prior to and during fence construction, tracks, sightings, and scats of wild dingoes, feral cats, and foxes were all observed within the dingo paddock area. However, spoor counts and spotlighting transects indicated that there were no foxes or dingoes present in the paddock when the fence was completed. Subsequent spoor counts and remote cameras detected two uncollared foxes and two uncollared cats that had climbed into the paddock at different times during the experiment.

Both control and dingo paddock transects exhibited similar trends of cat activity during the initial stages of the project (Figure 3). However, despite the presence of at least five cats in the paddock prior to dingoes being released in December 2008, as well as the addition of 4 cats in 2009 and another cat that was captured after climbing into the paddock, cat activity declined to zero by February 2010. All ten of these cats were radiocollared, and all died during the experiment. Cat activity fluctuated in the control area, but cats remained present throughout the experiment. Both areas experienced a decline in activity in 2009, possibly partly due to the dry conditions experienced during this time.

Fox spoor was recorded in the paddock when foxes were released in June but declined to zero by the end of the experiment. Fox activity in the control area was variable over the study period (Figure 4). The presence of rabbit spoor on transects followed similar trends at both dingo and control sites and were recorded on 50 to 85% of track transects in the dingo and control areas during 2009 (Figure 5). Inside the dingo paddock, dingo tracks were present on an average of 27% of track transects during the study period. The control area averaged dingo tracks on 4% of transects suggesting very low dingo activity.

5. Discussion

Many previous studies have suggested that dingoes suppress fox abundance [2, 4, 33–36, 46], but this is the first time that a direct negative interaction between dingoes and cats and foxes has been demonstrated. Small amounts of cat hair have been recorded in dingo scats [42], and some researchers have suggested that study cats were killed by dingoes [40]. However, other researchers have suggested that the presence of dingoes may assist cat survival by providing carrion [50]. Similar studies in North America have reported 25% of radiocollared cats killed by coyotes [23]. Both male and female, and large and small, animals were killed by dingoes in our study suggesting that all foxes and cats may be susceptible to dingo attack.

The primary mechanism for suppression of cats and foxes by dingoes in this study appeared to be direct physical attack rather than suppression of breeding or exclusion from resource points as has been suggested elsewhere [44]. The dingoes did not eat any of the carcasses, despite staying with and/or returning to them for extended periods, which suggests that they were killing due to interference competition rather than intraguild predation. Similar results were found by Molsher et al. [51] for red foxes and cats in Australia and Helldin et al. [52] for lynx (*Lynx lynx*) and red foxes in Sweden, and radiocollared animals were killed but rarely eaten by the dominant predator. However, intraguild predation has been previously recorded in dingoes. Marsack and Campbell [39] observed dingoes eating foxes in arid Western Australia, and both fox and cat remains have been found in dingo scats and stomach contents [39, 42–45]. Intraguild predation has also been recorded in the United States of America where cats are eaten by coyotes and can contribute up to 13.1% of coyote diet [53–55]. It is likely that mesopredator suppression mechanisms are influenced by resource availability, habitat type, breeding season, and

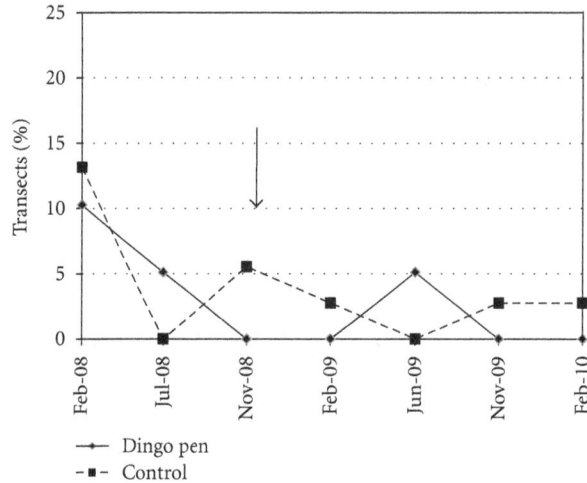

FIGURE 4: The percentage of transects (Dingo Pen $n = 29$, control $n = 38$) with fox tracks recorded at sites within the Dingo Pen and control area. Foxes were released into the Dingo Pen between June and October 2009 and into the control area between October and December 2009. The pen was completed in November 2008 and the arrow indicates when dingoes were released.

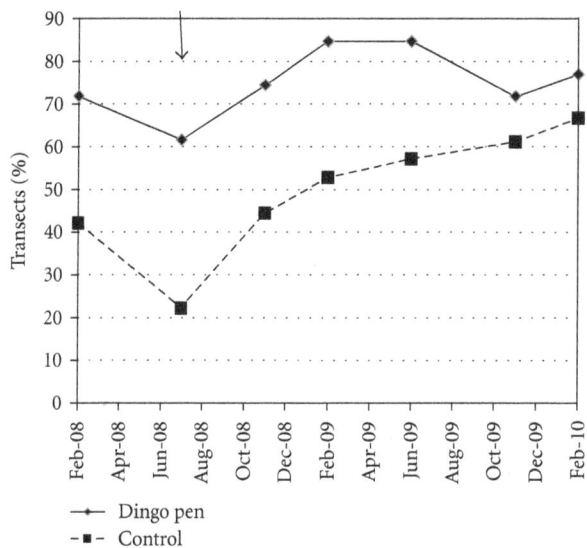

FIGURE 5: The percentage of transects (Dingo Pen $n = 39$, control $n = 38$) with rabbit tracks recorded within the Dingo Pen and control area. The arrow indicates when dingoes were released into the pen in December 2008.

intraspecific behavioural differences. Interestingly, most animals killed by dingoes showed very little external sign of injury suggesting that many "unexplained" deaths of radio-collared cats and foxes in other studies previously attributed to nutritional stress (e.g., [56]) may have been the result of dingo attack.

Although cats were subjected to direct dingo attack, other forms of suppression may also have been occurring in the paddock. Burrows et al. [15] found higher breeding success of cats in the area from which foxes and dingoes had been controlled. Despite lactating cats being captured within the paddock in our study, and kitten spoor being found briefly at 1 cat den site, no successful cat recruitment was recorded. Juvenile cat spoor was not recorded on any track transects

nor were any young uncollared cats photographed at the carcass dump. Other studies have suggested that dingoes may change cat's spatial behaviour, and both Edwards et al. [56] and Palmer (pers. obs) found cats used wooded or mulga habitat more than open habitats when dingoes were present, possibly due to predation risk by dingoes. One cat in the paddock was found to frequently shelter in a wedge-tailed eagle nest in a mulga tree, one of the few trees in the paddock that was above 2 m in height. Cat deaths were recorded in all habitat types suggesting that habitat may not have influenced predation risk in our study, but this result may not be consistent in wooded habitats.

Fox and cat deaths occurred at times when they were most active, foxes at night time and cats mainly at dusk. This is consistent with dingoes killing cats and foxes when they encounter them rather than digging them out of warrens or using olfactory cues to seek them out. Corroborating this assumption was the independent movement patterns of dingoes, cats, and foxes in the 24 hrs prior to death and deaths occurring when animals unknowingly moved into the path of the dingoes or vice versa. Therefore, it is likely that dingoes killed cats and foxes on an opportunistic basis, but they were probably aware of the foxes in the paddock and may have increased their daily movements to increase the chances of encountering them.

Containing all three animal species within a paddock, albeit a landscape scale one, may have influenced the outcomes of the study by restricting the movement of some cats and foxes and perhaps rendering them more likely to encounter or be cornered by a dingo. The home range of cats and foxes vary considerably in the arid zone with averages of between 20 and 30 km^2 commonly recorded [15, 20, 56]. The average cat and fox home range recorded in the study area during a previous study was 16-17 km^2 with a range of 0.5 to 123 km^2 [20]. However, several factors suggest that the paddock represented a realistic arid zone environment. Although cat home ranges can be large, they do overlap [49] and track counts in control, and paddock areas were similar at the start of the study, suggesting that the density of cats during the experiment was similar to that naturally recorded outside the paddock. Five cats were resident in the paddock when it was fenced, a similar or higher density to that maintained during the experiment. Only one fox was present in the paddock at a time, and the density of 0.027 foxes per km^2 is much lower than that recorded in other arid zone studies (0.46–0.52 [57], 0.6 [58]). Rabbit track counts suggested that food resources were similar in control and paddock areas, and the presence of a carcass dump provided supplementary food if required. Additionally, all animals used in the experiment were captured inside the paddock or from within 50 km in similar habitat.

The dingoes also appeared to have behaved typically, breeding in April/May and whelping in June, as recorded elsewhere in arid Australia [44]. The dingoes were recorded howling and scent marking and stopped trying to escape from the paddock after one week, also an indication that they were behaving like a dingo pack and maintaining a territory. Although dingo home ranges in the arid zone have been reported to be up to 77 km^2 [44] and even as large as 272 km^2 [59], other studies have reported arid zone dingo density between 1 and 22 per 100 km^2 (usually 5) [60], similar density to that recorded in the paddock during our study. Pack size of two dingoes is commonly recorded in the arid zone [60]. Dingoes are known to feed on rabbits, reptiles, kangaroos, and carrion [41], all of which were present during the study. Thus, although the paddock may have influenced the results, the large size, availability of different habitat types, densities of predators, and presence of suitable food and rabbit warrens for shelter should have minimised these influences.

Deaths were recorded at various times between April and October, before, during and, after the denning period. It is not known if dingoes will kill cats and foxes over the summer months. Although resident cats in the paddock when the dingoes were introduced in December were not killed until April or June, the dingo pair may have been more likely to influence other predators or competitors once they had formed a pack and started defending resources [61]. The female was only recorded killing animals on her own during her 6-week denning period when the male was guarding the den site. Dingoes may consider foxes and cats a threat to their pups and increase their intolerance of them during the breeding season. In North America, coyotes were also found to kill domestic cats at any time of year but with higher kill rates during puprearing [62].

Several animals appeared to have been chased around bushes or over short distances prior to death. The dingoes were travelling together when nine of the 12 foxes and cats were killed. It is impossible to determine if both dingoes assisted in the kills, but it appears likely, as dingoes regularly hunt and kill prey cooperatively [63]. Additionally, tracking observations of a fox killed by dingoes in a separate arid site in South Australia indicated that 2 dingoes chased and killed a fox, whose fresh carcass was located on a sand dune (J.L. Read pers obs). The female dingo killed a large 4 kg cat, and there was no indication that any particular size or sex of cat or fox was less susceptible to dingo attack by lone or pack dingoes. Cooperative dingo packs will more effectively hunt large mammals such as macropods, buffalo, feral horses, or cattle [43, 63, 64], but solitary dingoes can effectively hunt rabbits, small mammals, and sheep to achieve their daily energy requirements [41, 43, 44]. Grubbs and Krausman [62] documented coyotes killing domestic cats in the United States of America and found that single coyotes were just as effective at killing domestic cats as coyotes hunting in groups. However, dingoes may be more efficient hunters of cats as coyotes only killed cats in just over 50% of interactions [62].

Of those cats killed by dingoes, the resident cats survived longer than the cats placed in the paddock, possibly suggesting that the resident cats were more familiar with shelter sites and able to avoid interactions with the dingoes for longer. However, three of the five cats placed in the paddock do not appear to have been killed by dingoes, and their causes of death are unknown. Feral cats in the arid zone are thought to suffer from periods of nutritional stress leading to high natural mortality of more than 50% in less than 12 months [20, 56]. It appears unlikely that most cats were significantly nutritionally stressed as rabbit activity did not fluctuate significantly during the study period, and carcasses were regularly dumped at the carcass dump but rarely used by cats. Additionally, most recaptured cats either maintained or increased in weight. All of the cats and foxes placed in the paddock were adults and had been previously surviving unaided in the paddock or surrounding similar habitat. The death of two of these resident cats also suggests that it is unlikely that the translocation itself was responsible for other cat deaths. Two of these cats may have been preyed on by wedge-tailed eagles, but the third cat found down a rabbit warren may have died from natural causes.

Results from this study need to be extrapolated cautiously. Our experiment is a single replicate. Due to logistical constraints, we could only trial one pair of dingoes in a single paddock. Ideally, the experiment should be repeated using another dingo pair, and foxes and cats added in different seasons. It is likely that interactions between cats, foxes, and dingoes will vary depending on habitat types, breeding seasons, and food availability. The relatively open habitat in the paddock, despite numerous rabbit warrens for shelter, may have made it easier for dingoes to locate and catch cats and foxes. More wooded environments or areas with denser understorey may enable cats, foxes, and dingoes to coexist more readily. Despite similar habitat types in capture and release locations, for some animals, the paddock was an unfamiliar environment and may have influenced their susceptibility to dingo attack. Track searches of the paddock in early June 2010, 6 months after the experiment finished, located very low abundance of fox and cat tracks suggesting that these species had reinvaded the paddock. It is not known if dingoes permanently suppress cats and foxes over long periods or are more tolerant of cats and foxes outside the breeding period. Finally, drought conditions may have influenced results and increased dingo attacks due to competition for food resources.

Several studies have identified a loss in species biodiversity when a keystone or "apex" mammalian predator is removed [23, 28, 65]. The release of competitive restraints previously imposed on mesopredators can lead to changes in prey species' composition and diversity. Previous research has suggested that dingoes may suppress cat and fox abundance, but our trial is the first time that this has been proven experimentally. We found interference competition via direct attack to be the key suppression mechanism. However, the important question for threatened species conservation is whether the positive role that dingoes appear to play in suppressing cats and foxes will counteract dingo predation on these same threatened species and equate to a net benefit for native wildlife. We believe that there are several critical factors that will determine whether a native species may benefit from cat and fox suppression. Firstly, the size and behaviour of prey species may be important. Medium-sized native mammals that are preyed on by cats and foxes and dingoes may not benefit to the extent of smaller mammals, for which dingoes are less efficient predators. Although dingoes are known to prey on smaller mammals such as rodents [37, 41, 44], they are not preferred stable prey items and may only be targeted during natural irruptions when they are plentiful. Solitary, wide-ranging species such as the Greater Bilby (*Macrotis lagotis*) may benefit more than communal sedentary species such as the Burrowing Bettong (*Bettongia lesueur*). Sedentary, communal species are more conspicuous and easier to target by predators. Proposed continued monitoring of rabbit and small native rodent abundance inside and outside the Arid Recovery dingo paddock should elucidate the net ecological role of dingoes for these different-sized mammals. Furthermore, reintroducing threatened native mammal species with different social and movement systems into the dingo paddock will help determine whether positive suppression of cats and foxes outweighs any direct predation by dingoes.

Secondly, like other canids, foxes and dingoes both have a predisposition to kill several prey and consume only few or none of the total kill [66]. This behaviour, known as surplus killing, is why dingoes and foxes can pose a significant threat to native fauna and sheep populations; especially spatially restricted or threatened populations [67, 68]. There is some evidence to suggest that surplus killing in the dingo is not as common or devastating to native wildlife as the introduced red fox [66], but this is yet to be proven experimentally.

Thirdly, the relationship between dingo density and the magnitude of cat and fox suppression will have a major influence on whether a net benefit to prey species is realised. If low dingo density, particularly in concert with established breeding territories [61], is sufficient to significantly suppress cat and fox abundance then the net predation impact is likely to be low, leading to a net benefit to some wildlife species. However, the abundance of dingoes has increased significantly since European settlement due to the proliferation of stock watering points and plentiful rabbits. If the density of dingoes required to adequately suppress cats and foxes for the protection of wildlife is significantly higher than pre-European densities, then any benefit to wildlife may be offset by artificially high predation rates by dingoes.

Finally, unlike cats and foxes, dingoes are dependent upon water, at least during summer. Therefore, in desert areas dingo density and their predation and mesopredator suppression will be spatially and temporally patchy compared with cats and foxes. Many desert animals rely on restricted refugia areas for survival during drought [69], and unless these refugia areas coincide with areas of mesopredator suppression, long-term benefits to wildlife may not occur.

6. Management Implications

Although the ecological role of the dingo requires further verification in other environments, our study supports a growing body of evidence that the dingo plays an important role in ecosystem function. Therefore, we recommend that functional dingo populations in rangeland areas are maintained at landscape scales and that dingo control for calf protection is restricted to targeted control during exceptional circumstances. Research should now focus on whether dingoes provide a net benefit to threatened wildlife species by investigating the influence of prey size and behavioural traits, surplus killing, and dingo density. We predict that smaller, solitary, and wide-ranging native species close to permanent watering points will benefit the most from mesopredator suppression. Finally, the red fox, feral cat, and dingo all have catholic diets that can change rapidly depending on resource availability. Despite the dingo arriving in Australia several thousand years ago, all three species are relatively new arrivals in Australia. Researchers should consider that the mechanisms and benefits of mesopredator suppression in Australia may not mirror those recorded in North America and Europe where mesopredators are usually native and their diets more prey specific.

Acknowledgments

This study was conducted by Arid Recovery, a joint conservation initiative between BHP Billiton, the S.A. Department for Environment and Natural Resources, The University of Adelaide and the local community. Funding was provided by the South Australian Arid Lands Natural Resource Management Board, Olympic Dam Expansion Project, and Arid Recovery. Greg Kammermann and his assistants worked tirelessly on building the dingo paddock on time and within budget. Many other volunteers and staff have contributed to this work, particularly, A. Kilpatrick, A. Clarke, T. Moyle, D. Sandow, C. McGoldrick, G. Miller, and B. Arnold. Thanks are due to M. Lloyd for providing carcasses for our carcass dump and assisting with fox captures. Rob Savage gave permission to use Mulgaria Station as a control site, and Dr D. Paton provided useful comments on the manuscript. The authors are indebted to the veterinarians from Zoos South Australia, in particular I. Smith, W. Boardman, and D. McLelland, who conducted the necropsies, provided veterinary advice, and assisted with attempting to retrieve the dingo collars. Dr A. Melville-Smith from the Roxby Downs Veterinary Clinic assisted with anaesthesia dosages and training. This study was conducted under ethics approval (permit no. 6/2007-M3) from the South Australian Wildlife Ethics Committee.

References

[1] J. H. Seebeck, "Status of the barred bandicoot *Perameles gunnii*, in Victoria with a note on husbandry of a captive colony," *Australian Wildlife Research*, vol. 6, pp. 255–264, 1979.

[2] A. P. Smith and D. G. Quin, "Patterns and causes of extinction and decline in Australian conilurine rodents," *Biological Conservation*, vol. 77, no. 2-3, pp. 243–267, 1996.

[3] J. Short, "The extinction of rat-kangaroos (Marsupialia:Potoroidae) in New South Wales, Australia," *Biological Conservation*, vol. 86, no. 3, pp. 365–377, 1998.

[4] C. N. Johnson, J. L. Isaac, and D. O. Fisher, "Rarity of a top predator triggers continent-wide collapse of mammal prey: dingoes and marsupials in Australia," *Proceedings of the Royal Society B*, vol. 274, no. 1608, pp. 341–346, 2007.

[5] J. Short, S. D. Bradshaw, R. I. T. Prince, and G. R. Wilson, "Reintroduction of macropods (Marsupialia: Macropodoidea) in Australia-A review," *Biological Conservation*, vol. 62, no. 3, pp. 189–204, 1992.

[6] P. E. S. Christensen and N. D. Burrows, "Project desert dreaming: the reintroduction of mammals to the Gibson Desert," in *Reintroduction Biology of Australian and New Zealand Fauna*, M. Serena, Ed., pp. 199–208, Surrey Beatty and Sons, Chipping Norton, New Zealand, 1995.

[7] D. F. Gibson, G. Lundie-Jenkins, D. G. Langford, J. R. Cole, D. E. Clarke, and K. A. Johnson, "Predation by feral cats, *Felis catus*, on the rufous hare-wallaby, *Lagorchestes hirsutus*, in the Tanami Desert," *Australian Mammalogy*, vol. 17, pp. 103–107, 1994.

[8] R. Southgate and H. Possingham, "Modelling the reintroduction of the greater bilby *Microtus lagotis* using the metapopulation model analysis of the likelihood of extinction (ALEX)," *Biological Conservation*, vol. 73, no. 2, pp. 151–160, 1995.

[9] D. Priddel and R. Wheeler, "An experimental translocation of brush-tailed bettongs (*Bettongia penicillata*) to western New South Wales," *Wildlife Research*, vol. 31, no. 4, pp. 421–432, 2004.

[10] D. Algar and R. Smith, "Approaching Eden," *Landscape*, vol. 13, pp. 28–34, 1998.

[11] L. E. Twigg, G. R. Martin, and T. J. Lowe, "Evidence of pesticide resistance in medium-sized mammalian pests: a case study with 1080 poison and Australian rabbits," *Journal of Applied Ecology*, vol. 39, no. 4, pp. 549–560, 2002.

[12] D. A. Risbey, M. C. Calver, and J. Short, "The impact of cats and foxes on the small vertebrate fauna of Heirisson Prong, Western Australia. I. Exploring potential impact using diet analysis," *Wildlife Research*, vol. 26, no. 5, pp. 621–630, 1999.

[13] B. Catling and A. M. Reid, *Predator and Critical Weight Range Species 5. Results of Spring 2002 and Autumn 2003 Surveys*, CSIRO Sustainable Ecosystems, Canberra, Australia, 2003.

[14] D. A. Risbey, M. Calver, and J. Short, "Control of feral cats for nature conservation. I. Field tests of four baiting methods," *Wildlife Research*, vol. 24, no. 3, pp. 319–326, 1997.

[15] N. D. Burrows, D. Algar, A. D. Robinson, J. Sinagra, B. Ward, and G. Liddelow, "Controlling introduced predators in the Gibson Desert of Western Australia," *Journal of Arid Environments*, vol. 55, no. 4, pp. 691–713, 2003.

[16] D. Algar and N. D. Burrows, "Feral cat control research: Western Shield review—February 2003," *Conservation Science Western Australia*, vol. 5, no. 2, pp. 131–163, 2004.

[17] D. Hegglin, F. Bontadina, S. Gloor et al., "Baiting red foxes in an urban area: a camera trap study," *Journal of Wildlife Management*, vol. 68, no. 4, pp. 1010–1017, 2004.

[18] M. Olsson, E. Wapstra, G. Swan, E. Snaith, R. Clarke, and T. Madsen, "Effects of long-term fox baiting on species composition and abundance in an Australian lizard community," *Austral Ecology*, vol. 30, no. 8, pp. 899–905, 2005.

[19] D. Algar, G. J. Angus, M. R. Williams, and A. Mellican, "Influence of bait type, weather and prey abundance on bait uptake by feral cats (*felis catus*) on peron Peninsula, Western Australia," *Conservation Science Western Australia*, vol. 6, no. 1, pp. 109–149, 2007.

[20] K. E. Moseby, J. Stott, and H. Crisp, "Improving the effectiveness of poison baiting for the feral cat and European fox in northern South Australia: the influence of movement, habitat use and activity," *Wildlife Research*, vol. 36, pp. 1–14, 2009.

[21] E. A. Denny and C. R. Dickman, *Review of Cat Ecology and Management Strategies in Australia*, Invasive Animals Cooperative Research Centre, Canberra, Australia, 2010.

[22] F. Palomares and T. M. Caro, "Interspecific killing among mammalian carnivores," *American Naturalist*, vol. 153, no. 5, pp. 492–508, 1999.

[23] K. R. Crooks and M. E. Soulé, "Mesopredator release and avifaunal extinctions in a fragmented system," *Nature*, vol. 400, no. 6744, pp. 563–566, 1999.

[24] A. S. Glen and C. R. Dickman, "Complex interactions among mammalian carnivores in Australia, and their implications for wildlife management," *Biological Reviews of the Cambridge Philosophical Society*, vol. 80, no. 3, pp. 387–401, 2005.

[25] E. G. Ritchie and C. N. Johnson, "Predator interactions, mesopredator release and biodiversity conservation," *Ecology Letters*, vol. 12, no. 9, pp. 982–998, 2009.

[26] J. M. Goodrich and S. W. Buskirk, "Control of abundant native vertebrates for conservation of endangered species," *Conservation Biology*, vol. 9, no. 6, pp. 1357–1364, 1995.

[27] R. O. Peterson, "Wolves as interspecific competitors in canid ecology," in *Ecology and Conservation of Wolves in a Changing World*, L. N. Carbyn, Fritts , S.H., and D. R. Seip, Eds., pp. 315–323, Alberta, Canada, Canadian Circumpolar Institute, 1995.

[28] S. E. Henke and F. C. Bryant, "Effects of coyote removal on the faunal community in western Texas," *Journal of Wildlife Management*, vol. 63, no. 4, pp. 1066–1081, 1999.

[29] R. J. Hobbs, "Synergisms among habitat fragmentation, livestock grazing, and biotic invasions in southwestern Australia," *Conservation Biology*, vol. 15, no. 6, pp. 1522–1528, 2001.

[30] A. E. Newsome, "The biology and ecology of the dingo," in *A Symposium on the Dingo*, C. R. Dickman and D. Lunney, Eds., pp. 20–23, Royal Zoological Society of New South Wales, Sydney, Australia, 2001.

[31] A. E. Newsome, P. C. Catling, B. D. Cooke, and R. Smyth, "Two ecological universes separated by the dingo barrier fence in semi-arid Australia: interactions between landscapes, herbivory and carnivory, with and without dingoes," *Rangeland Journal*, vol. 23, no. 1, pp. 71–98, 2001.

[32] M. J. Daniels and L. Corbett, "Redefining introgressed protected mammals: when is a wildcat a wild cat and a dingo a wild dog?" *Wildlife Research*, vol. 30, no. 3, pp. 213–218, 2003.

[33] M. Letnic, F. Koch, C. Gordon, M. S. Crowther, and C. R. Dickman, "Keystone effects of an alien top-predator stem extinctions of native mammals," *Proceedings of the Royal Society B*, vol. 276, no. 1671, pp. 3249–3256, 2009.

[34] M. Letnic, M. S. Crowther, and F. Koch, "Does a top-predator provide an endangered rodent with refuge from an invasive mesopredator?" *Animal Conservation*, vol. 12, no. 4, pp. 302–312, 2009.

[35] N. G. Hairston, F. E. Smith, and L. B. Slobodkin, "Community structure, population control, and competition," *American Naturalist*, vol. 95, pp. 421–425, 1960.

[36] A. R. Pople, G. C. Grigg, S. C. Cairns, L. A. Beard, and P. Alexander, "Trends in the numbers of red kangaroos and emus on either side of the South Australian dingo fence: evidence for predator regulation?" *Wildlife Research*, vol. 27, no. 3, pp. 269–276, 2000.

[37] B. D. Mitchell and P. B. Banks, "Do wild dogs exclude foxes? Evidence for competition from dietary and spatial overlaps," *Austral Ecology*, vol. 30, no. 5, pp. 581–591, 2005.

[38] A. D. Wallach, C. N. Johnson, E. G. Ritchie, and A. J. O'Neill, "Predator control promotes invasive dominated ecological states," *Ecology Letters*, vol. 13, no. 8, pp. 1008–1012, 2010.

[39] P. Marsack and G. Campbell, "Feeding behaviour and diet of dingoes in the Nullarbor Region, Western Australia," *Australian Wildlife Research*, vol. 17, no. 4, pp. 349–357, 1990.

[40] R. Palmer, *Feral Pest Program Project 12, Cat Research and Management in Queensland, year 4; and Project 43. Dispersal and Spatial Organisation of Cats on the Diamantina Plains, year 3*, Interim Final Report to the Feral Pests Program, Australian Nature Conservation Agency, Canberra, Australia, 1996.

[41] P. Fleming, L. Corbett, R. Harden, and P. Thomson, *Managing the Impacts of Dingoes and other Wild Dogs*, Bureau of Rural Sciences, Canberra, Australia, 2001.

[42] D. Lunney, B. Triggs, P. Eby, and E. Ashby, "Analysis of scats of dogs *Canis* familiaris and foxes *Vulpes vulpes* (Canidae: Carnivora) in coastal forests near Bega, New South Wales," *Australian Wildlife Research*, vol. 17, no. 1, pp. 61–68, 1990.

[43] P. C. Thomson, "The behavioural ecology of dingoes in north-western Australia. III. Hunting and feeding behaviour, and diet," *Wildlife Research*, vol. 19, no. 5, pp. 531–541, 1992.

[44] L. Corbett, *The Dingo in Australia and Asia*, University of New South Wales Press, Sydney, Australia, 1995.

[45] R. Paltridge, "The diets of cats, foxes and dingoes in relation to prey availability in the Tanami Desert, Northern Territory," *Wildlife Research*, vol. 29, no. 4, pp. 389–403, 2002.

[46] C. N. Johnson and J. Vanderwal, "Evidence that dingoes limit abundance of a mesopredator in eastern Australian forests," *Journal of Applied Ecology*, vol. 46, no. 3, pp. 641–646, 2009.

[47] K. E. Moseby and J. L. Read, "The efficacy of feral cat, fox and rabbit exclusion fence designs for threatened species protection," *Biological Conservation*, vol. 127, no. 4, pp. 429–437, 2006.

[48] J. Read, "Recruitment characteristics of the white cypress pine (*Callitris glaucophylla*) in arid South Australia," *The Rangeland Journal*, vol. 17, no. 2, pp. 228–240, 1995.

[49] K. E. Moseby, B. M. Hill, and J. L. Read, "Arid Recovery—A comparison of reptile and small mammal populations inside and outside a large rabbit, cat and fox-proof exclosure in arid South Australia," *Austral Ecology*, vol. 34, no. 2, pp. 156–169, 2009.

[50] R. Paltridge, D. Gibson, and G. Edwards, "Diet of the feral cat (*Felis catus*) in Central Australia," *Wildlife Research*, vol. 24, no. 1, pp. 67–76, 1997.

[51] R. Molsher, A. Newsome, and C. Dickman, "Feeding ecology and population dynamics of the fetal cat (*Felis catus*) in relation to the availability of prey in central-eastern New South Wales," *Wildlife Research*, vol. 26, no. 5, pp. 593–607, 1999.

[52] J. O. Helldin, O. Liberg, and G. Glöersen, "Lynx (*Lynx lynx*) killing red foxes (*Vulpes vulpes*) in boreal Sweden—Frequency and population effects," *Journal of Zoology*, vol. 270, no. 4, pp. 657–663, 2006.

[53] J. G. MacCracken, "Coyote food in a Southern California suburb," *Wildlife Society Bulletin*, vol. 10, pp. 280–281, 1982.

[54] E. S. Shargo, *Home range, movement, and activity patterns of coyotes (Canis latrans) in Los Angeles suburbs*, Ph.D. thesis, University of California, Los Angeles, Calif, USA, 1988.

[55] T. Quinn, "Coyote (*Canis latrans*) food habits in three urban habitat types of Western Washington," *Northwest Science*, vol. 71, no. 1, pp. 1–5, 1997.

[56] G. P. Edwards, N. De Preu, B. J. Shakeshaft, I. V. Crealy, and R. M. Paltridge, "Home range and movements of male feral cats (*Felis catus*) in a semiarid woodland environment in central Australia," *Austral Ecology*, vol. 26, no. 1, pp. 93–101, 2001.

[57] N. J. Marlow, P. C. Thomson, D. Algar, K. Rose, N. E. Kok, and J. A. Sinagra, "Demographic characteristics and social organisation of a population of red foxes in a rangeland area in Western Australia," *Wildlife Research*, vol. 27, no. 5, pp. 457–464, 2000.

[58] J. Read and Z. Bowen, "Population dynamics, diet and aspects of the biology of feral cats and foxes in arid South Australia," *Wildlife Research*, vol. 28, no. 2, pp. 195–203, 2001.

[59] S. R. Eldridge, B. J. Shakeshaft, and T. J. Nano, "The impact of wild dog control on cattle, native and introduced herbivores and introduced predators in central Australia," Final Scientific Report to Bureau of Rural Sciences, Northern Territory, Australia, 2002.

[60] P. C. Thomson, "The behavioural ecology of dingoes in north-western Australia. IV. Social and spatial organisation, and movements," *Wildlife Research*, vol. 19, no. 5, pp. 543–563, 1992.

[61] A. D. Wallach, E. G. Ritchie, J. Read, and A. J. O'Neill, "More than mere numbers: the impact of lethal control on the social stability of a top-order predator," *PLoS One*, vol. 4, no. 9, Article ID e6861, 2009.

[62] S. E. Grubbs and P. R. Krausman, "Observations of coyote-cat interactions," *Journal of Wildlife Management*, vol. 73, no. 5, pp. 683–685, 2009.

[63] L. K. Corbett and A. E. Newsome, "The feeding ecology of the dingo—III. Dietary relationships with widely fluctuating prey populations in arid Australia: an hypothesis of alternation of predation," *Oecologia*, vol. 74, no. 2, pp. 215–227, 1987.

[64] A. E. Newsome and B. J. Coman, "Canids," in *Fauna of Australia. Mammalia*, D. W. Walron and B. J. Richardson, Eds., vol. 1B, pp. 993–1005, Australian Government Publishing Service, Canberra, Australia, 1989.

[65] M. E. Soulé, J. A. Estes, B. Miller, and D. L. Honnold, "Strongly interacting species: conservation policy, management, and ethics," *BioScience*, vol. 55, no. 2, pp. 168–176, 2005.

[66] J. Short, J. E. Kinnear, and A. Robley, "Surplus killing by introduced predators in Australia—Evidence for ineffective anti-predator adaptations in native prey species?" *Biological Conservation*, vol. 103, no. 3, pp. 283–301, 2002.

[67] J. E. Kinnear, M. L. Onus, and R. N. Bromilow, "Fox control and rock-wallaby population dynamics," *Australian Wildlife Research*, vol. 15, no. 4, pp. 435–450, 1988.

[68] J. E. Kinnear, M. L. Onus, and N. R. Sumner, "Fox control and rock-wallaby population dynamics—II. An update," *Wildlife Research*, vol. 25, no. 1, pp. 81–88, 1998.

[69] S. R. Morton, "The impact of European settlement on the vertebrate animals of arid Australia: a conceptual model," *Proceedings of the Ecological Society of Australia*, vol. 16, pp. 201–213, 1990.

18

Underappreciated Consequences of Phenotypic Plasticity for Ecological Speciation

Benjamin M. Fitzpatrick

Ecology and Evolutionary Biology, University of Tennessee, Knoxville, TN 37996, USA

Correspondence should be addressed to Benjamin M. Fitzpatrick, benfitz@utk.edu

Academic Editor: Andrew Hendry

Phenotypic plasticity was once seen primarily as a constraint on adaptive evolution or merely a nuisance by geneticists. However, some biologists promote plasticity as a source of novelty and a factor in evolution on par with mutation, drift, gene flow, and selection. These claims are controversial and largely untested, but progress has been made on more modest questions about effects of plasticity on local adaptation (the first component of ecological speciation). Adaptive phenotypic plasticity can be a buffer against divergent selection. It can also facilitate colonization of new niches and rapid divergent evolution. The influence of non-adaptive plasticity has been underappreciated. Non-adaptive plasticity, too can interact with selection to promote or inhibit genetic differentiation. Finally, phenotypic plasticity of reproductive characters might directly influence evolution of reproductive isolation (the second component of ecological speciation). Plasticity can cause assortative mating, but its influence on gene flow ultimately depends on maintenance of environmental similarity between parents and offspring. Examples of plasticity influencing mating and habitat choice suggest that this, too, might be an underappreciated factor in speciation. Plasticity is an important consideration for studies of speciation in nature, and this topic promises fertile ground for integrating developmental biology with ecology and evolution.

1. Introduction

Phenotypic plasticity has often been seen primarily as an alternative to genetic divergence and a feature making populations less responsive to natural selection [1–3]. For example, those studying adaptation and speciation have often used phrases like "merely plastic" to contrast environmentally induced variation against geographic or species differences with strong genetic bases [4–8]. However, others suggested that phenotypically plastic traits can promote adaptive evolution and the origin of species [9–16]. The general issue of plasticity and adaptation has been reviewed extensively in the last decade (e.g., [15, 17–26]). Adaptive plasticity's impact on speciation was recently reviewed by Pfennig et al. [27], and I do not attempt to duplicate their efforts. Instead I make a few points that have not been emphasized in the recent literature. In particular, nonadaptive plasticity and environmentally induced barriers to gene flow deserve greater attention.

After making explicit my working definitions of key terms, I argue that the "developmental plasticity hypothesis of speciation" [13–15] is a special case of ecological speciation, and I review the subject by breaking down the effects of plasticity on the two components of ecological speciation: adaptive divergence and the evolution of reproductive isolation [28]. I close with a few suggestions for future work.

By any definition, speciation requires genetic divergence. Therefore, integration of ecological developmental biology with the well-developed body of fact and theory on the genetics of speciation [29–31] will be more productive than attempting to replace this population genetic foundation. Recent reviews and models support this perspective [27, 32–37].

2. Definitions

Understanding the relationship between environmental induction and speciation requires a set of consistent definitions. Terms like "environment," "speciation," "plasticity," and "natural selection" are sometimes assumed by different workers to have different definitions, and this can affect communication [38]. The definitions that follow are intended to clarify what I mean by certain words and phrases within this paper; they are not intended to challenge or replace alternative definitions. I believe I have followed recent convention in all cases [39, 40].

Adaptation. Genetic change in response to natural selection and resulting in organisms with improved performance with respect to some function or feature of the environment.

Countergradient Variation. Pattern of geographic variation in which genetic differences between populations affect their phenotypes in the opposite way from environmental differences between populations. For example, if the mean phenotype of population i is given by the sum of genetic and environmental effects $P_i = G_i + E_i$, and the effect of environment 2 tends to increase the phenotypic value relative to environment 1 ($E_2 > E_1$), countergradient variation would exist if $G_2 < G_1$. Without genetic differentiation the expected difference in phenotype would be $P_2 - P_1 = E_2 - E_1$, but countergradient variation reduces that difference and might even make $P_2 < P_1$. More formally, countergradient variation is negative covariance between genetic and environmental effects on phenotype [41].

Environment. Here, I consider environment to include anything external to a given individual organism. Environment includes other organisms (siblings, mates, competitors, predators, prey, etc.) in addition to the physical and chemical surroundings. Given this definition, different individuals in the same place might experience different environments. Neither ecological speciation nor environmentally induced variation require environmental differences to be associated with geography. To put it another way, the effects of environment on fitness and on development might differ among co-occurring individuals for a variety of reasons, often involving feedbacks between phenotype and environment [42]. For example, small tadpoles in a pond might experience food shortages or attacks from predators while large tadpoles in the same pond have access to more food and experience less predation (or a different set of predators).

Environmental Induction. Any effect of environment on trait expression is environmental induction. Environmental induction usually refers to an event (developmental outcome or process) caused by an environmental condition, whereas *plasticity* (see below) refers to the *propensity* of an organism or trait to respond to environmental change.

Genetic Assimilation. Evolutionary reduction in the degree of plasticity such that a character state or trait value that was once conditionally expressed depending on the environment becomes expressed constitutively (unconditionally, regardless of environment).

Natural Selection. Natural selection refers to any nonrandom difference between entities in survival or reproduction.

To put it another way, natural selection exists whenever phenotypic variation *causes* covariance between phenotype and fitness [43, 44]. *Fitness* is metaphorical shorthand for the ability to survive and reproduce. It is important to emphasize that natural selection, under this definition, can exist without genetic variation and can recur over many generations without causing evolution [15, 17].

Plasticity. The ability of a single genotype to express different phenotypic values or states under different environmental conditions, that is, in response to environmental induction. Plasticity can include developmental plasticity, physiological acclimation, or behavioral flexibility. Plasticity might be adaptive or not. *Adaptive plasticity* is a tendency for a genotype to express a phenotype that enhances its ability to survive and reproduce in each environment. *Nonadaptive plasticity* includes any response to environmental induction that does not enhance fitness (including maladaptive responses). *Noisy plasticity* is effectively unpredictable phenotypic variation owing, for example, to developmental instability or random perturbations within environments [45]. Phenotypic plasticity and environmental induction are twin concepts; plasticity emphasizes an organismal property (the propensity to express different phenotypes in different environments), and environmental induction emphasizes the action of the environment. Phenotypic plasticity might *exist* even in a homogeneous population in a homogeneous environment. Environmental induction *happens* when environmental heterogeneity causes phenotypic heterogeneity.

Polyphenism. Expression of more than one discrete phenotypic state (alternative phenotypes) by a single genotype (a special case of phenotypic plasticity).

Reaction Norm [or Norm of Reaction]. The set of expected phenotypic states or values expressed by a genotype over a range of environments.

Speciation. Speciation is any process in which an ancestral species gives rise to two or more distinct descendant species. There is some disagreement about whether "speciation" should be synonymous with the evolution of reproductive isolation [30, 46, 47] or broadened to include anagenesis or phyletic speciation [48–50]. In any case, speciation is usually a gradual, continuous process of genetic divergence resulting in a discontinuous pattern of variation (species taxa). *Ecological speciation* is the evolution of reproductive isolation as a consequence of divergent ecological adaptation [28, 51, 52]. Without reproductive isolation, this is local adaptation. This definition is based on the biological species concept [53], which emphasizes genetically based reproductive isolation as the primary explanation for the existence of distinct kinds of organisms (i.e., those recognized as species taxa). I am not making a recommendation about taxonomic practice. Rather, from the perspective of evolutionary biology, the evolution of reproductive isolation is what distinguishes speciation from more general phenomena of genetic divergence [30, 40].

3. The Developmental Plasticity Hypothesis of Speciation

Matsuda [54] hypothesized that phenotypic plasticity was a crucial first step in the adaptive evolution of distinct, ecologically specialized lineages. As an example, Matsuda [54] suggested that major differences in life history, such as presence or absence of metamorphosis prior to reproduction in salamanders (Figure 1), likely began as nongenetic polyphenisms and evolved via genetic assimilation in habitat specialists. Widespread generalists such as *Ambystoma tigrinum* and *A. velasci* show conditional expression of metamorphosis from aquatic larva to terrestrial adult in small temporary ponds versus a fully aquatic life cycle with no metamorphosis in more permanent water bodies. This polyphenism likely represents the ancestral state of the tiger salamander clade [55]. In several isolated lakes in Mexico, permanently aquatic endemics such as *A. mexicanum* and *A. dumerilii* no longer express metamorphosis in nature owing to genetic changes in the thyroid hormone system [54–56].

Along the same lines, West-Eberhard [13–15] proposed a generalized "developmental plasticity hypothesis of speciation" in which the evolution of ecologically distinct forms in different environments depends on the initial appearance of those distinct forms as alternative phenotypes in a phenotypically plastic ancestor. She argued that when adaptive phenotypic plasticity results in strong associations between phenotypes and environments, rapid speciation could occur in three steps. First, alternative phenotypes become fixed in different populations owing to environmental differences, but with little or no genetic change. Then, genetic assimilation and/or other adaptive modifications of each phenotype occur owing to divergent selection. Finally, reproductive isolation evolves as a byproduct of adaptive divergence or via reinforcement if there is contact between the diverging populations.

Clearly, West-Eberhard's [13–15] hypothesis is a kind of ecological speciation in which developmental plasticity promotes genetic divergence in response to ecologically based selection. In the absence of plasticity, divergence might be prevented entirely if the single expressed phenotype cannot establish a viable population in the alternative environment [22, 27, 36] or might be much slower if phenotypic divergence must await new mutations and their gradual fixation [13–15].

West-Eberhard's developmental plasticity hypothesis of speciation is focused on adaptive phenotypic plasticity and its influence on one component of ecological speciation: the evolutionary response to divergent selection. However, nonadaptive plasticity might be equally if not more influential in promoting an evolutionary response [18]. Further, the other component of ecological speciation, the evolution of reproductive isolation [28, 62, 63], also can be directly influenced by phenotypic plasticity. In the next sections, I examine how plasticity can interact with these two components of ecological speciation.

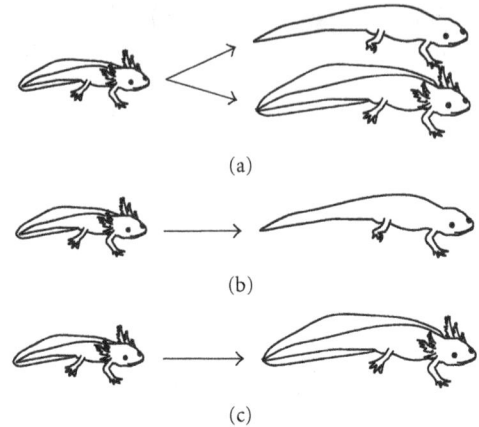

(a)

(b)

(c)

FIGURE 1: The tiger salamander radiation exemplifies the developmental plasticity hypothesis of ecological speciation [13–15, 54]. (a) Environmentally induced alternative phenotypes of the tiger salamander (*Ambystoma tigrinum*) include terrestrial metamorphs and aquatic paedomorphs. (b) The California tiger salamander (*A. californiense*) is an obligate metamorph derived from a developmentally plastic common ancestor with *A. tigrinum* [55]. (c) The Mexican axolotl (*A. mexicanum*) is one of several obligate paedomorphs, independently derived from plastic ancestors within the last few million years [55, 57, 58].

4. Phenotypic Plasticity and Ecological Divergence

Phenotypic plasticity can slow or enhance genetic divergence. How plasticity affects divergence depends to some extent on whether plasticity is adaptive or not.

4.1. Adaptive Plasticity. Adaptive plasticity can dampen or eliminate divergent selection. If any individual can express a nearly optimal phenotype in whatever environment it finds itself, then there is little or no variation in the ability to survive and reproduce, hence little or no divergent selection [64]. This has long been an intuitive reason to regard plasticity as a constraint on genetic evolution and to discount the evolutionary potential of environmentally induced variation [1]. Models have supported the prediction that adaptive plasticity can effectively take the place of genetic divergence between environments [65–67]. And a large number of empirical studies are consistent with increased plasticity in species with high dispersal rates [68]. However, the extent to which plasticity prevents or slows genetic divergence depends on several factors explored by Thibert-Plante and Hendry [36] in individual-based simulations.

First, is development sufficiently flexible that an individual can express traits near either environmental optimum? Given alternative environments or niches with divergent fitness functions, the only way environmental induction can completely eliminate divergent selection is to cause the mean expressed trait of a single gene pool to match the optimum in each environment [22] (Figure 2). If adaptive plasticity is less than perfect, divergent selection might still exist. Then the question is whether plasticity quantitatively dampens

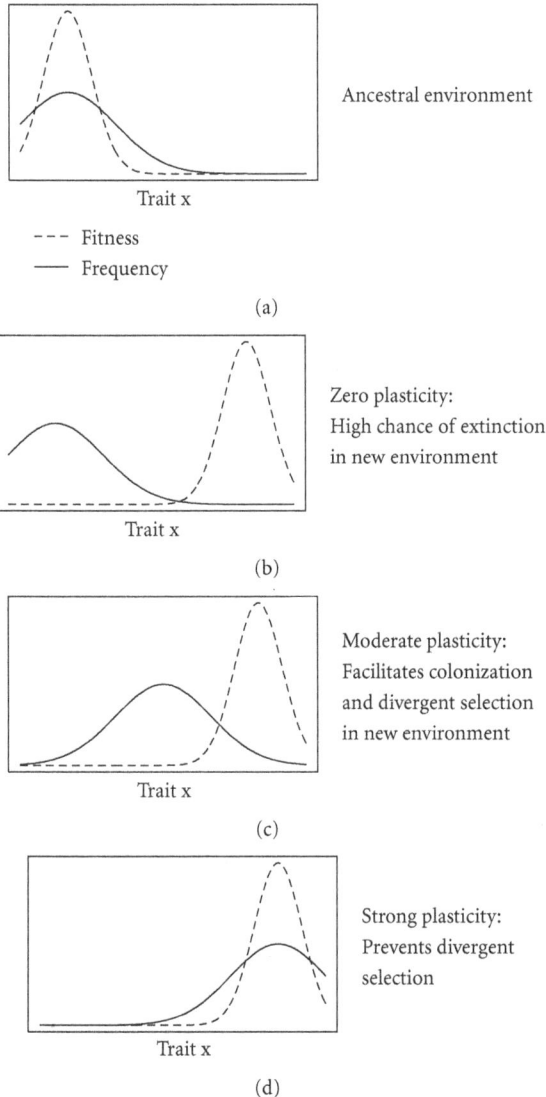

--- Fitness
—— Frequency

(a)

Ancestral environment

(b)

Zero plasticity:
High chance of extinction
in new environment

(c)

Moderate plasticity:
Facilitates colonization
and divergent selection
in new environment

(d)

Strong plasticity:
Prevents divergent
selection

FIGURE 2: Effects of adaptive plasticity on colonization and adaptation to a new niche. A population well adapted to its niche is illustrated in (a) by coincidence of the mean trait value (black line) with the peak of the fitness function (dashed line). If the fitness function is dramatically different in a new environment, a population with trait values favored in the old environment (b) might have such low fitness as to have little chance of survival. If environmental induction produces a shift in trait values toward higher fitness phenotypes (c), the population might persist but still experience selection. If phenotypic plasticity results in a perfect match between mean trait value and fitness optimum (d), then there is no effect of selection on the population mean.

the fitness tradeoff enough to substantially slow or prevent divergent evolution [36].

Second, how much dispersal occurs between environments? In the absence of gene flow, almost any amount of divergent selection will eventually cause evolutionary divergence. When there is gene flow between populations, the effect of divergent selection depends on the relative magnitudes of selection and gene flow [1, 69–73], in addition

to demographic factors [74–76]. Therefore the impact of adaptive plasticity on the potential for genetic divergence depends on how it affects the tension between divergent selection and gene flow [36, 77, 78]. Moreover, plasticity itself is adaptive only to the extent that individuals have a reasonable chance of experiencing alternative environments. If the populations expressing alternative phenotypes are isolated in their respective environments, the ability to express the alternative phenotype is likely to be lost owing to selection for efficient development or simply because loss of function mutations are likely to accumulate neutrally in genes that are never expressed [79]. This process of a conditionally expressed trait becoming constitutively expressed is genetic assimilation [11, 16, 80].

Third, are the systems sensing the environment and regulating trait expression sufficiently accurate that the best phenotype is reliably expressed in each environmental context? There are two components to this, first is simply the question of how well developmental or behavioral systems are able to sense and react to environmental stimuli [81]. Again, if adaptive plasticity is less than perfect, divergent selection can exist. Second is the question of whether the timing of key developmental and life history events is such that future environmental conditions can be correctly predicted [36]. For example, if individuals disperse and settle before completing development (e.g., seeds or planktonic larvae), they might be able to accurately tune their adult phenotypes to the environment in which they settle. However, if a substantial number of individuals disperse after completing development (e.g., animals with extended parental care [82]), then developmental plasticity would do little to help them accommodate new environmental challenges because their phenotypes are adjusted to their natal habitat rather than their new habitat. In this case, there might be strong selection against immigrants before any genetic differences arise between populations [36].

Finally, is there any cost to plasticity? Several modeling studies have confirmed the idea that plasticity is less likely to evolve if there are fitness costs to maintaining multiple developmental pathways or changing expression during development [36]. Empirical tests for costs of plasticity itself are rare [83], but it is conceivable that some pathways might have inherent tradeoffs between efficiency and plasticity [84, 85], and adaptive plasticity probably always comes with some potential for error; that is, the best developmental "decision" might not be made every time [64, 81]. When plasticity is costly enough to outweigh its fitness benefits, possible alternative outcomes are the evolution of a single "compromise" or generalist phenotype, evolution of a simple genetic "switch" enabling coexistence of alternative specialist phenotypes [14, 15, 86], or divergent evolution of specialist populations (local adaptation) [45, 67, 87]. In general we know very little about the prevalence or influence of costs of plasticity in nature.

The potential for adaptive plasticity to evolve as a response to ecological tradeoffs instead of genetic divergence is well supported. The dampening effect of plasticity is reduced but not eliminated by reduction in the extent and precision of plasticity, reduction in gene flow between environments,

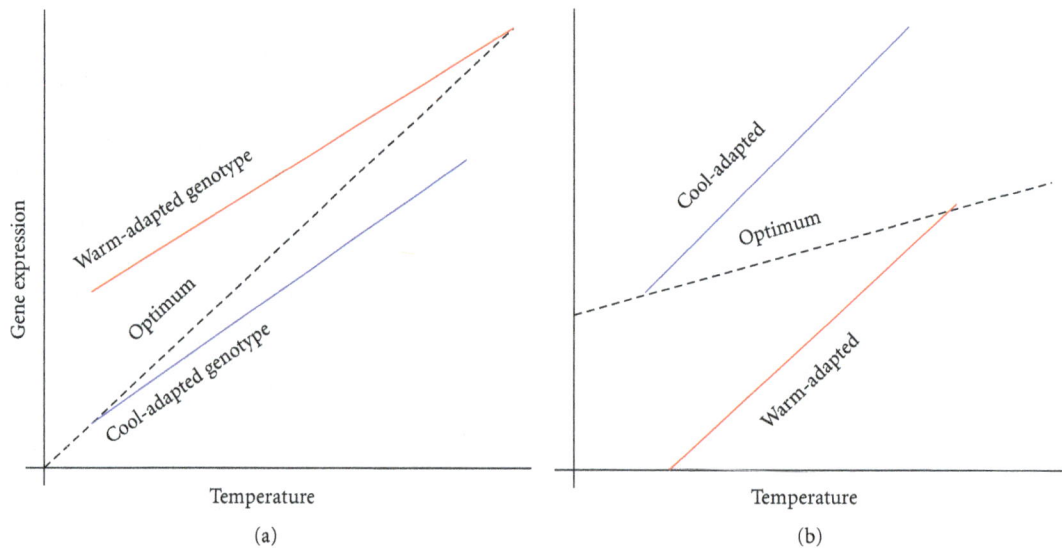

FIGURE 3: Adaptive and nonadaptive reaction norms. These hypothetical examples suppose that the optimum expression level for some gene increases with temperature and that increased temperature induces increased expression of the gene. (a) Plasticity is adaptive when it keeps expression closer to the optimum than it would be if expression, were constant across temperatures. (b) Plasticity is nonadaptive (maladaptive, in this case) when induced changes in a new environment take expression further from the optimum than it would be if it had remained constant. (b) is an example of countergradient variation, in which genetic differences cause cool-adapted genotypes to have higher gene expression than warm-adapted genotypes at the same temperature. The result is overexpression by cool-adapted genotypes transplanted to warm environments.

and increased costs of plasticity [36]. However, plasticity can actually promote genetic divergence under some conditions. In particular, when development is completed after dispersal (e.g., sessile organisms), adaptive plasticity might make successful colonization of new environments more likely (Figure 2). In West-Eberhard's [14, 15] conceptual model and Thibert-Plante and Hendry's [36] mathematical model, individuals are able to colonize a radically new environment by adjusting developmentally, behaviorally, and/or physiologically. This adaptive plasticity allows a population to persist in the new environment, continually exposed to divergent selection. If instead all individuals entering the new environment die or leave, there is no divergent selection. Thus, without adaptive plasticity, there might simply be suitable and unsuitable environments, with little opportunity for divergent evolution. Successful colonization of a new environment can initiate divergent selection, not only on the plastic trait, but possibly also on other traits, which might then cause ecological speciation. This to some extent reconciles the conflicting effects of plasticity. In other circumstances, when dispersal occurs after development (e.g., animals with extended parental care), individuals settling in new environments are especially likely to express suboptimal phenotypes, which might accentuate the effects of selection and spatial separation once a new habitat has been colonized [36]. This effect is similar to effects of nonadaptive plasticity discussed below.

4.2. Nonadaptive Plasticity. The potential for adaptive plasticity to promote colonization and adaptation to alternative environments has been promoted by advocates of

developmental evolutionary biology [14, 15, 17, 88] and treated extensively in recent reviews and models [22, 27, 32, 36]. The effects of nonadaptive phenotypic plasticity have received less attention. However, any environmental effect on phenotypes can affect the strength and direction of selection in addition to the genetic variances and covariances of important traits [18, 22, 42]. Some kinds of nonadaptive environmental induction might affect the probability of ecological speciation. In particular, suboptimal development or noisy plasticity [45] in stressful environments could inhibit adaptation by decreasing the fitness of local relative to immigrant individuals. However, it would also increase the strength of selection and potentially result in cryptic adaptive divergence (countergradient variation) [18, 41].

Countergradient variation is negative covariance between genetic and environmental effects on phenotype [41]. Classic examples are poikilotherms such as fish and molluscs [89], flies [90], or frogs [91] that grow more slowly in cold climates, but cold-adapted populations have higher growth rates than warm-adapted populations when raised at the same temperature. The best explanation for this pattern is that genetic differences have evolved to compensate for divergent effects of environmental induction, resulting in populations that appear similar when measured each in their native habitat but show maladaptive plasticity when transplanted (Figure 3). Note that negative covariance between genetic and environmental effects is not necessarily maladaptive, but environmental effects will tend to be maladaptive if the optimum phenotype is roughly constant across the environment range.

When countergradient variation exists, we expect immigrants to have a fitness disadvantage owing to under- or overexpression of an environmentally sensitive trait relative to a local optimum (Figure 3). This might well promote evolution of restricted or nonrandom dispersal; hence intrinsic barriers to gene exchange as a result of divergent ecological selection. However, countergradient variation will evolve only if selection is stronger than gene flow and if reaction norms are genetically constrained (otherwise, we might expect adaptive evolution to flatten the reaction norm). At least in the early stages of colonization of a challenging new habitat, nonadaptive plasticity (such as stunted growth or suboptimal metabolic rates) might make resident individuals less viable and fecund than healthy immigrants from a less stressful habitat [22]. This potential fitness asymmetry could offset effects of adaptive genetic changes on the relative fitness of immigrants and residents. That is, offspring of immigrants might have lower fitness than offspring of native genotypes with locally adaptive alleles. However, offspring of immigrants might nevertheless outnumber offspring of natives if immigrants come into the stressful habitat with substantial viability and fertility advantages from being raised in a higher quality habitat. Parental care and physiological maternal effects could further extend those environmentally induced advantages to the offspring. The net effect could be a tendency for locally adapted genotypes to be replaced ("swamped") by immigrants owing to a negative covariance between environmental and genetic effects. This effect of nonadaptive plasticity is synergistic with the potential for demographic swamping, a well-known constraint on local adaptation to novel habitats [74, 76, 87, 92].

For now, it appears that the impact of nonadaptive plasticity on the probability of ecological speciation cannot be predicted without additional detailed knowledge. Just as gene flow can constrain or facilitate local adaptation [74, 76], and adaptive plasticity can inhibit or promote adaptive genetic divergence [22, 27, 36], nonadaptive plasticity might impede genetic divergence by accentuating fitness advantages of immigrants and/or promote divergence by increasing the intensity of selection.

5. Phenotypic Plasticity and the Evolution of Reproductive Isolation

Plasticity of a different sort might directly affect reproductive compatibility between populations developing in different environments. Environmental induction might generate differences in preference, reproductive phenology, or expression of secondary sexual characteristics. Coincidence of environmentally induced reproductive barriers and potentially divergent selection can be genetically equivalent to a geographic barrier between divergent environments [59, 93, 94]. A key element of the developmental plasticity model of ecological speciation is the establishment of a consistent relationship between the environment of parents and that of their offspring. Similarity of parent and offspring environments maintains shared environmental effects on phenotype, consistency of selection, and reduces gene flow.

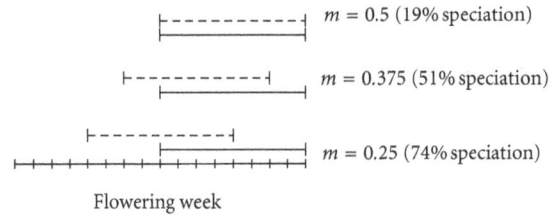

FIGURE 4: Effects of phenotypic plasticity of flowering time on ecological speciation. The lines illustrate flowering periods for plants on two soil types (dashed and solid) based on initial conditions in the simulations of Gavrilets and Vose [59] for ecological speciation based on the *Howea* palm tree study [60]. I illustrate the case where flowering time is affected by 8 loci, and the environmental effect of soil type is either 0, 2, or 4 weeks difference. In the simulations [59], changing the number of loci has confounding effects on the initial environmental variance component and the fitness effects of mutations, so a single genetic scenario is illustrated here for simplicity. I calculated the initial rate of gene flow between soil types as $m = 0.5$ (% time overlap), based on the assumption that any tree is equally likely to receive pollen from any other currently flowering tree. %speciation is based on the number of simulations ending in speciation, given in Table 2 of Gavrilets and Vose [59].

Environmental similarity can be a byproduct of geographic isolation or might be promoted by environmentally induced variation in habitat choice, phenology, or other aspects of mating behavior. Whether or not ecological speciation ensues depends on whether genetic reproductive barriers evolve and whether that evolution can be attributed to divergent selection.

A seemingly commonplace example of environmentally induced barriers to gene flow is flowering time in plants. Flowering time is often accelerated or delayed when a given genotype is grown on different substrates [34]. For example, grasses and monkey flowers colonizing contaminated soils around mines show environmentally induced shifts in flowering [95, 96], and palm trees on Lord Howe Island show soil-dependent flowering times [60]. These examples are also widely recognized cases of ecological speciation. Gavrilets and Vose [59] used simulations to explore the impact of environmentally induced shifts in flowering time on the probability of adaptive divergence and ecological speciation and confirmed that this instantaneous barrier to gene flow between habitats can markedly increase the probability and rate of divergence (Figure 4).

A similar effect arises owing to behavioural imprinting [97, 98]. Many animals, such as birds and anadromous fishes, are known for imprinting on their natal habitat [99–101]. When this is a direct matter of memorizing where home is (as might be the case in *Ficedula* flycatchers [101]), it simply accentuates the relationship between geography and gene flow. Slightly different implications emerge from imprinting on a *kind* of habitat, host, or resource because then the environmental influence on gene flow is independent of geography. Some phytophagous insects imprint on their host plant species based on chemical cues [102], and nest-parasitic indigobirds imprint on their hosts [103]. In these examples, phenotypic plasticity helps maintain similarity between maternal and offspring environments, which is important both for maintaining phenotypic similarity and consistency of selection on parents and offspring. However, it is not always obvious that habitat or resource imprinting should directly affect mating preferences. In the *Vidua* indigobirds, there is an imprinting effect of host song, such that nest parasites raised by the same host species tend to mate assortatively owing to learned elements of their own songs and preferences [103, 104]. In many phytophagous insects, mating occurs on or near the host plant [102, 105], making plant choice a "magic trait" simultaneously effecting ecological and sexual differentiation [31]. As with genetically determined traits, phenotypic plasticity of traits directly linked to both ecological adaptation and assortative mating is most likely to contribute to ecological speciation.

Environmental effects on traits directly involved in sexual selection are not unusual. Expression of pigments, pheromones, and other displays can depend on diet, condition, or experience [19, 106]. For example, premating isolation is induced by larval host plant differences in *Drosophila mojavensis*, because the chemical properties of their cuticular hydrocarbons (important contact pheromones) are strongly influenced by diet [107, 108]. Sharon et al. [109] recently showed that mate choice in *D. melanogaster* can be modified by symbiotic gut bacteria. Flies raised on a high-starch medium had microbiota dominated by *Lactobacillus plantarum*, which was only a minor constituent of the microbiota of flies raised on a standard cornmeal-molasses medium. Sharon et al. [109] found that the differences in bacterial composition can affect cuticular hydrocarbon levels, providing a probable mechanism affecting mate choice.

In Sockeye Salmon (*Oncorhynchus nerka*), postmating isolation might be caused by sexual selection on diet-derived coloration. Anadromous sockeye sequester carotenoids from crustaceans consumed in the ocean and use them to express their brilliant red mating colors. The nonanadromous morph (kokanee) expresses equally bright red mating color on a diet with much less carotenoids, an example of countergradient variation [110]. Anadromous morphs and hybrids raised in the freshwater habitat of the kokanee (low-carotenoid diet) underexpress the red pigment and probably suffer reduced mating success as a consequence [111, 112].

The actual effect of phenotypic plasticity on gene flow depends on environmental similarity between parents and offspring. Environmentally induced mate discrimination will have little or no hindrance on gene flow unless it also affects the phenotypes and/or environments of the offspring. For example, imagine a phytophagous insect with environmentally induced contact pheromones causing perfect assortative mating between individuals raised on the same host plant. If there are no differences in host choice, then the offspring of each mating type are equally likely to grow up on each plant and therefore have no tendency to develop the same pheromone profile as their parents (Figure 5). In this hypothetical case, there is free gene flow despite assortative mating of phenotypes.

Environmentally induced differences in habitat choice reduce gene flow when individuals are more likely to mate with other individuals using the same habitat, and offspring are more likely to grow up in habitats similar to those of their parents. If habitat choice is entirely determined by the individual's environment (i.e., if there is no tendency for the offspring of immigrants to return to their parent's original habitat), then the effect is genetically identical to a geographic barrier. Either way we can describe the system in terms of populations of individuals or gametes with some probability (m) of "moving" from their natal population to breed in a different population. In the simple case of two environments and nonoverlapping generations, the expected frequency of an allele in generation t in environment i is a weighted average of the frequencies in environments i and j in generation $t - 1$ [113]:

$$p_{i(t)} = (1 - m)p_{i(t-1)} + mp_{j(t-1)}. \qquad (1)$$

It makes no difference whether m is determined by geography or environmental induction as long as there are not heritable differences in m among individuals within a habitat. More generally, geographic or spatial covariance is a special case of environmental similarity, and the extensive knowledge from decades of conceptual and mathematical modeling of gene flow's effects on adaptation and speciation [30, 31, 87, 114, 115] can be extended directly to include this kind of phenotypic plasticity. In particular, we might expect environmentally induced restrictions on gene flow to facilitate the evolution of postzygotic incompatibilities (both environment-dependent and -independent selection against hybrids) and genetic assimilation of behavioral reproductive barriers (habitat and mate choice), but also to lessen the potential for selective reinforcement of assortative mating (just as adaptive plasticity lessens divergent selection on ecological phenotypes).

6. Conclusions and Future Directions

Opinions still seem to outnumber data about the impact of environmental induction and plasticity on evolution, but substantial progress has been made in the last 20 years [23, 64]. Plasticity appears to be a common if not universal feature of developmental systems and should not be ignored. Plasticity and environmental effects were once black boxes, ignored by some, uncritically promoted as threats to evolutionary theory by others. But theoretical and empirical investigations have increasingly shed light on how plasticity evolves and interacts with natural selection. Whether developmental plasticity rivals mutation as a source of quantitative or qualitative change (the evolution of "novelty") remains contentious [17, 80, 116]. However, to the extent that speciation is defined by genetic divergence, genes will not be displaced from their central role in the study of speciation. Phenotypic plasticity can promote or constrain adaptive evolution and ecological speciation. The effects of plasticity in a particular case, and whether there is an overall trend, are empirical questions.

Important theoretical challenges for understanding the importance of plasticity for adaptation and speciation

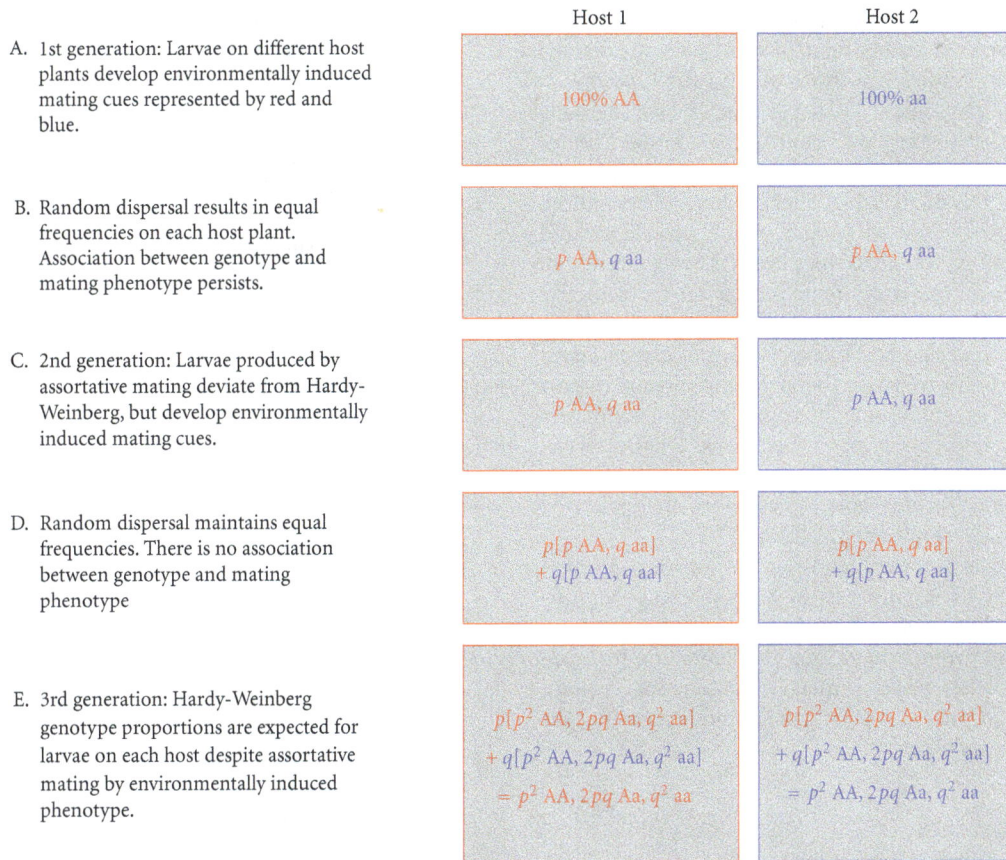

FIGURE 5: Free gene flow despite assortative mating. If mating cues are environmentally induced but there is no habitat choice, host-associated populations will not be genetically differentiated, and it takes two generations to establish Hardy-Weinberg equilibrium. If the host densities are p and $1 - p = q$, a locus that is fixed for different alleles on different hosts has allele frequencies p and q. At first, genotypes are perfectly associated with the host-induced mating cue, but random dispersal eliminates that association in one generation if there are no other factors maintaining covariance between parent and offspring phenotype (e.g., maternal effects or divergent selection). Once genotype frequencies are equalized between phenotypes, mating within phenotypes establishes Hardy-Weinberg genotype proportions [61]. Although imagining a locus with complete differentiation makes for the simplest illustration, the result is completely general for any allele frequency. If we somehow know what alleles have ancestors in each habitat in generation 0, the result is that it takes just two generations to completely randomize that ancestry, regardless of whether or not the alleles are actually different by state.

include the extension of models (such as [36]) to incorporate nonadaptive plasticity and countergradient variation, further investigation of how costs of plasticity affect genetic assimilation and reproductive isolation, and careful examination of what kinds of environmental effects on phenotype can be considered equivalent to geographic restrictions of gene flow (environmental effects on dispersal) [59, 93, 94, 114]. Not all environmental effects on mate choice will affect gene flow (Figure 5). However, maternal effects and factors that promote environmental similarity between relatives seem particularly likely to reduce gene flow and promote adaptive divergence. Incorporation of such effects into the classic models of local adaptation [87] and ecological speciation [31] might bring substantial clarification to the subject. Moreover, how environmentally induced barriers to gene flow might affect the evolution of constitutive genetic barriers has not been explored. Should we expect environmental effects on assortative mating to slow the evolution of genetic barriers to gene flow? Finally, virtually no attention has been

given to the relationship between plasticity and postzygotic isolation. Key questions include how might plasticity affect ecologically based selection on hybrids, and are highly plastic developmental pathways good or bad candidates for involvement in postzygotic developmental incompatibilities?

Addressing empirical challenges might best begin by establishing criteria for recognizing plasticity as a causal factor in speciation or adaptive radiation. West-Eberhard [15] and Pfennig and McGee [117] suggested that associations between intraspecific plasticity and species diversity support a role for plasticity in promoting adaptive radiation. Examples like the tiger salamander radiation (Figure 1) and others [117], where an intraspecific polyphenism parallels a repeated pattern of interspecific divergence, are consistent with adaptive plasticity as an origin of ecological divergence between species. More confirmatory evidence for a causal role of plasticity in adaptive divergence (the first component of ecological speciation) would come from testing the strengths of tradeoffs, costs of plasticity, and selection against immigrants and hybrids. Effects of plasticity on reproductive isolation (the second component of ecological speciation) are illustrated in studies like those of parasitic indigobirds [103, 104] and *Drosophila* [107–109] on environmentally induced barriers to gene flow. We presently know little of the prevalence of this phenomenon or whether it is often strong enough to facilitate substantial genetic differentiation. Field studies are needed to document the prevalence and strength, in nature, of costs of plasticity, evolvability of reaction norms, and environmental effects on reproductive isolation. The principle illustrated in Figure 5 should be considered in the design and interpretation of experiments. In order for plasticity of mating behaviour to directly impact gene flow, there must be some factor maintaining similarity of environmental effects across generations. Finally, developmental genetic studies are needed to assess whether genes causing postzygotic or prezygotic isolation are often involved in highly plastic developmental pathways.

The theoretical and empirical foundations of speciation research are very strong in terms of genetics and geography [29–31, 115]. The roles of ecology, environment, and development are prominent among the remaining frontiers [27, 36, 52]. Further research integrating genetics, ecology, and development promises great gains in understanding the origins of biological diversity.

Acknowledgments

The author thanks A. Hendry and two reviewers for their very helpful comments on an earlier draft of this paper. His knowledge and ideas on the subject, and their presentation herein, were greatly improved by discussions with J. Fordyce and X. Thibert-Plante. Research relevant to this paper has been supported by the United States National Science Foundation (DEB-0516475 and DEB-1011216).

References

[1] S. Wright, "Evolution in Mendelian populations," *Genetics*, vol. 16, no. 2, pp. 97–159, 1931.

[2] S. C. Stearns, "The role of development in the evolution of life histories," in *Evolution and Development*, J. T. Bonner, Ed., pp. 237–258, Springer, Berlin, Germany, 1982.

[3] D. A. Levin, "Plasticity, canalization and evolutionary stasis in plants," in *Plant Population Ecology*, A. J. Davy, M. J. Hutchings, and A. R. Watkinson, Eds., pp. 35–45, Blackwell Scientific, 1988.

[4] J. Clausen, "Principles for a joint attack on evolutionary problems," in *Proceedings of the 6th International Congress of Genetics*, D. F. Jones, Ed., vol. 2, pp. 21–23, Brooklyn Botanic Garden, Ithaca, NY, USA, 1932.

[5] R. K. Clements, J. M. Baskin, and C. C. Baskin, "The comparative biology of the two closely-related species *Penstemon tenuiflorus* Pennell and *P. hirsutus* (L.) Willd. (Scrophulariaceae, section *Graciles*): I. Taxonomy and geographical distribution," *Castanea*, vol. 63, no. 2, pp. 138–153, 1998.

[6] P. B. Reich, I. J. Wright, J. Cavender-Bares et al., "The evolution of plant functional variation: traits, spectra, and strategies," *International Journal of Plant Sciences*, vol. 164, no. 3, pp. S143–S164, 2003.

[7] U. Dieckmann and M. Heino, "Probabilistic maturation reaction norms: their history, strengths, and limitations," *Marine Ecology-Progress Series*, vol. 335, pp. 253–269, 2007.

[8] M. G. Palacios, A. M. Sparkman, and A. M. Bronikowski, "Developmental plasticity of immune defence in two life-history ecotypes of the garter snake, thamnophis elegans - a common-environment experiment," *Journal of Animal Ecology*, vol. 80, no. 2, pp. 431–437, 2011.

[9] J. M. Baldwin, "A new factor in evolution," *American Naturalist*, vol. 30, pp. 441–451, 1896.

[10] C. H. Waddington, "Canalization of development and the inheritance of acquired characters," *Nature*, vol. 150, no. 3811, pp. 563–565, 1942.

[11] C. H. Waddington, "Genetic assimilation of acquired characters," *Evolution*, vol. 7, no. 4676, pp. 118–126, 1953.

[12] Schmalhausen II, *Factors of Evolution*, Blakiston, Philadelphia, Pa, USA, 1949.

[13] M. J. West-Eberhard, "Alternative adaptations, speciation, and phylogeny," *Proceedings of the National Academy of Sciences of the United States of America*, vol. 83, no. 5, pp. 1388–1392, 1986.

[14] M. J. West-Eberhard, "Phenotypic plasticity and the origins of diversity," *Annual review of Ecology and Systematics*, vol. 20, pp. 249–278, 1989.

[15] M. J. West-Eberhard, *Developmental Plasticity and Evolution*, Oxford University Press, Oxford, UK, 2003.

[16] C. D. Schlichting, "The role of phenotypic plasticity in diversification," in *Phenotypic Plasticity: Functional and Conceptual Approaches*, T. J. DeWitt and S. M. Scheiner, Eds., pp. 191–200, Oxford University Press, 2004.

[17] M. J. West-Eberhard, "Developmental plasticity and the origin of species differences," *Proceedings of the National Academy of Sciences of the United States of America*, vol. 102, no. 1, pp. 6543–6549, 2005.

[18] G. F. Grether, "Environmental change, phenotypic plasticity, and genetic compensation," *The American Naturalist*, vol. 166, no. 4, pp. E115–123, 2005.

[19] T. D. Price, "Phenotypic plasticity, sexual selection and the evolution of colour patterns," *Journal of Experimental Biology*, vol. 209, no. 12, pp. 2368–2376, 2006.

[20] E. Crispo, "The Baldwin effect and genetic assimilation: revisiting two mechanisms of evolutionary change mediated by phenotypic plasticity," *Evolution*, vol. 61, no. 11, pp. 2469–2479, 2007.

[21] E. Crispo, "Modifying effects of phenotypic plasticity on interactions among natural selection, adaptation and gene flow," *Journal of Evolutionary Biology*, vol. 21, no. 6, pp. 1460–1469, 2008.

[22] C. K. Ghalambor, J. K. McKay, S. P. Carroll, and D. N. Reznick, "Adaptive versus non-adaptive phenotypic plasticity and the potential for contemporary adaptation in new environments," *Functional Ecology*, vol. 21, no. 3, pp. 394–407, 2007.

[23] S. F. Gilbert and D. Epel, *Ecological Developmental Biology: Integrating Epigenetics, Medicine, and Evolution*, Sinauer Associates, 2009.

[24] G. Fusco and A. Minelli, "Phenotypic plasticity in development and evolution: facts and concepts," *Philosophical Transactions of the Royal Society B*, vol. 365, no. 1540, pp. 547–556, 2010.

[25] T. Schwander and O. Leimar, "Genes as leaders and followers in evolution," *Trends in Ecology and Evolution*, vol. 26, no. 3, pp. 143–151, 2011.

[26] T. Uller and H. Helantera, "When are genes 'leaders' or 'followers' in evolution?" *Trends in Ecology and Evolution*, vol. 26, no. 9, pp. 435–436, 2011.

[27] D. W. Pfennig, M. A. Wund, E. C. Snell-Rood, T. Cruickshank, C. D. Schlichting, and A. P. Moczek, "Phenotypic plasticity's impacts on diversification and speciation," *Trends in Ecology and Evolution*, vol. 25, no. 8, pp. 459–467, 2010.

[28] H. D. Rundle and P. Nosil, "Ecological speciation," *Ecology Letters*, vol. 8, no. 3, pp. 336–352, 2005.

[29] M. Turelli, N. H. Barton, and J. A. Coyne, "Theory and speciation," *Trends in Ecology and Evolution*, vol. 16, no. 7, pp. 330–343, 2001.

[30] J. A. Coyne and H. A. Orr, *Speciation*, Sinauer Associates, Sunderland, Mass, USA, 2004.

[31] S. Gavrilets, *Fitness Landscapes and the Origin of Species*, Princeton University Press, Princeton, NJ, USA, 2004.

[32] T. D. Price, A. Qvarnstrom, and D. E. Irwin, "The role of phenotypic plasticity in driving genetic evolution," *Proceedings of the Royal Society B*, vol. 270, no. 1523, pp. 1433–1440, 2003.

[33] R. Lande, "Adaptation to an extraordinary environment by evolution of phenotypic plasticity and genetic assimilation," *Journal of Evolutionary Biology*, vol. 22, no. 7, pp. 1435–1446, 2009.

[34] D. A. Levin, "Flowering-time plasticity facilitates niche shifts in adjacent populations," *New Phytologist*, vol. 183, no. 3, pp. 661–666, 2009.

[35] R. Svanbäck, M. Pineda-Krch, and M. Doebeli, "Fluctuating population dynamics promotes the evolution of phenotypic plasticity," *American Naturalist*, vol. 174, no. 2, pp. 176–189, 2009.

[36] X. Thibert-Plante and A. P. Hendry, "The consequences of phenotypic plasticity for ecological speciation," *Journal of Evolutionary Biology*, vol. 24, no. 2, pp. 326–342, 2011.

[37] M. D. Herron and M. Doebeli, "Adaptive diversification of a plastic trait in a predictably fluctuating environment," *Journal of Theoretical Biology*, vol. 285, pp. 58–68, 2011.

[38] D. J. Funk, "Of 'host forms' and host races: terminological issues in ecological speciation," *International Journal of Ecology*, vol. 2012, Article ID 506957, 8 pages, 2012.

[39] J. K. Conner and D. L. Hartl, *A Primer of Ecological Genetics*, Sinauer Associates, Sunderland, Mass, USA, 2004.

[40] D. J. Futuyma, *Evolution*, Sinauer Associates, Sunderland, Mass, USA, 2nd edition, 2009.

[41] D. O. Conover and E. T. Schultz, "Phenotypic similarity and the evolutionary significance of countergradient variation," *Trends in Ecology and Evolution*, vol. 10, no. 6, pp. 248–252, 1995.

[42] J. A. Fordyce, "The evolutionary consequences of ecological interactions mediated through phenotypic plasticity," *Journal of Experimental Biology*, vol. 209, no. 12, pp. 2377–2383, 2006.

[43] G. R. Price, "Selection and covariance," *Nature*, vol. 227, no. 5257, pp. 520–521, 1970.

[44] J. A. Endler, *Natural Selection in the Wild*, Princeton University Press, Princeton, NJ, USA, 1986.

[45] S. Via, "The evolution phenotypic plasticity: what do we really know?" in *Ecological Genetics*, L. A. Real, Ed., pp. 35–57, Princeton University Press, 1994.

[46] G. L. Bush, "Reply [to M. Claridge] from G. L. Bush," *Trends in Ecology & Evolution*, vol. 10, no. 1, p. 38, 1995.

[47] M. Claridge, "Species and speciation," *Trends in Ecology & Evolution*, vol. 10, no. 1, p. 38, 1995.

[48] G. G. Simpson, *Tempo and Mode in Evolution*, Columbia University Press, New York, NY, USA, 1944.

[49] S. J. Gould, *The Structure of Evolutionary Theory*, Harvard University Press, Cambridge, Mass, USA, 2002.

[50] T. F. Stuessy, G. Jakubowsky, R. S. Gomez et al., "Anagenetic evolution in island plants," *Journal of Biogeography*, vol. 33, no. 7, pp. 1259–1265, 2006.

[51] A. P. Hendry, "Ecological speciation! or the lack thereof?" *Canadian Journal of Fisheries and Aquatic Sciences*, vol. 66, no. 8, pp. 1383–1398, 2009.

[52] D. Schluter, "Evidence for ecological speciation and its alternative," *Science*, vol. 323, no. 5915, pp. 737–741, 2009.

[53] E. Mayr, *Systematics and the Origin of Species from the Viewpoint of a Zoologist*, Columbia University Press, New York, NY, USA, 1942.

[54] R. Matsuda, "The evolutionary process in talitrid amphipods and salamanders in changing environments, with a discussion of "genetic assimilation" and some other evolutionary concepts," *Canadian Journal of Zoology*, vol. 60, no. 5, pp. 733–749, 1982.

[55] H. B. Shaffer and S. R. Voss, "Phylogenetic and mechanistic analysis of a developmentally integrated character complex: alternate life history modes in ambystomatid salamanders," *American Zoologist*, vol. 36, no. 1, pp. 24–35, 1996.

[56] S. R. Voss and H. B. Shaffer, "Adaptive evolution via a major gene effect: paedomorphosis in the Mexican axolotl," *Proceedings of the National Academy of Sciences of the United States of America*, vol. 94, no. 25, pp. 14185–14189, 1997.

[57] H. B. Shaffer, "Evolution in a paedomorphic lineage. I. An electrophoretic analysis of the Mexican ambystomatid salamanders," *Evolution*, vol. 38, pp. 1194–1206, 1984.

[58] H. B. Shaffer and M. L. Mcknight, "The polytypic species revisited: genetic differentiation and molecular phylogenetics of the tiger salamander Ambystoma tigrinum (Amphibia: Caudata) complex," *Evolution*, vol. 50, no. 1, pp. 417–433, 1996.

[59] S. Gavrilets and A. Vose, "Case studies and mathematical models of ecological speciation. 2. Palms on an oceanic island," *Molecular Ecology*, vol. 16, no. 14, pp. 2910–2921, 2007.

[60] V. Savolainen, M. C. Anstett, C. Lexer et al., "Sympatric speciation in palms on an oceanic island," *Nature*, vol. 441, no. 7090, pp. 210–213, 2006.

[61] G. H. Hardy, "Mendelian proportions in a mixed population," *Science*, vol. 28, no. 706, pp. 49–50, 1908.

[62] D. Schluter, "Ecology and the origin of species," *Trends in Ecology and Evolution*, vol. 16, no. 7, pp. 372–380, 2001.

[63] K. Räsänen and A. P. Hendry, "Disentangling interactions between adaptive divergence and gene flow when ecology drives diversification," *Ecology Letters*, vol. 11, no. 6, pp. 624–636, 2008.

[64] T. J. DeWitt and S. M. Scheiner, "Phenotypic variation from single genotypes: a primer," in *Phenotypic Plasticity: Functional and Conceptual Approaches*, T. J. DeWitt and S. M. Scheiner, Eds., pp. 1–9, Oxford University Press, New York, NY, USA, 2004.

[65] S. Via and R. Lande, "Genotype-environment interaction and the evolution of phenotypic plasticity," *Evolution*, vol. 39, no. 3, pp. 505–522, 1985.

[66] L. Zhivotovsky, M. Feldman, and A. Bergman, "On the evolution of phenotypic plasticity in a spatially heterogeneous environment," *Evolution*, vol. 50, no. 2, pp. 547–558, 1996.

[67] S. Sultan and H. G. Spencer, "Metapopulation structure favors plasticity over local adaptation," *American Naturalist*, vol. 160, no. 2, pp. 271–283, 2002.

[68] J. Hollander, "Testing the grain-size model for the evolution of phenotypic plasticity," *Evolution*, vol. 62, no. 6, pp. 1381–1389, 2008.

[69] J. B. S. Haldane, "A mathematical theory of natural and artificial selection—part VI. Isolation," *Proceedings of the Cambridge Philosophical Society*, vol. 26, pp. 220–230, 1930.

[70] M. G. Bulmer, "Multiple niche polymorphism," *American Naturalist*, vol. 106, pp. 254–257, 1972.

[71] M. Slatkin, "Gene flow and the geographic structure of natural populations," *Science*, vol. 236, no. 4803, pp. 787–792, 1987.

[72] W. R. Rice and E. E. Hostert, "Laboratory experiments on speciation: what have we learned in 40 years?" *Evolution*, vol. 47, no. 6, pp. 1637–1653, 1993.

[73] S. Gavrilets, "Models of speciation: what have we learned in 40 years?" *Evolution*, vol. 57, no. 10, pp. 2197–2215, 2003.

[74] R. D. Holt, "On the evolutionary ecology of species' ranges," *Evolutionary Ecology Research*, vol. 5, no. 2, pp. 159–178, 2003.

[75] D. Garant, S. E. Forde, and A. P. Hendry, "The multifarious effects of dispersal and gene flow on contemporary adaptation," *Functional Ecology*, vol. 21, no. 3, pp. 434–443, 2007.

[76] T. J. Kawecki, "Adaptation to marginal habitats," *Annual Review of Ecology, Evolution, and Systematics*, vol. 39, pp. 321–342, 2008.

[77] S. Via, "Genetic constraints on the evolution of phenotypic plasticity," in *Genetic Constraints on Adaptive Evolution*, V. Loeschcke, Ed., pp. 47–71, Springer, Boston, Mass, USA, 1987.

[78] S. Via, R. Gomulkiewicz, G. de Jong, S. M. Scheiner, C. D. Schlichting, and P. H. Van Tienderen, "Adaptive phenotypic plasticity: consensus and controversy," *Trends in Ecology and Evolution*, vol. 10, no. 5, pp. 212–217, 1995.

[79] A. Romero and S. M. Green, "The end of regressive evolution: examining and interpreting the evidence from cave fishes," *Journal of Fish Biology*, vol. 67, no. 1, pp. 3–32, 2005.

[80] M. Pigliucci, C. J. Murren, and C. D. Schlichting, "Phenotypic plasticity and evolution by genetic assimilation," *Journal of Experimental Biology*, vol. 209, no. 12, pp. 2362–2367, 2006.

[81] R. B. Langerhans and T. J. DeWitt, "Plasticity constrained: over-generalized induction cues cause maladaptive phenotypes," *Evolutionary Ecology Research*, vol. 4, no. 6, pp. 857–870, 2002.

[82] F. C. James, "Environmental component of morphological differentiation in birds," *Science*, vol. 221, no. 4606, pp. 184–186, 1983.

[83] T. Steinger, B. A. Roy, and M. L. Stanton, "Evolution in stressful environments II: adaptive value and costs of plasticity in response to low light in *Sinapis arvensis*," *Journal of Evolutionary Biology*, vol. 16, no. 2, pp. 313–323, 2003.

[84] P. H. Van Tienderen, "Evolution of generalists and specialists in spatially heterogeneous environments," *Evolution*, vol. 45, no. 6, pp. 1317–1331, 1991.

[85] T. J. DeWitt, A. Sih, and D. S. Wilson, "Costs and limits of phenotypic plasticity," *Trends in Ecology and Evolution*, vol. 13, no. 2, pp. 77–81, 1998.

[86] T. B. Smith and S. Skulason, "Evolutionary significance of resource polymorphisms in fishes, amphibians, and birds," *Annual Review of Ecology and Systematics*, vol. 27, pp. 111–133, 1996.

[87] T. Lenormand, "Gene flow and the limits to natural selection," *Trends in Ecology and Evolution*, vol. 17, no. 4, pp. 183–189, 2002.

[88] C. D. Schlichting and M. Pigliucci, *Phenotypic Evolution: A Reaction Norm Perspective*, Sinauer Associates, 1998.

[89] T. H. Bullock, "Compensation for temperature in the metabolism and activity of poikilotherms," *Biological Reviews*, vol. 30, pp. 311–342, 1955.

[90] R. Levins, "Thermal acclimation and heat resistance in *Drosophila* species," *American Naturalist*, vol. 103, pp. 483–499, 1969.

[91] K. A. Berven, D. E. Gill, and S. J. Smithgill, "Countergradient selection in the green frog, rana clamitans," *Evolution*, vol. 33, pp. 609–623, 1979.

[92] M. Kirkpatrick and N. H. Barton, "Evolution of a species' range," *American Naturalist*, vol. 150, no. 1, pp. 1–23, 1997.

[93] B. M. Fitzpatrick, J. A. Fordyce, and S. Gavrilets, "What, if anything, is sympatric speciation?" *Journal of Evolutionary Biology*, vol. 21, no. 6, pp. 1452–1459, 2008.

[94] B. M. Fitzpatrick, J. A. Fordyce, and S. Gavrilets, "Pattern, process and geographic modes of speciation," *Journal of Evolutionary Biology*, vol. 22, no. 11, pp. 2342–2347, 2009.

[95] J. Antonovics, "Evolution in closely adjacent plant populations. X. Long-term persistence of prereproductive isolation at a mine boundary," *Heredity*, vol. 97, no. 1, pp. 33–37, 2006.

[96] M. C. Hall and J. H. Willis, "Divergent selection on flowering time contributes to local adaptation in *Mimulus guttatus* populations," *Evolution*, vol. 60, no. 12, pp. 2466–2477, 2006.

[97] J. B. Beltman and P. Haccou, "Speciation through the learning of habitat features," *Theoretical Population Biology*, vol. 67, no. 3, pp. 189–202, 2005.

[98] J. Beltman and J. A. Metz, "Speciation: more likely through a genetic or through a learned habitat preference?" *Proceedings of the Royal Society B*, vol. 272, no. 1571, pp. 1455–1463, 2005.

[99] J. M. Davis and J. A. Stamps, "The effect of natal experience on habitat preferences," *Trends in Ecology and Evolution*, vol. 19, no. 8, pp. 411–416, 2004.

[100] A. P. Hendry, V. Castric, M. T. Kinnison, and T. P. Quinn, "The evolution of philopatry and dispersal: homing versus straying in salmonids," in *Evolution illuminated: Salmon and their relatives*, A. P. Hendry and S. C. Stearns, Eds., pp. 52–91, Oxford University Press, Oxford, UK, 2004.

[101] N. Vallin and A. Qvarnstrom, "Learning the hard way: imprinting can enhance enforced shifts in habitat choice," *International Journal of Ecology*, vol. 2011, Article ID 287532, 7 pages, 2011.

[102] E. A. Bernays and R. F. Chapman, *Host-Plant Selection by Phytophagous Insects*, Chapman and Hall, New York, NY, USA, 1994.

[103] M. D. Sorenson, K. M. Sefc, and R. B. Payne, "Speciation by host switch in brood parasitic indigobirds," *Nature*, vol. 424, no. 6951, pp. 928–931, 2003.

[104] C. N. Balakrishnan, K. M. Sefc, and M. D. Sorenson, "Incomplete reproductive isolation following host shift in brood parasitic indigobirds," *Proceedings of the Royal Society B*, vol. 276, no. 1655, pp. 219–228, 2009.

[105] S. H. Berlocher and J. L. Feder, "Sympatric speciation in phytophagous insects: moving beyond controversy?" *Annual Review of Entomology*, vol. 47, pp. 773–815, 2002.

[106] H. Knüttel and K. Fiedler, "Host-plant-derived variation in ultraviolet wing patterns influences mate selection by male butterflies," *Journal of Experimental Biology*, vol. 204, no. 14, pp. 2447–2459, 2001.

[107] M. D. Stennett and W. J. Etges, "Premating isolation is determined by larval rearing substrates in cactophilic *Drosophila mojavensis*. III. Epicuticular hydrocarbon variation is determined by use of different host plants in *Drosophila mojavensis*

and *Drosophila arizonae*," *Journal of Chemical Ecology*, vol. 23, no. 12, pp. 2803–2824, 1997.

[108] W. J. Etges, C. L. Veenstra, and L. L. Jackson, "Premating isolation is determined by larval rearing substrates in cactophilic *Drosophila mojavensis*. VII. Effects of larval dietary fatty acids on adult epicuticular hydrocarbons," *Journal of Chemical Ecology*, vol. 32, pp. 2629–2646, 2006.

[109] G. Sharon, D. Segal, J. M. Ringo, A. Hefetz, I. Zilber-Rosenberg, and E. Rosenberg, "Commensal bacteria play a role in mating preference of *Drosophila melanogaster*," *Proceedings of the National Academy of Sciences of the United States of America*, vol. 107, no. 46, pp. 20051–20056, 2010.

[110] J. K. Craig and C. J. Foote, "Countergradient variation and secondary sexual color: phenotypic convergence promotes genetic divergence in carotenoid use between sympatric anadromous and nonanadromous morphs of sockeye salmon (*Onchorhynchus nerka*)," *Evolution*, vol. 55, pp. 380–391, 2001.

[111] C. J. Foote, G. S. Brown, and C. W. Hawryshyn, "Female colour and male choice in sockeye salmon: implications for the phenotypic convergence of anadromous and nonanadromous morphs," *Animal Behaviour*, vol. 67, no. 1, pp. 69–83, 2004.

[112] J. K. Craig, C. J. Foote, and C. C. Wood, "Countergradient variation in carotenoid use between sympatric morphs of sockeye salmon (*Oncorhynchus nerka*) exposes nonanadromous hybrids in the wild by their mismatched spawning colour," *Biological Journal of the Linnean Society*, vol. 84, no. 2, pp. 287–305, 2005.

[113] D. L. Hartl and A. G. Clark, *Principles of Population Genetics*, Sinauer Associates, Sunderland, Mass, USA, 3rd edition, 1997.

[114] M. Kirkpatrick and V. Ravigné, "Speciation by natural and sexual selection: models and experiments," *American Naturalist*, vol. 159, pp. S22–S35, 2002.

[115] D. I. Bolnick and B. M. Fitzpatrick, "Sympatric speciation: models and empirical evidence," *Annual Review of Ecology, Evolution, and Systematics*, vol. 38, pp. 459–487, 2007.

[116] G. de Jong, "Evolution of phenotypic plasticity: patterns of plasticity and the emergence of ecotypes," *New Phytologist*, vol. 166, no. 1, pp. 101–117, 2005.

[117] D. W. Pfennig and M. McGee, "Resource polyphenism increases species richness: a test of the hypothesis," *Philosophical Transactions of the Royal Society B*, vol. 365, no. 1540, pp. 577–591, 2010.

19

Synergy between Allopatry and Ecology in Population Differentiation and Speciation

Yann Surget-Groba,[1,2] Helena Johansson,[1,3] and Roger S. Thorpe[1,4]

[1] School of Biological Sciences, Bangor University, Bangor LL57 2UW, UK
[2] Department of Ecology and Evolutionary Biology, University of California Santa Cruz, Santa Cruz, CA 95064, USA
[3] Department of Biosciences, University of Helsinki, P.O. Box 65, 00014 Helsinki, Finland
[4] School of Biological Sciences, College of Natural Sciences, Bangor University, Deiniol Road, Bangor, Gwynedd LL57 2UW, UK

Correspondence should be addressed to Roger S. Thorpe, r.s.thorpe@bangor.ac.uk

Academic Editor: Andrew Hendry

The general diversity pattern of the Caribbean anole radiation has been described in detail; however, the actual mechanisms at the origin of their diversification remain controversial. In particular, the role of ecological speciation, and the relative importance of divergence in allopatry and in parapatry, is debated. We describe the genetic structure of anole populations across lineage contact zones and ecotones to investigate the effect of allopatric divergence, natural selection, and the combination of both factors on population differentiation. Allopatric divergence had no significant impact on differentiation across the lineage boundary, while a clear bimodality in genetic and morphological characters was observed across an ecotone within a single lineage. Critically, the strongest differentiation was observed when allopatry and ecology act together, leading to a sharp reduction in gene flow between two lineages inhabiting different habitats. We suggest that, for Caribbean anoles to reach full speciation, a synergistic combination of several historical and ecological factors may be requisite.

1. Introduction

Speciation, the mechanism at the origin of species diversification, is one of the most studied subjects in evolutionary biology. Despite this enormous interest, very little is known about the factors needed for speciation to occur. For instance, the relative importance of ecological versus purely historical factors, as well as the geographic context of speciation, is still debated, and mechanisms that are sometimes invoked to explain lack of speciation, such as observation of a species-area relationship [1], are themselves not fully understood.

The most widely recognised speciation model is the allopatric model, where different populations that are geographically isolated develop genetic incompatibilities, purely by genetic drift or founder effects [2–4], by adapting to different habitats (by-product speciation [5]), by sexual selection [6], or by fixation of incompatible mutations through adaptation to a similar habitat (mutation-order speciation [7]). Because there is no gene flow to counter the differentiation of populations, this model is the easiest to explain speciation, especially if the isolated populations are exposed to distinct selective environments (reviewed in [8]).

The possibility of speciation in the presence of recurrent gene flow (sympatric or parapatric speciation) is much more debated. In this model, the divergence between populations exchanging migrants is driven by ecological differentiation (ecological speciation [9]), and/or sexual differentiation (speciation by sexual selection [10]). Because even a limited amount of gene flow is expected to counter differentiation, early theoretical studies rejected this speciation model [11]. However, other theoretical and empirical work suggested that nonallopatric speciation is possible under particular circumstances (reviewed in [12]).

A group in which both the geographic context of speciation, and the role of ecological speciation, is debated is the Caribbean anoles. The anole radiation is a highly diverse species group useful for understanding the mechanisms at the origin of diversification. *Anolis* is one of the most speciose vertebrate genera with more than 400 described species. Of these, ca. 150 are found in the Caribbean islands, and are thought to originate from just two independent colonizations from the mainland. This species diversity is accompanied by a very high morphological and ecological diversity. For these reasons, the anole radiation in the Caribbean has been the focus of intense studies in the last decades (reviewed in [13]).

One of the striking patterns in *Anolis* diversity is the contrast between the Greater and the Lesser Antilles. Whereas numerous species coexist within the large islands of the Greater Antilles, islands of the Lesser Antilles have at most two naturally occurring species with most of the islands having only one. Furthermore, if most of the diversification in the Greater Antilles occurred within island, most of the species pairs in the Lesser Antilles are not sister species, suggesting that these species pairs did not diverge within island but rather came together after dispersal from another island, with one possible exception in Saint Vincent [13].

The Caribbean anoles show a well-documented species-area relationship [14, 15]; however, the reason why so many speciation events happened in the Greater Antilles while almost none happened in the Lesser Antilles is still speculative. Losos and Schluter [14] proposed two explanations for this observation. First, larger islands could offer more opportunities of geographic fragmentation and hence lead to the formation of more species by allopatric speciation. Second, larger islands may have more habitat diversity, and hence populations submitted to divergent selective pressure in different habitats could lead to the formation of more species by ecological speciation. The observation that some of the largest islands in the Lesser Antilles, like Dominica, Guadeloupe, or Martinique, have a very high habitat diversity and hence should be able to support several species if ecological speciation was the driving force in anole diversification led Losos [13] to suggest that nonallopatric modes of speciation driven by ecological speciation are not supported in anoles.

Several studies suggest the opposite. For instance, because of its complex geologic history [16], the island of Martinique has offered plenty of opportunities for allopatric speciation to occur in its endemic anole, *Anolis roquet*. Present day Martinique was once formed of separate proto-islands where distinct mtDNA lineages of *A. roquet* evolved in allopatry for millions of years before coming back into secondary contact [17]. However, high gene flow is observed between previously allopatric lineages showing that this long geographic isolation did not lead to complete speciation [18]. Similarly, deep mtDNA lineages are observed in several species, both in the Lesser Antilles [19–23], and in the Greater Antilles [24–31]. In the cases where it has been studied, no restriction of gene flow has been detected between these previously allopatric lineages [18, 32, 33], with one possible exception in North-Eastern Martinique [18]. All this suggests that even if allopatric speciation probably occurred in some situations, geographic isolation alone is not sufficient to reach speciation in anoles.

Instead, ecological gradients seem to be driving population differentiation in Martinique anoles. This island is very heterogeneous, both topographically and ecologically. The mountains in the North are exposed to the trade winds and receive a very high amount of precipitation all year round. At the opposite, the northern Caribbean coast is in the "rain shadow" of these mountains and is much drier and seasonal. Hence, the habitat changes dramatically from a cool montane rainforest to a hot xeric scrubland in just a few kilometres. Previous studies [18] have shown that the divergent selective forces along this habitat gradient lead to significant morphological and genetic differentiation between coastal and mountain populations of *A. roquet*. On the neighbouring island of Dominica, a similar situation appears to occur with its endemic anole, *Anolis oculatus* [32], and indeed, in the single case to date where contact zones have been studied in the Greater Antilles, a significant reduction in gene flow between divergent lineages is only observed in an area with a steep ecological gradient [33].

In this paper we reanalysed data from a previous study [18] and added a new transect of *A. roquet* populations where the ecotone and the lineage boundary overlap. We compared the population structure of this new transect to the two previously published ones, one between two lineages within a single habitat, and the other between two habitats within a single lineage. We studied the population structure and admixture rates along these different transects to investigate the effects of geographic isolation, ecological isolation, and the combination of these two factors on anole population differentiation and speciation. We observe that population differentiation is at its highest when both factors act simultaneously and suggest that a possible explanation of the species-area relationship observed in Caribbean anoles is that the probability of both allopatric and ecological factors acting in synergy increases with island size.

2. Material and Methods

2.1. Sampling. The distribution of anoles in Martinique is more or less continuous. Hereafter we use the term "population" to refer to a discrete sampling site, not to genetically or ecologically differentiated entities. We sampled three distinct transects in northern Martinique (Figure 1). First, the "lineage transect" (transect II in [18]) consisted of eight populations sampled in similar habitat (transitional forest) across the lineage boundary between the North-West (NW) and the Central (C) lineages. Second, the "habitat transect" (populations 1 to 6 from transect IV in [18]) consisted of six populations sampled within the NW lineage and across an ecotone between coastal scrubland and montane rainforest. Finally, the "combined transect" (new data) consisted of eight populations sampled across the lineage boundary between the NW and the C lineages and across the ecotone between coastal scrubland and montane rainforest. For each population, tail tips from 48 individuals were sampled, and quantitative traits measurements (see below) were collected on ten adult males.

2.2. Genetic Analyses. For each individual, genotypes were scored at nine microsatellite loci (AAE-P2F9, ABO-P4A9,

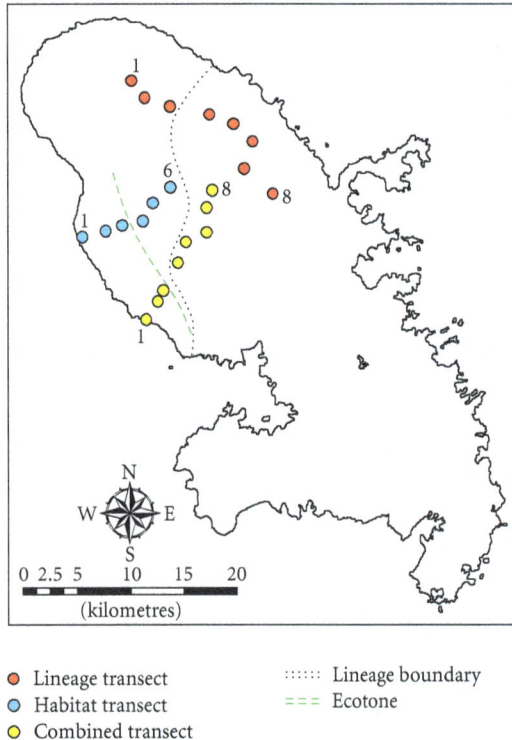

FIGURE 1: Map of the sampled populations. The lineage transect is indicated in red, the habitat transect in blue, and the combined transect in yellow. The first and last sites of each transect are numbered on the map.

AEX-P1H11, ARO-HJ2 [34], ARO-035, ARO-062, ARO-065, ARO-120 [35], and ALU-MS06 [36]. Mitochondrial DNA lineage was inferred by amplifying the complete *Cyt-b* gene, digesting the PCR product with the restriction enzymes SspI and DraI (New England Biolabs) whose cutting pattern allows to distinguish the different lineages present in this species. More details on these molecular techniques can be found in [18]. Only individuals successfully genotyped at least at four loci were kept for analyses. Three sets of analyses were conducted.

(i) Admixture Analysis. We estimated the genetic structure within each transect using the Bayesian clustering method implemented in STRUCTURE v2.3.2 [37]. First, we estimated the most likely number of clusters (k) by running the analysis with k ranging from 1 to the real number of populations. For each k value, the analysis was run 10 times using the admixture model, with a burn-in of 100,000 steps for a total run length of 500,000 steps. The optimal number of clusters (K) was inferred following the method outlined in [38], choosing the number of clusters corresponding to the highest rate of change of the log probability of data between successive K values. We considered an individual as admixed if less than 80% of its genome was assigned to a single cluster. We also estimated the standardised pairwise Fst (Fst' [39]) between adjacent populations along each transect using RecodeData v0.1 [39] and FSTAT v2.9.3 [40].

(ii) We then estimated the influence of different potentially important factors on the genetic structure using the

method implemented in GESTE v2.0 [41]. GESTE employs a hierarchical Bayesian framework to estimate population specific Fst (representing the differentiation of a given population relative to all other populations) and uses a generalized linear model in order to test the contribution of biotic or nonbiotic factors to genetic structuring. For our analyses we used the default settings (sample size of 10000, thinning interval 20), and in accordance with guidelines we allowed ten pilot runs to estimate means and variances for the required input parameters [41]. The analysis was repeated five times for each transect to ensure that the results were consistent. Three different factors were considered. First 19 bioclimatic variables were obtained for each population from the WorldClim database. These variables were annual mean temperature, mean diurnal range, isothermality, temperature seasonality, maximum temperature warmest month, minimum temperature coldest month, temperature annual range, mean temperature wettest quarter, mean temperature driest quarter, mean temperature warmest quarter, mean temperature coldest quarter, annual precipitation, precipitation wettest month, precipitation driest month, precipitation seasonality, precipitation wettest quarter, precipitation driest quarter, precipitation warmest quarter, and precipitation coldest quarter. A principal component analysis (PCA) of the log transformed variables was performed using the ade4 package [42] in the R environment [43]. The first component explained more than 75% of the variation and was used as a composite bioclimatic variable to describe the environment among each site. This first axis described the variation from hot and seasonally dry coastal sites to cool and wet montane sites. Second, the geographic connectivity of the populations was estimated as described in [44] to include a measure of geographic isolation for each individual population (mean geographical distance between a given population and all other populations). Third, each site was assigned to a lineage (lineage transect), a habitat (habitat transect), or both (combined transect) as follows. For the lineage transect, sites 1–3 were assigned to the NW lineage, and sites 4–8 were assigned to the C lineage. For the habitat transect, sites 1–3 were assigned to coastal habitat, and sites 4–6 were assigned to montane habitat. For the combined transect, sites 1–3 were assigned to NW/coastal group, while sites 4–8 were assigned to C/montane group.

(iii) Transects Comparison. To determine if the genetic structure was different among transects, we computed and plotted the pairwise Fst' values between populations on each side of the contact zone within each transect. This allowed us to compare the level of across habitat/lineage differentiation among transects. Because the observations are nonindependent and hence violate the assumption of both parametric and nonparametric tests, these plots provide qualitative information that has not been tested statistically.

2.3. Combined Analysis of Genetic and Quantitative Data. We conducted a modified version of the Discriminant Analysis of Principal Components (DAPC, [45]) to describe the global structure of populations within each transect using both genetics and morphology.

First different morphological characters were recorded as previously described [22, 46]. These included body dimensions (jaw length, head length, head depth, head width, upper leg length, lower leg length, dewlap length), scalation (number of postmental scales, scales between supraorbitals, ventral scales, and dorsal scales), colour pattern (number of dorsal chevrons, chevron intensity, occipital "A" mark, black dorsal reticulation, white spots), and trunk background colour (percentage red, green and blue hue on the posterior trunk). These were combined to six independent hues (UV 330–380 nm, UV/violet 380–430 nm, blue 430–490 nm, green 520–590 nm, yellow/orange 590–640 nm, and red 640–710 nm) extracted from the spectrum of the anterior and posterior dewlap by a multiple-group eigenvector procedure [23]. This combined data set was subjected to a principal component analysis (PCA) using the ade4 package in R.

Then, the microsatellite data were also subjected to a PCA using the adegenet package [47] in R. Finally, the components from the genetic (14 components) and quantitative (4 components) datasets were combined and subjected to a linear discriminant analysis using the MASS package [48] in R, using the population of origin as the grouping factor.

3. Results

3.1. Lineage Transect. Two genetic clusters were identified in this transect, with individuals from populations 1 to 3 being assigned in majority to the first cluster and individuals from populations 4 to 8 to the second cluster (Figure 2(a)). This separation corresponds to what was observed with mitochondrial DNA, populations 1 to 3 being in majority from the NW lineage while populations 4 to 8 are in majority from the C lineage [18, 46]. However, the proportion of admixed individuals is very high and relatively constant all along this transect (between 35 and 56%), and there is no trend in the pairwise Fst' that vary between 0.072 and 0.101 (Figure 2(a)). Morphological and genetic data analysed separately (Figure S1) show the same trend that is magnified in the combined analyses. We observe a slight differentiation between populations 1 to 3 and populations 4 to 8, but there is an overlap between these groups (Figure 3(a)). The population specific Fst estimated with GESTE (Table 1) that represents the level of differentiation of one population relative to all other [41] are very low (ranging from 0.008 to 0.013, Table 1) and do not correlate with any of the factors investigated (geographic connectivity, environment, lineage), underlining again the lack of genetic structure along this transect (Table 2).

3.2. Habitat Transect. Two genetic clusters were also identified in this transect (cluster 1: populations 1–3; cluster 2: populations 4–6, Figure 2(b)). This genetic division corresponds to the habitat division, the ecotone being situated between populations 3 and 4 [18]. The admixture rates are globally lower than on the lineage transect (range: 15–55%). It is relatively low at the two opposite sides of the transect (29% in population 1 and 15% in population 6) and gets higher in the centre of the transect, around the ecotone (55% in population 3 and 45% in population 4). Here again, there

is no obvious trend in the pairwise Fst' that vary between 0.059 and 0.112 (Figure 2(b)). The combined dataset shows a much higher level of variation than on the lineage transect (almost twice as high), and a marked differentiation between populations 1-2 and 4–6, with population 3 being intermediate (Figure 3(b)). The population specific Fst are higher than in any site of the lineage transect (ranging from 0.0148 to 0.0522, Table 1) confirming the higher genetic structuring along this transect. The best model to explain this genetic structure only incorporates a constant, but the second and third best models have a nonnegligible probability and incorporate, respectively, the habitat type and the bioclimatic variable (Table 1); these two factors have a combined probability of 0.175 and 0.134, respectively (Table 2).

3.3. Combined Transect. Here again two genetic clusters were identified (cluster 1: populations 1–3; cluster 2: populations 4–8, Figure 2(c)), but with a much higher differentiation. This genetic division corresponds both to the habitat and lineage boundaries, the ecotone being situated between populations 3 and 4, as well as the lineage boundary (Figure 4). On this transect, the admixture rate is much lower (range 4–27%) than on the other transects. It is somewhat higher in population 4, situated at the lineage boundary and at the ecotone, but even in this population, it is lower than in any of the lineage transect's populations and than in most of the habitat transect's populations (Figure 2(c)). There is also a marked increase of the pairwise Fst' at the contact zone, with a value of 0.169 between populations 3 and 4, while the other values are much lower (range: 0.052–0.0760, Figure 2(b)), suggesting the existence of a barrier to gene flow at the lineage/habitat boundary. The combined dataset shows a similar level of variation than in the habitat transect, with a strong differentiation between coastal (1-2) and montane (5–8) populations, with the populations 3 and 4 being intermediate. Along this transect, the population specific Fst are similar or higher to what is observed in the habitat transect (Table 1). The genetic structure is significantly associated with the habitat type/lineage factor, and the second best model includes the bioclimatic variable with a nonnegligible probability (Table 2).

4. Discussion

In this paper, we describe the differentiation between two lineages of *Anolis roquet* living in contrasting habitats and coming into secondary contact at the ecotone between these habitats. As a comparison, we reanalysed two previously published transects, a "lineage transect" where two lineages meet within a same habitat, and a "habitat transect" where different populations of the same lineage live in contrasting environments and are in contact at the ecotone between these habitats. This design allowed to investigate the effects of allopatry, habitat, and of the combination of these two factors on anole population differentiation and speciation.

A marked difference could be observed in the population structure along these three transects (Figure 5), suggesting

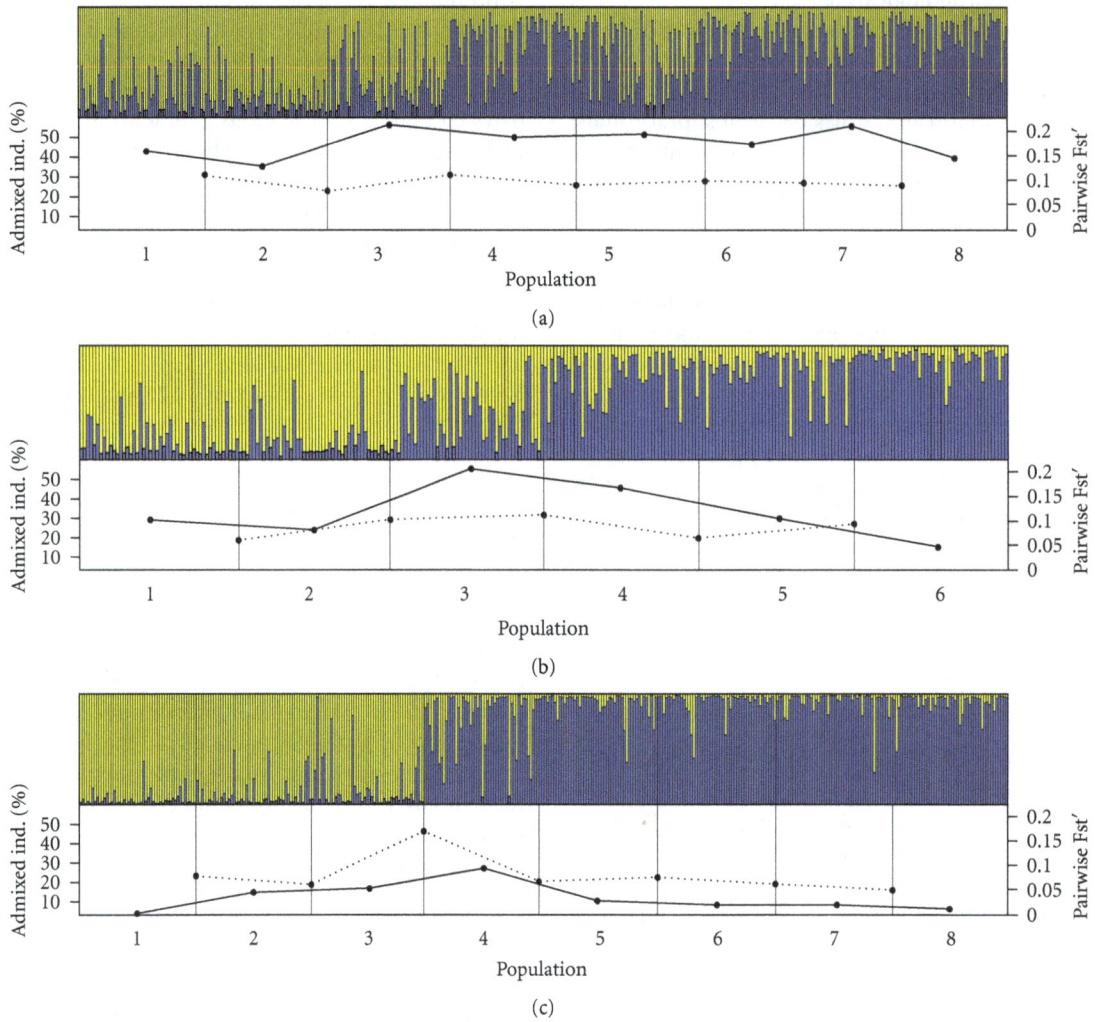

FIGURE 2: Population structure and admixture rates along the three transects. Solid lines represent admixture rates, and broken lines represent pairwise Fst values.

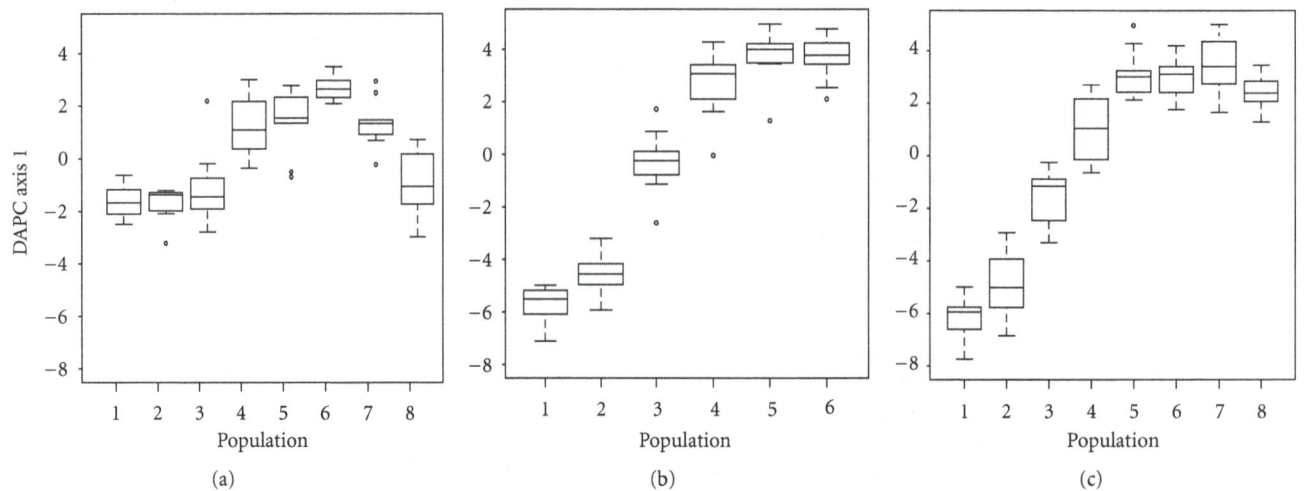

FIGURE 3: DAPC (discriminant analysis of principal components) based on the combination of genetic and quantitative trait characters (see supplementary Figure S1 available online at doi: 10.1155/2012/273413 for the separate analyses).

TABLE 1: Bayesian estimates of Fst values for each population. The mean value and the lower and upper limits of the 95% highest probability density interval are indicated. These Fst values measure the differentiation of a given population relative to all other populations.

Population	Lineage transect	Habitat transect	Combined transect
1	0.0078 [0.0038; 0.0123]	0.0522 [0.0357; 0.0695]	0.0721 [0.0524; 0.0934]
2	0.0104 [0.0055; 0.0155]	0.0464 [0.0301; 0.0627]	0.0513 [0.0357; 0.0674]
3	0.0054 [0.0019; 0.0093]	0.0330 [0.0209; 0.0468]	0.0507 [0.0350; 0.0671]
4	0.0097 [0.0049; 0.0151]	0.0148 [0.0071; 0.0228]	0.0148 [0.0075; 0.0223]
5	0.0061 [0.0022; 0.0101]	0.0219 [0.0122; 0.0323]	0.0131 [0.0064; 0.0206]
6	0.0132 [0.0075; 0.0193]	0.0233 [0.0142; 0.0339]	0.0134 [0.0071; 0.0205]
7	0.0090 [0.0044; 0.0139]		0.0122 [0.0060; 0.0190]
8	0.0101 [0.0052; 0.0153]		0.0136 [0.0066; 0.0210]

TABLE 2: Environmental factors determining the genetic structure of populations. (a) Sum of posterior probabilities of models including a given factor. (b) Posterior probability of the 8 models considered. The best model is indicated in bold. (c) Estimates of the regression parameters for the best model (mean value and lower and upper limits of the 95% highest probability density interval are indicated).

	Lineage transect	Habitat transect	Combined transect
(a)			
Connectivity (G1)	0.069	0.107	0.081
Environment (G2)	0.070	0.184	0.458
Lineage/habitat (G3)	0.079	0.228	0.572
(b)			
Constant	**0.800**	**0.554**	0.068
Constant, G1	0.057	0.067	0.012
Constant, G2	0.058	0.134	0.319
Constant, G3	0.006	0.175	**0.431**
Constant, G1, G2	0.067	0.017	0.029
Constant, G1, G3	0.006	0.020	0.031
Constant, G2, G3	0.006	0.030	0.100
Constant, G1, G2, G3	0.001	0.003	0.010
(c)			
$\alpha 0$	-3.80 $[-4.28; -3.34]$	-3.49 $[-4.17; -2.78]$	-3.77 $[-4.24; -3.24]$
$\alpha 1$			0.75 $[-1.22; -0.29]$
$\sigma 2$	0.39 [0.10; 0.84]	0.70 [0.13; 1.63]	0.40 [0.10; 0.89]

that they are at different stages of the speciation continuum [49, 50]. According to Hendry et. al [49], four stages can be distinguished along this continuum: "(1) continuous variation within panmictic populations, (2) partially discontinuous variation with minor reproductive isolation, (3) strongly discontinuous variation with strong but reversible reproductive isolation and (4) complete and irreversible reproductive isolation." The weak structure observed in the lineage transect, and the high admixture level correspond the State 1 of the continuum. This pattern is similar to what was observed in a previous study [18] for all but one (transect I in [18]) transects sampled across lineages (transects II, III, IV, V, VI, VII, VIII in [18]). In the habitat transect, the variation is clearly bimodal, but the high level of admixture in the contact zone between the two habitats suggests that the reproductive isolation is still limited between the ecotypes, corresponding to a stage between State 2 and State 3 of the continuum. Here again, this pattern is similar to the other transect sampled across habitats (transect III in [18]). Finally, the combined transects

present strongly discontinuous variation, but there is still a significant level of admixture at the contact zone suggesting that the reproductive isolation is not complete despite a strong reduction in gene flow. This would correspond to State 3 of the speciation continuum. The situation of this last transect, where the ecotone and the lineage boundary overlap, is unique on this island, and hence this result could not be replicated.

In terms of association between observed genetic structure and biotic and abiotic factors, no significant factors were detected on the lineage transect, all the models other than the one incorporating only a constant having a very low probability. To the contrary, for the combined transect both the habitat/lineage category and the bioclimatic composite variable are the best at explaining the genetic structure, suggesting the important driving force of the environment into population differentiation. The situation in the habitat transect is not so clear. The best model only incorporates a constant, but the combined probability of the habitat type and the environment factor are not negligible as they were on

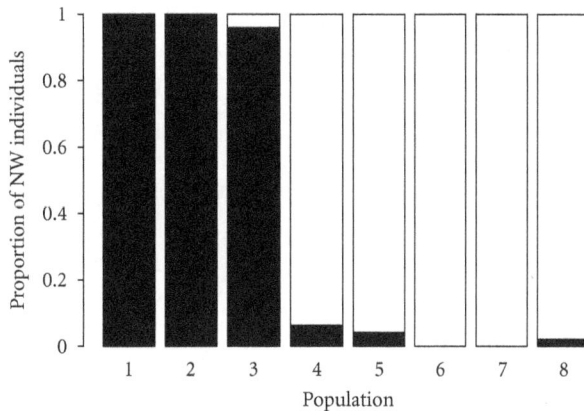

FIGURE 4: Proportion of individuals from the NW (black) and C (white) mitochondrial lineages along the combined transect.

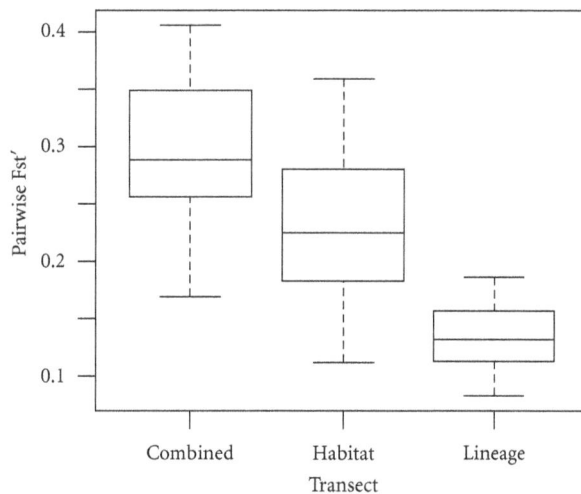

FIGURE 5: Box plot of the pairwise Fst' values between populations across the lineage and/or habitat contact zone for the three transects.

the lineage transect. The lack of significance of these factors could be due to the lack of power of this method when the number of populations analysed is low. Indeed, simulations showed that this method failed to identify the true model with five or seven populations [41]. However, the observed trend suggests that environment factors may play a role in the genetic structuring of the populations along this transect too.

As expected, we find the strongest signal of divergence in morphological data that are known to react quickly to selection [51–54] (Figure S1). The genetic data is similar in pattern but smaller in magnitude. Recent simulation studies suggest that neutral markers, that is, markers that are not linked with the traits under selection, are not very sensitive to detect ecological speciation [55, 56]. The authors conclude that this would lead to false negatives (failure to detect ecological speciation) rather than false positives. Taking this into account, and the fact that we found a clear signal of gene flow reduction both in the habitat and the combined transect

in accordance with an extreme environment gradient, the divergence we demonstrate in this study with these neutral markers is undoubtedly considerably less than the divergence of the traits and loci under selection.

Since the habitat along the lineage transect is very homogeneous we do not expect ecological speciation to currently play a role in this area. However, this does not mean it has always been the case. When the populations from the two lineages were isolated on different proto-islands, it is possible that they were submitted to different environmental pressures. In such a situation, several studies have demonstrated that isolated populations can rapidly evolve partial reproductive isolation as a byproduct of local adaptation [57–60]. Such a mechanism could also explain the differences observed between the three transects. When populations from the C and NW lineages came back into secondary contact, they did so in two very different contexts: either along a very sharp environmental gradient (combined transect), where any preexisting reproductive isolation could be maintained or strengthened by current disruptive selection, or within an homogeneous habitat (lineage transect) where any preexisting reproductive isolation may have been lost. Breakdowns of reproductive barriers associated with ecological changes have been recently described in various species [61, 62].

The relative role of geography and ecology in speciation remains a subject of debate. The main discussion relates to whether or not ecological factors can drive speciation in the presence of gene flow. Several convincing empirical studies suggest that it is indeed possible to reach full speciation in sympatry by ecological speciation (e.g., [63–67]), while others emphasize the combined role of historical and ecological factors in shaping species diversity [68–70]. For instance, in Trinidadian guppies, ecological speciation played a role in premating isolation either in allopatry (byproduct speciation) or in parapatry (to avoid maladaptive matings) [69]. For Martinique anoles, it is clear that for the populations that reached secondary contact very little evidence of the role of geographic isolation exists, while ecological factors seem to play a more important role. However, without conducting mate-choice experiments, it is not possible to determine conclusively the role of ecological speciation in allopatry.

Despite the large number of studies on the Caribbean anole radiation, very little is known about the factors at the origin of their diversity. Several papers have described the diversity patterns, demonstrating a correlation between island size and species diversity in large islands, while no speciation events were recorded on islands below a threshold size [13]. Recent work demonstrated that within island diversity could be the result of ecological opportunities and that net speciation rate decreased with time as opportunities decreased to reach an equilibrium at the island carrying capacity [15]. However, these studies do not explain the mechanisms at the origin of these diversity patterns and only propose several hypotheses to explain these observations.

In line with previous studies, we demonstrate that indeed environmental factors have a strong effect on the genetic structure of Martinique anole populations, with a reduction of gene flow at the ecotone between coastal and montane habitats and that it does not seem sufficient to lead to full reproductive isolation. Furthermore, we refine our understanding of divergence in these anoles by demonstrating that when allopatric lineages come into secondary contact on an ecotone, the differentiation is much stronger, with a significant reduction in gene flow. The absence of replication of the combined transect does not allow the generalisation of these findings, but we hypothesize that to reach full speciation anoles need first to evolve in allopatry and then come into secondary contact in an area where divergent natural selection will allow them to stay separate and further reinforce their divergence. Because this combination of factors is more likely to be found on large islands, it could be the mechanism at the origin of the species-area relationship observed in Caribbean anoles.

Acknowledgments

The authors wish to thank the DIREN Martinique for providing permits to collect samples, and W. Grail for assistance with molecular work. This work was funded by a Marie Curie Intra-European Fellowship to Y. Surget-Groba, a BBSRC grant to R. S. Thorpe, and a NERC studentship to H. Johansson.

References

[1] Y. Kisel and T. G. Timothy, "Speciation has a spatial scale that depends on levels of gene flow," *American Naturalist*, vol. 175, no. 3, pp. 316–334, 2010.

[2] E. Mayr, *Systematics and the Origin of Species*, Columbia University Press, New York, NY, USA, 1942.

[3] E. Mayr, *Animal Species and Speciation*, Harvard University Press, Cambridge, Mass, USA, 1963.

[4] M. Nei, T. Maruyama, and C. Wu, "Models of evolution of reproductive isolation," *Genetics*, vol. 103, no. 3, pp. 557–579, 1983.

[5] E. Mayr, "Ecological factors in speciation," *Evolution*, vol. 1, pp. 263–288, 1947.

[6] T. M. Panhuis, R. Butlin, M. Zuk, and T. Tregenza, "Sexual selection and speciation," *Trends in Ecology and Evolution*, vol. 16, no. 7, pp. 364–371, 2001.

[7] P. Nosil and S. M. Flaxman, "Conditions for mutation-order speciation," *Proceedings of the Royal Society B: Biological Sciences*, vol. 278, no. 1704, pp. 399–407, 2011.

[8] J. A. Coyne and H. A. Orr, *Speciation*, Sinauer Associates, Sunderland, Mass, USA, 2004.

[9] H. D. Rundle and P. Nosil, "Ecological speciation," *Ecology Letters*, vol. 8, no. 3, pp. 336–352, 2005.

[10] G. F. Turner and M. T. Burrows, "A model of sympatric speciation by sexual selection," *Proceedings of the Royal Society B: Biological Sciences*, vol. 260, no. 1359, pp. 287–292, 1995.

[11] J. Felsenstein, "Skepticism towards Santa Rosalia, or why are there so few kinds of animals," *Evolution*, vol. 35, pp. 124–138, 1981.

[12] M. R. Servedio, G. S.V. Doorn, M. Kopp, A. M. Frame, and P. Nosil, "Magic traits in speciation: "magic" but not rare?" *Trends in Ecology and Evolution*, vol. 26, no. 8, pp. 389–397, 2011.

[13] J. B. Losos, *Lizards in an Evolutionary Tree*, University of California Press, Berkeley, Calif, USA, 2009.

[14] J. B. Losos and D. Schluter, "Analysis of an evolutionary species-area relationship," *Nature*, vol. 408, no. 6814, pp. 847–850, 2000.

[15] D. L. Rabosky and R. E. Glor, "Equilibrium speciation dynamics in a model adaptive radiation of island lizards," *Proceedings of the National Academy of Sciences of the United States of America*, vol. 107, no. 51, pp. 22178–22183, 2010.

[16] D. Westercamp, P. Andreieff, P. Bouysse, S. Cottez, and R. Battistini, *Notice Explicative, Carte Géol. France (1/50 000), Feuille MARTINIQUE*, Bureau de Recherches Géologiques et Minières, Orléans, France, 1989.

[17] R. S. Thorpe and A. G. Stenson, "Phylogeny, paraphyly and ecological adaptation of the colour and pattern in the *Anolis roquet* complex on Martinique," *Molecular Ecology*, vol. 12, no. 1, pp. 117–132, 2003.

[18] R. S. Thorpe, Y. Surget-Groba, and H. Johansson, "Genetic tests for ecological and allopatric speciation in *Anoles* on an Island archipelago," *PLoS Genetics*, vol. 6, no. 4, Article ID e1000929, 2010.

[19] A. G. Stenson, R. S. Thorpe, and A. Malhotra, "Evolutionary differentiation of bimaculatus group anoles based on analyses of mtDNA and microsatellite data," *Molecular Phylogenetics and Evolution*, vol. 32, no. 1, pp. 1–10, 2004.

[20] R. S. Thorpe, A. G. Jones, A. Malhotra, and Y. Surget-Groba, "Adaptive radiation in Lesser Antillean lizards: molecular phylogenetics and species recognition in the Lesser Antillean dwarf gecko complex, *Sphaerodactylus fantasticus*," *Molecular Ecology*, vol. 17, no. 6, pp. 1489–1504, 2008.

[21] R. S. Thorpe and A. Malhotra, "Molecular and morphological evolution within small islands," *Philosophical Transactions of the Royal Society B: Biological Sciences*, vol. 351, no. 1341, pp. 815–822, 1996.

[22] R. S. Thorpe, Y. Surget-Groba, and H. Johansson, "The relative importance of ecology and geographic isolation for speciation in anoles," *Philosophical Transactions of the Royal Society B: Biological Sciences*, vol. 363, no. 1506, pp. 3071–3081, 2008.

[23] R. S. Thorpe, "Analysis of color spectra in comparative evolutionary studies: molecular phylogeny and habitat adaptation in the St. Vincent anole (*Anolis trinitatis*)," *Systematic Biology*, vol. 51, no. 4, pp. 554–569, 2002.

[24] T. Jezkova, M. Leal, and J. A. Rodríguez-Robles, "Living together but remaining apart: comparative phylogeography of *Anolis poncensis* and *A. cooki*, two lizards endemic to the aridlands of Puerto Rico," *Biological Journal of the Linnean Society*, vol. 96, no. 3, pp. 617–634, 2009.

[25] J. A. Rodríguez-Robles, T. Jezkova, and M. Leal, "Climatic stability and genetic divergence in the tropical insular lizard *Anolis krugi*, the Puerto Rican "Lagartijo Jardinero de la Montaña"," *Molecular Ecology*, vol. 19, no. 9, pp. 1860–1876, 2010.

[26] T. R. Jackman, D. J. Irschick, K. De Queiroz, J. B. Losos, and A. Larson, "Molecular phylogenetic perspective on evolution of lizards of the *Anolis grahami* series," *Journal of Experimental Zoology*, vol. 294, no. 1, pp. 1–16, 2002.

[27] R. E. Glor, J. J. Kolbe, R. Powell, A. Larson, and J. B. Lossos, "Phylogenetic analysis of ecological and morphological diversification in hispaniolan trunk-ground anoles (*Anolis cybotes* group)," *Evolution*, vol. 57, no. 10, pp. 2383–2397, 2003.

[28] R. E. Glor, J. B. Losos, and A. Larson, "Out of Cuba: overwater dispersal and speciation among lizards in the *Anolis carolinensis* subgroup," *Molecular Ecology*, vol. 14, no. 8, pp. 2419–2432, 2005.

[29] J. J. Kolbe, R. E. Glor, L. R. Schettino, A. C. Lara, A. Larson, and J. B. Losos, "Genetic variation increases during biological invasion by a Cuban lizard," *Nature*, vol. 431, no. 7005, pp. 177–181, 2004.

[30] J. A. Rodríguez-Robles, T. Jezkova, and M. A. García, "Evolutionary relationships and historical biogeography of *Anolis desechensis* and *Anolis monensis*, two lizards endemic to small islands in the eastern Caribbean Sea," *Journal of Biogeography*, vol. 34, no. 9, pp. 1546–1558, 2007.

[31] R. E. Glor, M. E. Gifford, A. Larson et al., "Partial island submergence and speciation in an adaptive radiation: a multilocus analysis of the Cuban green anoles," *Proceedings of the Royal Society B: Biological Sciences*, vol. 271, no. 1554, pp. 2257–2265, 2004.

[32] A. G. Stenson, A. Malhotra, and R. S. Thorpe, "Population differentiation and nuclear gene flow in the Dominican anole (*Anolis oculatus*)," *Molecular Ecology*, vol. 11, no. 9, pp. 1679–1688, 2002.

[33] J. Ng and R. E. Glor, "Genetic differentiation among populations of a Hispaniolan trunk anole that exhibit geographical variation in dewlap colour," *Molecular Ecology*, vol. 20, no. 20, pp. 4302–4317, 2011.

[34] J. L. Gow, H. Johansson, Y. Surget-Groba, and R. S. Thorpe, "Ten polymorphic tetranucleotide microsatellite markers isolated from the *Anolis roquet* series of Caribbean lizards," *Molecular Ecology Notes*, vol. 6, no. 3, pp. 873–876, 2006.

[35] R. Ogden, T. J. Griffiths, and R. S. Thorpe, "Eight microsatellite loci in the Caribbean lizard, *Anolis roquet*," *Conservation Genetics*, vol. 3, no. 3, pp. 345–346, 2002.

[36] H. Johansson, Y. Surget-Groba, J. L. Gow, and R. S. Thorpe, "Development of microsatellite markers in the St Lucia anole, *Anolis luciae*," *Molecular Ecology Resources*, vol. 8, no. 6, pp. 1408–1410, 2008.

[37] J. K. Pritchard, M. Stephens, and P. Donnelly, "Inference of population structure using multilocus genotype data," *Genetics*, vol. 155, no. 2, pp. 945–959, 2000.

[38] G. Evanno, S. Regnaut, and J. Goudet, "Detecting the number of clusters of individuals using the software STRUCTURE: a simulation study," *Molecular Ecology*, vol. 14, no. 8, pp. 2611–2620, 2005.

[39] P. G. Meirmans, "Using the amova framework to estimate a standardized genetic differentiation measure," *Evolution*, vol. 60, no. 11, pp. 2399–2402, 2006.

[40] J. Goudet, "FSTAT (Version 1.2): a computer program to calculate F-statistics," *Journal of Heredity*, vol. 86, pp. 485–486, 1995.

[41] M. Foll and O. Gaggiotti, "Identifying the environmental factors that determine the genetic structure of populations," *Genetics*, vol. 174, no. 2, pp. 875–891, 2006.

[42] J. Thioulouse, D. Chessel, S. Dolédec, and J. M. Olivier, "ADE-4: a multivariate analysis and graphical display software," *Statistics and Computing*, vol. 7, no. 1, pp. 75–83, 1997.

[43] R Development Core Team, *R: A Language and Environment for Statistical Computing*, R Foundation for Statistical Computing, Vienna, Austria, 2011.

[44] O. E. Gaggiotti, D. Bekkevold, H. B. H. Jørgensen et al., "Disentangling the effects of evolutionary, demographic, and environmental factors influencing genetic structure of natural populations: atlantic herring as a case study," *Evolution*, vol. 63, no. 11, pp. 2939–2951, 2009.

[45] T. Jombart, S. Devillard, and F. Balloux, "Discriminant analysis of principal components: a new method for the analysis of genetically structured populations," *BMC Genetics*, vol. 11, article 94, 2010.

[46] H. Johansson, Y. Surget-Groba, and R. S. Thorpe, "The roles of allopatric divergence and natural selection in quantitative trait variation across a secondary contact zone in the lizard *Anolis roquet*," *Molecular Ecology*, vol. 17, no. 23, pp. 5146–5156, 2008.

[47] T. Jombart, "Adegenet: a R package for the multivariate analysis of genetic markers," *Bioinformatics*, vol. 24, no. 11, pp. 1403–1405, 2008.

[48] W. Venables and B. Ripley, *Modern Applied Statistics with S*, Springer, New York, NY, USA, 2002.

[49] A. P. Hendry, D. I. Bolnick, D. Berner, and C. L. Peichel, "Along the speciation continuum in sticklebacks," *Journal of Fish Biology*, vol. 75, no. 8, pp. 2000–2036, 2009.

[50] P. Nosil, L. J. Harmon, and O. Seehausen, "Ecological explanations for (incomplete) speciation," *Trends in Ecology and Evolution*, vol. 24, no. 3, pp. 145–156, 2009.

[51] A. Malhotra and R. S. Thorpe, "Experimental detection of rapid evolutionary response in natural lizard populations," *Nature*, vol. 353, no. 6342, pp. 347–351, 1991.

[52] M. Leal and L. J. Fleishman, "Differences in visual signal design and detectability between allopatric populations of *Anolis* lizards," *American Naturalist*, vol. 163, no. 1, pp. 26–39, 2004.

[53] R. S. Thorpe, J. T. Reardon, and A. Malhotra, "Common garden and natural selection experiments support ecotypic differentiation in the Dominican anole (*Anolis oculatus*)," *American Naturalist*, vol. 165, no. 4, pp. 495–504, 2005.

[54] R. Calsbeek, J. H. Knouft, and T. B. Smith, "Variation in scale numbers is consistent with ecologically based natural selection acting within and between lizard species," *Evolutionary Ecology*, vol. 20, no. 4, pp. 377–394, 2006.

[55] X. Thibert-Plante and A. P. Hendry, "Five questions on ecological speciation addressed with individual-based simulations," *Journal of Evolutionary Biology*, vol. 22, no. 1, pp. 109–123, 2009.

[56] X. Thibert-Plante and A. P. Hendry, "When can ecological speciation be detected with neutral loci?" *Molecular Ecology*, vol. 19, no. 11, pp. 2301–2314, 2010.

[57] D. J. Funk, "Isolating a role for natural selection in speciation: host adaptation and sexual isolation in *Neochlamisus bebbianae* leaf beetles," *Evolution*, vol. 52, no. 6, pp. 1744–1759, 1998.

[58] T. H. Vines and D. Schluter, "Strong assortative mating between allopatric sticklebacks as a by-product of adaptation to different environments," *Proceedings of the Royal Society B: Biological Sciences*, vol. 273, no. 1589, pp. 911–916, 2006.

[59] R. B. Langerhans, M. E. Gifford, and E. O. Joseph, "Ecological speciation in *Gambusia* fishes," *Evolution*, vol. 61, no. 9, pp. 2056–2074, 2007.

[60] A. P. Hendry, P. Nosil, and L. H. Rieseberg, "The speed of ecological speciation," *Functional Ecology*, vol. 21, no. 3, pp. 455–464, 2007.

[61] E. B. Taylor, J. W. Boughman, M. Groenenboom, M. Sniatynski, D. Schluter, and J. L. Gow, "Speciation in reverse: morphological and genetic evidence of the collapse of a three-spined stickleback (*Gasterosteus aculeatus*) species pair," *Molecular Ecology*, vol. 15, no. 2, pp. 343–355, 2006.

[62] O. Seehausen, G. Takimoto, D. Roy, and J. Jokela, "Speciation reversal and biodiversity dynamics with hybridization in changing environments," *Molecular Ecology*, vol. 17, no. 1, pp. 30–44, 2008.

[63] E. Rolan-Alvarez, M. Carballo, J. Galindo et al., "Nonallopatric and parallel origin of local reproductive barriers between two

snail ecotypes," *Molecular Ecology*, vol. 13, no. 11, pp. 3415–3424, 2004.

[64] V. Savolainen, M. C. Anstett, C. Lexer et al., "Sympatric speciation in palms on an oceanic island," *Nature*, vol. 441, no. 7090, pp. 210–213, 2006.

[65] M. Barluenga, K. N. Stölting, W. Salzburger, M. Muschick, and A. Meyer, "Sympatric speciation in Nicaraguan crater lake cichlid fish," *Nature*, vol. 439, no. 7077, pp. 719–723, 2006.

[66] A. S.T. Papadopulos, W. J. Baker, D. Crayn et al., "Speciation with gene flow on Lord Howe Island," *Proceedings of the National Academy of Sciences of the United States of America*, vol. 108, no. 32, pp. 13188–13193, 2011.

[67] E. B. Rosenblum and L. J. Harmon, ""same same but different": replicated ecological speciation at white sands," *Evolution*, vol. 65, no. 4, pp. 946–960, 2011.

[68] P. Nosil, "Ernst Mayr and the integration of geographic and ecological factors in speciation," *Biological Journal of the Linnean Society*, vol. 95, no. 1, pp. 26–46, 2008.

[69] A. K. Schwartz, D. J. Weese, P. Bentzen, M. T. Kinnison, and A. P. Hendry, "Both geography and ecology contribute to mating isolation in guppies," *PLoS ONE*, vol. 5, no. 12, Article ID e15659, 2010.

[70] M. A. Frey, "The relative importance of geography and ecology in species diversification: evidence from a tropical marine intertidal snail (*Nerita*)," *Journal of Biogeography*, vol. 37, no. 8, pp. 1515–1528, 2010.

Permissions

The contributors of this book come from diverse backgrounds, making this book a truly international effort. This book will bring forth new frontiers with its revolutionizing research information and detailed analysis of the nascent developments around the world.

We would like to thank all the contributing authors for lending their expertise to make the book truly unique. They have played a crucial role in the development of this book. Without their invaluable contributions this book wouldn't have been possible. They have made vital efforts to compile up to date information on the varied aspects of this subject to make this book a valuable addition to the collection of many professionals and students.

This book was conceptualized with the vision of imparting up-to-date information and advanced data in this field. To ensure the same, a matchless editorial board was set up. Every individual on the board went through rigorous rounds of assessment to prove their worth. After which they invested a large part of their time researching and compiling the most relevant data for our readers. Conferences and sessions were held from time to time between the editorial board and the contributing authors to present the data in the most comprehensible form. The editorial team has worked tirelessly to provide valuable and valid information to help people across the globe.

Every chapter published in this book has been scrutinized by our experts. Their significance has been extensively debated. The topics covered herein carry significant findings which will fuel the growth of the discipline. They may even be implemented as practical applications or may be referred to as a beginning point for another development. Chapters in this book were first published by Hindawi Publishing Corporation; hereby published with permission under the Creative Commons Attribution License or equivalent.

The editorial board has been involved in producing this book since its inception. They have spent rigorous hours researching and exploring the diverse topics which have resulted in the successful publishing of this book. They have passed on their knowledge of decades through this book. To expedite this challenging task, the publisher supported the team at every step. A small team of assistant editors was also appointed to further simplify the editing procedure and attain best results for the readers.

Our editorial team has been hand-picked from every corner of the world. Their multi-ethnicity adds dynamic inputs to the discussions which result in innovative outcomes. These outcomes are then further discussed with the researchers and contributors who give their valuable feedback and opinion regarding the same. The feedback is then collaborated with the researches and they are edited in a comprehensive manner to aid the understanding of the subject.

Apart from the editorial board, the designing team has also invested a significant amount of their time in understanding the subject and creating the most relevant covers. They scrutinized every image to scout for the most suitable representation of the subject and create an appropriate cover for the book.

The publishing team has been involved in this book since its early stages. They were actively engaged in every process, be it collecting the data, connecting with the contributors or procuring relevant information. The team has been an ardent support to the editorial, designing and production team. Their endless efforts to recruit the best for this project, has resulted in the accomplishment of this book. They are a veteran in the field of academics and their pool of knowledge is as vast as their experience in printing. Their expertise and guidance has proved useful at every step. Their uncompromising quality standards have made this book an exceptional effort. Their encouragement from time to time has been an inspiration for everyone.

The publisher and the editorial board hope that this book will prove to be a valuable piece of knowledge for researchers, students, practitioners and scholars across the globe.

List of Contributors

Fernando Rubino and Manuela Belmonte
Laboratory of Plankton Ecology, IAMC CNR, UOS Talassografico "A. Cerruti," 74123 Taranto, Italy

Salvatore Moscatello, Manuela Belmonte, Gianmarco Ingrosso and Genuario Belmonte
Laboratory of Zoogeography and Fauna, CoNISMa U.O. Lecce, DiSTeBA University of the Salento, 73100 Lecce, Italy

Melissa Songer, Melanie Delion and Alex Biggs
Conservation Ecology Center, Smithsonian Conservation Biology Institute, National Zoological Park, Front Royal, VA 22630, USA

Qiongyu Huang
Geography Department, University of Maryland, College Park, MD 20742, USA

Marcia S. Meixler
Department of Ecology, Evolution and Natural Resources, Rutgers University, 14 College Farm Road, New Brunswick, NJ 08901, USA

Mark B. Bain
Department of Natural Resources, Cornell University, Ithaca, NY 14850, USA

Matthew D. Trager
USDA Forest Service, National Forests of Florida, 325 John Knox Road, Suite F-100, Tallahassee, FL 32303, USA

Matthew D. Trager, Matthew D. Thom and Jaret C. Daniels
Florida Museum of Natural History, 3215 Hull Road, Gainesville, FL 32611, USA

Matthew D. Thom and Jaret C. Daniels
Department of Entomology and Nematology, University of Florida, P.O. Box 110620, Gainesville, FL 32611, USA

Sophie E. Webster, Juan Galindo and Roger K. Butlin
Department of Animal and Plant Sciences, The University of Sheffield, Western Bank, Sheffield S10 2TN, UK

Juan Galindo
Departamento de Bioquimica, Genetica e Inmunolog´ia, Facultad de Biologia, Universidad de Vigo, 36310 Vigo, Spain

John W. Grahame
Institute of Integrative and Comparative Biology, Faculty of Biological Sciences, University of Leeds, Leeds LS2 9JT, UK

Tommy Lennartsson, Jörgen Wissman, and Hanna-Märtha Bergström
Swedish Biodiversity Centre, Swedish University of Agricultural Sciences, P.O. Box 7007, 750 07 Uppsala, Sweden

Gavin Ferris and Christopher K. Williams
Department of Entomology & Wildlife Ecology, University of Delaware, Newark, DE 19716, USA

Vincent D'Amico
US Forest Service, Newark, DE 19716, USA

Matsui Naohiro
Department of Environment, Kanso Technos Co., Ltd., Osaka 541-0052, Japan

Songsangjinda Putth
Fishery Department, Trang Coastal Aquaculture Station, Trang 92150, Thailand

Morimune Keiyo
Power Engineering R&D Center, The Kansai Electric Power Co., Inc., Kyoto 609-0237, Japan

Shimane W. Makhabu
Department of Basic Sciences, Botswana College of Agriculture, Private Bag 0027, Gaborone, Botswana

Balisana Marotsi
Department of Wildlife and National Parks, Ministry of Environment, Wild life and Tourism, P.O. Box 4, Tsabong, Botswana

Brian F. Kuhn
Institute for Human Evolution, School of Geosciences, University of the Witwatersrand (WITS), Johannesburg 2050, South Africa

Aaron E. Maxwell
Natural Resource Analysis Center, West Virginia University, Morgantown, WV 26506, USA

Michael P. Strager and Charles B. Yuill
Division of Resource Management, West Virginia University, Morgantown, WV 26506, USA

J. Todd Petty
Division of Forestry and Natural Resources, West Virginia University, Morgantown, WV 26506, USA

Stephen B. Heard
Department of Biology, University of New Brunswick, P.O. Box 4400, Fredericton, NB, Canada E3B 5A3

Alexander J. Werth
Department of Biology, Hampden-Sydney College, Hampden-Sydney, VA 23943-0162, USA

Marianne Elias
CNRS, UMR 7205, Mus´eum National d'Histoire Naturelle, 45 Rue Buffon, CP50, 75005 Paris, France

Rui Faria
CIBIO/UP—Centro de Investigac¸˜ao em Biodiversidade e Recursos Gen´eticos, Universidade do Porto, Campus Agr´ario de Vair˜ao, R. Monte-Crasto, 4485-661 Vair˜ao, Portugal
IBE—Institut de Biologia Evolutiva (UPF-CSIC), Universitat Pompeu Fabra, PRBB, Avenue Doctor Aiguader N88, 08003 Barcelona, Spain

Zachariah Gompert
Department of Botany, 3165, University of Wyoming, 1000 East University Avenue Laramie, WY 82071, USA

Andrew Hendry
Redpath Museum and Department of Biology, McGill University, 859 Sherbrooke Street West Montreal, QC, Canada H3A 2K6

L. A. Scriven, M. J. Sweet and G. R. Port
School of Biology, Ridley Building, Newcastle University, Newcastle upon Tyne NE1 7RU, UK

Cynthia F. Scholl and Matthew L. Forister
Department of Biology, University of Nevada, Reno, NV 89557, USA

Chris C. Nice
Department of Biology, Population and Conservation Biology Program, Texas State University, San Marcos, TX 78666, USA

James A. Fordyce
Department of Ecology and Evolutionary Biology, University of Tennessee, Knoxville, TN 37996, USA

Zachariah Gompert
Department of Botany, Program in Ecology, University of Wyoming, Laramie, WY 82071, USA

Katherine E. Moseby and John L. Read
School of Earth and Environmental Sciences, The University of Adelaide, SA 5005. Arid Recovery, P.O. Box 147, Roxby Downs, South Australia 5725, Australia

Katherine E. Moseby, Heather Neilly, John L. Read and Helen A. Crisp
Arid Recovery, P.O. Box 147, Roxby Downs, SA 5725, Australia

Benjamin M. Fitzpatrick
Ecology and Evolutionary Biology, University of Tennessee, Knoxville, TN 37996, USA

Yann Surget-Groba, Helena Johansson and Roger S. Thorpe
School of Biological Sciences, Bangor University, Bangor LL57 2UW, UK

Yann Surget-Groba
Department of Ecology and Evolutionary Biology, University of California Santa Cruz, Santa Cruz, CA 95064, USA

Helena Johansson
Department of Biosciences, University of Helsinki, P.O. Box 65, 00014 Helsinki, Finland

Roger S. Thorpe
School of Biological Sciences, College of Natural Sciences, Bangor University, Deiniol Road, Bangor, Gwynedd LL57 2UW, UK